Bioterrorism

FRONTIERS IN APPLIED MATHEMATICS

The SIAM series on Frontiers in Applied Mathematics publishes monographs dealing with creative work in a substantive field involving applied mathematics or scientific computation. All works focus on emerging or rapidly developing research areas that report on new techniques to solve mainstream problems in science or engineering.

The goal of the series is to promote, through short, inexpensive, expertly written monographs, cutting edge research poised to have a substantial impact on the solutions of problems that advance science and technology. The volumes encompass a broad spectrum of topics important to the applied mathematical areas of education, government, and industry.

BOOKS PUBLISHED IN FRONTIERS IN APPLIED MATHEMATICS

Banks, H. T. and Castillo-Chavez, Carlos, editors, *Bioterrorism: Mathematical Modeling Applications in Homeland Security*

Smith, Ralph C. and Demetriou, Michael, editors, *Research Directions in Distributed Parameter Systems*

Höllig, Klaus, *Finite Element Methods with B-Splines*

Stanley, Lisa G. and Stewart, Dawn L., *Design Sensitivity Analysis: Computational Issues of Sensitivity Equation Methods*

Vogel, Curtis R., *Computational Methods for Inverse Problems*

Lewis, F. L.; Campos, J.; and Selmic, R., *Neuro-Fuzzy Control of Industrial Systems with Actuator Nonlinearities*

Bao, Gang; Cowsar, Lawrence; and Masters, Wen, editors, *Mathematical Modeling in Optical Science*

Banks, H. T.; Buksas, M. W.; and Lin, T., *Electromagnetic Material Interrogation Using Conductive Interfaces and Acoustic Wavefronts*

Oostveen, Job, *Strongly Stabilizable Distributed Parameter Systems*

Griewank, Andreas, *Evaluating Derivatives: Principles and Techniques of Algorithmic Differentiation*

Kelley, C. T., *Iterative Methods for Optimization*

Greenbaum, Anne, *Iterative Methods for Solving Linear Systems*

Kelley, C. T., *Iterative Methods for Linear and Nonlinear Equations*

Bank, Randolph E., *PLTMG: A Software Package for Solving Elliptic Partial Differential Equations. Users' Guide 7.0*

Moré, Jorge J. and Wright, Stephen J., *Optimization Software Guide*

Rüde, Ulrich, *Mathematical and Computational Techniques for Multilevel Adaptive Methods*

Cook, L. Pamela, *Transonic Aerodynamics: Problems in Asymptotic Theory*

Banks, H. T. , *Control and Estimation in Distributed Parameter Systems*

Van Loan, Charles, *Computational Frameworks for the Fast Fourier Transform*

Van Huffel, Sabine and Vandewalle, Joos, *The Total Least Squares Problem: Computational Aspects and Analysis*

Castillo, José E., *Mathematical Aspects of Numerical Grid Generation*

Bank, R. E., *PLTMG: A Software Package for Solving Elliptic Partial Differential Equations. Users' Guide 6.0*

McCormick, Stephen F., *Multilevel Adaptive Methods for Partial Differential Equations*

Grossman, Robert, *Symbolic Computation: Applications to Scientific Computing*

Coleman, Thomas F. and Van Loan, Charles, *Handbook for Matrix Computations*

McCormick, Stephen F., *Multigrid Methods*

Buckmaster, John D., *The Mathematics of Combustion*

Ewing, Richard E., *The Mathematics of Reservoir Simulation*

Bioterrorism

Mathematical Modeling Applications in Homeland Security

Edited by

H. T. Banks
North Carolina State University
Raleigh, North Carolina

Carlos Castillo-Chavez
Cornell University
Ithaca, New York

siam

Society for Industrial and Applied Mathematics
Philadelphia

10 9 8 7 6 5 4 3 2 1

Library of Congress Cataloging-in-Publication Data

Bioterrorism : mathematical modeling applications in homeland security / edited by H.T. Banks, Carlos Castillo-Chavez.

p. cm. — (Frontiers in applied mathematics)

Includes bibliographical references and index.

ISBN 0-89871-549-0 (pbk.)

1. Biological warfare—Mathematical models. 2. Bioterrorism—Mathematical models. 3. Civil defense—United States—Mathematical models. 4. United States —Defenses—Mathematical models. I. Banks, H. Thomas. II. Castillo-Chavez, Carlos. III. Frontiers in applied mathematics (Unnumbered)

UG447.8.B585 2003

363.34'97—dc22

2003060646

Contents

Preface

The Centers for Disease Control established and developed intelligence epidemiological services in the early 1950s. This decision, driven by national concerns on the potential use of biological agents as a source of terror, was one of the first systematic responses to bioterrorism. The horror of September 11 and the events that followed have shown that the delivery of biological agents can be carried out by the systematic use of humans as well as by nontraditional means (such as mail).

Recent acts using anthrax have highlighted the use of biological and toxic agents as weapons of mass destruction as well as psychological agents of terror. Speculative discussions on the possible impact of the deliberate release of viruses such as smallpox into unsuspecting human populations have taken place from time to time over the years. The possible genetic manipulation of highly variable viruses such as influenza, for which society might not have an effective vaccine in storage, and their deliberate release are sources of great concern. The current SARS epidemic and its social and economic impact have revived the fears and concerns that were experienced during the anthrax scare of 2001. The avian flu epidemic (in April 2003) in the Netherlands has sent a strong reminder that we must now be prepared to live in a world where the impact of local "perturbations" is felt almost instantly everywhere.

Globalization and the possibility of bioterrorist acts have increased the demand for the development of theoretical and practical frameworks that can anticipate and predict the effects of initiation and considered response to acts of destabilization. Theoretical frameworks and the development of models to respond to specific focused questions will be useful to identify key pressure points in the system, to test for robust system features, and to look at the importance of system modularity and redundancy in addressing threats to various systems. Modeling and system interrogation in the presence of uncertainty have also become key areas of investigation, in which much work remains to be done.

The use of models is not limited to the biological sciences but in fact must be deeply connected to the social, behavioral, and economic sciences. For example, impact of bioterrorist acts on national and cultural behavioral norms has to be of great concern to those in charge of our national and homeland security as well as to economic and financial leaders. Current efforts to understand the mechanisms behind the spread of SARS, approaches to develop models that identify response strategies that minimize the impact associated with the potential deliberate release of biological agents in various topologies, the use of epidemiological approaches to model the spread of fanatic ideologies, and the development of mathematical and statistical approaches that can help in the building of biological and epidemiological "sensors" are but some of the currently hot areas of application that have benefited from recent advances in the fields of theoretical, mathematical, and computational epidemiology.

This volume brings the contributions of a selected group of experts from various fields who have begun to develop models to tackle some of the current challenges raised in biosurveillance, agroterrorism, bioterror response logistics, and assessment of the impact of the deliberate release of biological agents and the social forces that maintain groups of terror. Most of the contributors met first at a DIMACS workshop in March 2002 sponsored by NSF and DIMACS that led to a report co-authored by Fred Roberts and Carlos Castillo-Chavez (see http://dimacs.rutgers.edu/Workshops/WGDeliberate/). The idea of bringing the contributions together in book form came to fruition later in special sessions organized for the SIAM 50th Anniversary Annual Meeting in Philadelphia in July 2002.

Naturally, this volume gives a limited view of the challenges, possibilities, and opportunities that may be tackled with the use of mathematical and computational approaches. Our hope is that the contributions in its 10 chapters may inspire and stimulate research at the interface of homeland security and the mathematical and statistical sciences.

The editors would like to thank April Schilpp and the SIAM publication staff for their help and extreme efficiency in putting this volume together in such a short time. Their efforts and this publication epitomize the goals of the Frontiers series in bringing thoughtful and well-written considerations of emerging scientific issues to the community through rapid publication.

H. T. Banks
Center for Research in Scientific Computation
North Carolina State University

Carlos Castillo-Chavez
Departments of Biological Statistics and Computational Biology
Theoretical and Applied Mechanics
Cornell University
and
Center for Nonlinear Studies
Los Alamos National Laboratory

Chapter 1

Challenges for Discrete Mathematics and Theoretical Computer Science in the Defense against Bioterrorism

*Fred S. Roberts**

1.1 Introduction

The recent acts of bioterrorism using anthrax have raised great concern about the use of biological agents as weapons of mass destruction. This has led to major national debate and discussion—for example, about the possible impact of viruses such as smallpox as a biological weapon and about the possibility that bioterrorists would take advantage of genetic manipulation of highly variable viruses such as influenza. The great concern about the deliberate introduction of diseases by bioterrorists has raised new challenges for mathematical scientists. Dealing with bioterrorism requires detailed planning of preventive measures and responses: vaccination; vaccine dilution; vaccine stockpiling; quarantine of individuals, buildings, or regions; and systematic surveillance of emergency rooms, food supply, and water supply, to name just a few examples. Both planning and response require precise reasoning and extensive analysis of the type that mathematical scientists are very good at. Discrete mathematics and theoretical computer science, broadly defined, seem very relevant. In this chapter, we describe challenges for discrete mathematics and theoretical computer science raised by the need to plan for and defend against bioterrorist attacks. To begin, however, we talk briefly about a variety of mathematical sciences approaches to the defense against bioterrorism.

Understanding infectious systems consisting of parasites, vectors, and hosts requires being able to reason about highly complex biological systems, sometimes with hundreds of demographic and epidemiological variables. Obtaining such understanding is crucial in the defense against bioterrorism. Components of host-pathogen systems are sufficiently numerous and their interactions sufficiently complex that intuition alone is insufficient to fully understand the dynamics of such interactions. Yet experimentation or field trials are often prohibitively expensive or unethical and do not always lead to understanding at a fundamental level. Therefore, mathematical modeling becomes an important experimental and analytical tool. Mathematical models have become important tools in analyzing the spread

*DIMACS and Department of Mathematics, Rutgers University, 96 Frelinghuysen Road, Piscataway, NJ 08854.

and control of infectious diseases, especially when combined with the use of powerful, modern computer methods for analyzing and/or simulating the models. (See, for instance, [174].) Mathematical approaches can also be used to compare alternative intervention programs, to prepare responses, as guides for training exercises and scenario development, to help with risk assessment, as an aid in forensic analysis, and to predict future trends. All of these issues are central to the extensive planning required to deal with bioterrorist attacks.[1]

1.2 Methods of Mathematical and Computational Epidemiology

Many of the research challenges facing us in the defense against bioterrorism apply more generally to the defense against disease. Thus, we shall speak rather generally here about methods of "mathematical and computational epidemiology." Mathematical models of infectious diseases go back to Daniel Bernoulli's mathematical analysis of smallpox in 1760 and have been developed extensively since the early 1900s. Hundreds of mathematical models have been published since, exploring the effects of bacterial, parasitic, and viral pathogens on human populations. The results have highlighted and formalized such concepts as the core population in sexually transmitted diseases [113] and made explicit other concepts such as herd immunity for vaccination policies [3]. Key pathogens that have been studied are malaria [6], *Neisseria gonorrheae* [113], *Mycobacterium tuberculosis* [30, 31, 44, 166], HIV [189, 4], and *Treponema pallidum* [12]. Mathematical modeling has provided new insights on important issues such as drug-resistance, rate of spread of infection, epidemic trends, and the effects of treatment and vaccination. (For instance, for modeling work on vaccination strategies, see [98, 105, 112, 210, 213].) Yet, for many infectious diseases, we are far from understanding the mechanisms of disease dynamics. The modeling process can lend insight and clarification to data and theories. Mathematical models, with the aid of computer simulations, are useful theoretical and experimental tools for building and testing theories about complex biological systems involving disease, assessing quantitative conjectures, determining sensitivities to changes in parameter values, estimating key parameters from data, and preparing intervention strategies against newly emerging diseases or deliberate biological attacks. The size of modern epidemiological problems and the large data sets that arise call out for the use of powerful computational methods for studying these large models. As Levin et al. [144] point out, "imaginative and efficient computational approaches are essential in dealing with the overwhelming complexity of [such] biological systems." New mathematical and computational methods are needed to deal with the dynamics of multiple interacting strains of viruses through the construction and simulation of dynamic models, the problems of spatial spread of disease through pattern analysis and simulation, and the early detection of emerging diseases or bioterrorist acts through rapidly responding surveillance systems. An important challenge for the mathematical sciences is to investigate and seek to stimulate the development of such new methods and to use the new methods in a variety of new applications.[2]

Various methods from the mathematical sciences have been used in epidemiology. It is important to try to involve a larger number of mathematical scientists in partnership with epidemiologists by exposing the mathematical sciences community to the models and methods that have already been applied to epidemiology and by exploring new variants on these models and methods for dealing with new and complex issues as yet unexplored, under the guidance of leading mathematical and biological scientists already involved in the

[1] I would like to thank Denise Kirschner for some of the ideas in this paragraph.

[2] I would like to thank Don Hoover, Denise Kirschner, and David Ozonoff for some of the ideas in this paragraph.

enterprise. Statistical methods have long been used in mainstream epidemiology largely for the purpose of evaluating the role of chance and confounding induced associations [229]. Epidemiologists seek to ferret out sources of systematic error ("bias and confounding") in the observations, evaluate the role of uncontrollable error (using statistical methods) in producing the results, and interpret the results using correlative information from the medical and biological sciences.[3]

A smaller but venerable tradition within epidemiology has considered the spread of infectious disease as a dynamical system and applied difference equations and differential equations to that end [35, 62, 112]. Dynamical systems, both continuous and discrete, are relevant to many epidemiological models (classical examples are in host-parasitoid models such as Nicholson–Bailey or SIR or SIRS). Examples of relevant work on host-parasitoid and host-pathogen models are the models of urinary tract infections by Gordon and Riley [104], mycoparasite-immune dynamics by Antia, Levin, and May [8] and Antia, Koella, and Perrott [7], antimicrobial chemotherapy by Lipsitch and Levin [147, 148], and the recent work on tuberculosis by Wigginton and Kirschner [245]. Critical phenomena models (see, for example, [222]) can be used to model the sudden phase transition that occurs when a disease becomes epidemic. Threshold phenomena are well known in dynamical systems models of infectious disease [35]. These newer phase-transition models can be important tools to understand how sporadic disease occurrence can erupt into sudden epidemics. (Eastern equine encephalitis and west Nile virus are two important examples where understanding the phenomenon is critically needed.) Yet some dynamic models of spread of viruses in nonbiological systems [84, 2] do not exhibit threshold effects, and it will be important to understand why and what insight this can give us about spread of disease. In spite of this large body of work, little systematic effort has been made to apply today's powerful computational methods to these dynamical systems models, and relatively few mathematical scientists have been involved in the process. It is an important challenge to the mathematical sciences to change this situation.[4]

Probabilistic methods, in particular stochastic processes, have also played an important role in infectious disease modeling (see, e.g., [160, 75]). Random walk models have been used to study a variety of diseases (see, e.g., [108]). Percolation theory has been used to model spread of contagious phenomena like forest fires and epidemics [222] and related research on interacting particle systems and probabilistic cellular automata (see, e.g., [160, 197, 143]) has also become useful in epidemic modeling. Markov chain Monte Carlo methods are promising tools used to fit data to models of numerous diseases (see, e.g., [100, 178]). Moment closure methods have been useful in modeling sexually transmitted diseases, tuberculosis, and other diseases (see, e.g., [127]). Computational methods for simulating stochastic processes in complex spatial environments or on large networks have started to enable us to simulate more and more complex biological interactions. However, here again, few mathematical scientists have been involved in efforts to bring the power of modern computational methods to bear and it is a challenge to try to remedy this situation.

1.3 Discrete Mathematics and Theoretical Computer Science

A variety of other potentially useful approaches to epidemiological issues have not yet attracted the attention of many in the mathematical sciences community and some relevant

[3]Thanks to David Ozonoff for some of these ideas.

[4]Thanks to Denise Kirschner and David Ozonoff for some of the ideas in this paragraph.

methods of computer science and mathematics are not widely known among epidemiologists and biologists. Specifically, many fields of science, and in particular molecular biology, have made extensive use of the methods of discrete mathematics (DM), broadly defined, especially those that exploit the power of modern computational tools. These are guided by the algorithms, models, and concepts of theoretical computer science (TCS) that make these tools more available than they have been in the past. Yet these methods remain largely unused in the study of the transmission dynamics of infectious diseases.

One major development in epidemiology that makes the tools of DM and TCS especially relevant is the use of geographic information systems (GIS). These systems allow analytic approaches to spatial information not used previously (see, e.g., [53, 101, 123, 129, 175, 230, 234, 239, 240, 248]). Another related development is the availability of large and disparate computerized databases on subjects related to disease. This calls for use of modern methods of data mining. Data mining methods of cluster analysis, visualization, and learning, grounded in TCS and statistics, are key to detection and surveillance of bioterrorist attacks and are extremely relevant to spatial-temporal patterning, the recognition that a disease has reached epidemic stage, and the construction of exposure categories. (See, e.g., [43, 111].)

DM and TCS are also relevant to the increasing emphasis in infectious disease modeling of an evolutionary point of view. To fully understand issues such as immune responses of hosts; coevolution of hosts, parasites, and vectors; drug response; and antibiotic resistance, biologists are increasingly taking approaches that model the impact of mutation, selection, population structure, selective breeding, and genetic drift on the evolution of infectious organisms and their various hosts. (See, e.g., [174].) Moreover, recent advances in genomics provide useful tools that could be used to fight bioterrorism. For example, DNA sequencing is routinely used to characterize pathogens' strain phylogenies (see, e.g., [167]), a critical step in the identification of potential sources of supply of an agent. Epidemiologists are only beginning to become aware of some of the mathematical sciences tools available to analyze these complex problems, e.g., new methods of classification and phylogenetic tree reconstruction grounded in concepts and algorithms of TCS and developed through the explosion in "computational biology." Many recent methods of phylogenetic tree reconstruction resulted from partnerships between mathematical and computer scientists and biological scientists through the DIMACS Special Focus on Mathematical Support for Molecular Biology (see "Mathematical Support for Molecular Biology—Achievements—August 1994 to August 2000," http://dimacs.rutgers.edu/About/specialfocusreports.html). Yet a great deal more needs to be done. Many of the phylogenetic techniques were developed for more traditional, well-behaved evolutionary problems (see, e.g., [40, 41]). However, the traditional model of a binary tree with a small number of species as the leaves is insufficient to capture the "quasi species" nature of many viruses and the very high substitution rates of retroviruses. Collaboration between mathematical scientists and epidemiologists involving the development and use of phylogenetic methods in epidemiology is likely to take both fields of epidemiology and mathematical science in new and fruitful directions.[5]

Mathematical models that capture the process of one protein interfering with the folding of another seem relevant to diseases such as bovine spongiform encephalopathy ("mad cow" disease, or bovine spongiform encephalopathy), Alzheimer's disease, Huntington's disease, and amyloidosis [5, 149, 173]. These models might shed light on important epidemiological questions involving crossing species barriers and dose-response relationships. Models of protein folding are often based on global minimization of energy functions [182]

[5]I would like to thank Eddie Holmes and Mike Steel for some of the ideas in this paragraph.

but increasingly on methods blending DM and TCS [26, 220, 221].

The study of spread processes on graphs and networks has a long history in the mathematical analysis of social networks and opinion formation and in the developing theory of distributed computing. These processes, whether deterministic or stochastic, have had some application in the modeling of diseases, but there is a significant body of mathematical and modeling work that seems very relevant to epidemiology that needs exploration. (Examples of relevant work are [191, 83, 187, 168, 69].)

In the remainder of this chapter, we go into more detail about some of the challenges for DM and TCS, with special emphasis on bioterrorism. Many of the challenges are of interest in and of themselves and new results and methods stimulated by these challenges should be of use in a wide variety of fields.

1.4 Detailed Challenges for DM and TCS

DM deals with arrangements, designs, codes, patterns, schedules, and assignments. These are all very relevant to aspects of the defense against bioterrorism, as we shall point out. TCS deals with the theory of computer algorithms. During the first 40 years of the computer age, TCS, aided by powerful mathematical methods—in particular DM, but also logic and probability—had a direct impact on technology by developing models, data structures, algorithms, and lower bounds that are now at the core of computing. DM and TCS have found extensive use in many areas of science and public policy, with computational molecular biology being a specific and important case in point. These tools, which seem very relevant to epidemiological problems and in particular to the defense against bioterrorism, are not very well known among those working on public health problems.

1.4.1 Detection and Surveillance

A key to the prevention of bioterrorist attacks and to the successful response to them if they cannot be prevented is early detection that an attack is planned or has taken place. A basic requirement is the development of efficient and effective surveillance systems. Many of the most important challenges for DM and TCS arise from this requirement. Surveillance has long been a major function of the public health system. Concern about terrorist-instigated disease outbreaks has renewed interest in surveillance as one of the basic lines of defense.

Disease or event or symptom reporting and surveillance systems represent a primary epidemiological data source for the study of, and alert to, adverse reactions to medication, emerging diseases, or bioterrorist attacks. These systems synthesize data from millions of reports. There are many challenges for mathematical scientists to investigate current major issues confronting adverse event/disease reporting, surveillance, and analysis. A key issue is speed. Quick detection of a new disease outbreak is critical in determining the ultimate impact of the outbreak. In the US, two major reporting/surveillance systems are AERS and VAERS. Neither is specifically relevant to bioterrorism defense, but issues relating to these databases are. AERS, the Adverse Event Reporting System (http://www.fda.gov/cder/aers/) is a database of drug adverse reactions reported by health professionals and others. AERS is administered by the Food and Drug Administration (FDA). The system contains adverse reactions detected and reported after marketing of a drug for a specified time period. AERS contains over two million cases. VAERS, the Vaccine Adverse Event Reporting System (http://www.vaers.org), is a Cooperative Program for Vaccine Safety of the FDA and the Centers for Disease Control and Prevention (CDC). VAERS is a post-marketing safety surveillance program, collecting information about adverse events that occur after the administration of US licensed vaccines. Reports are provided by all concerned individuals: patients, parents, health care providers, pharmacists, and vaccine manufacturers. The VAERS database is publicly available and contains over 100,000 reports. It is important to develop new databases that are more specifically geared to bioterrorism defense. The appropriateness of different databases is a research issue of some significance. Other databases being developed and explored in connection with bioterrorism defense include emergency department reports, prescription drug information, and HMO data. Other potential sources of surveillance data under discussion include calls to doctors' offices and nurses' hotlines, hits to certain websites, etc. Even data from mobile or wearable sensors might be useful.[6]

Analyses of surveillance data must confront several difficulties. These include adverse event recognition, underreporting, biases, estimation of population exposure, report quality, and, most importantly, no denominator or control group. In many cases it is difficult to discern whether or not a reported symptom or adverse reaction was from a terrorist event or instead was a consequence of some underlying condition (see, for example, [130, 198, 227]). Similar considerations would apply if we were trying to understand whether symptoms came from a natural or terrorist-caused disease. Several methodological issues relating to these reporting mechanisms and others emphasizing disease and symptom reporting form the basis for a research agenda. These include application of computational and statistical methods for early detection of emerging trends; modification of algorithms of streaming data analysis designed to set off early warning alarms; application of data mining methods; development of causal inferential methods in the absence of controls; study of ways to eliminate bias; and design of verification methodology. This last issue is especially pressing since large-scale medical record databases that now exist in certain subpopulations (e.g., HMOs, the military) can provide a basis both for assessing the quality of surveillance data and for validating analyses. Other issues arise from taking reports supplied in "natural language" and translating them into a form amenable for analysis using formal or semiformal methods.[7]

New computational methods are needed to deal with a variety of large, complex data sets of various kinds arising in epidemiology, not just the disease and adverse event reporting systems. Notifiable infectious diseases provide a huge testbed of data for surveillance, for planning and evaluation of intervention programs, and for hypothesis generation in etiologic studies [144]. Data sets connecting environmental factors and disease can be used in much the same way [184, 236]. While there is a long history of use of such data sets in epidemiology, much work will be involved in applying or modifying known methods to bioterrorist applications. Some examples of large data sets arising in epidemiological studies involve fat-free body mass [46], linkage scans [193], the health of Gulf War veterans [192], and alcoholism [212]. A long list of links to large health-related data sets can be found at the website http://www.ehdp.com/vitalnet/datasets.htm. Such data, while often massive in quantity, is uneven in quality and completeness and heterogeneous in nature, and is typical of data we might expect in bioterrorist surveillance. Work will be required to build on relevant data mining studies in the epidemiological literature (see, e.g., [36, 37, 73, 91, 115, 179, 188, 201]), in particular the development of new algorithmic methods for data mining in epidemiology involving clustering methods. Automatic environmental monitoring and risk evaluation for cancer provide sample motivations even if cancer is not considered a bioterrorist threat. The issues are similar. In the US, there is a rich history of cancer mapping, highlighted by the release of the first US Cancer Mortality Atlas in 1975, the recent development of the US Cause

[6]I would like to thank Don Hoover and David Madigan for some of the ideas in this paragraph.

[7]Thanks to David Madigan for some of these ideas.

of Death Atlas, and the National Cancer Institute's data set of about 10 million US cancer cases. Additional data comes from questionnaires based on individual patient and resident information; public registries with cancer incidences aggregated by county; population-based cancer registries aggregated by city and town; birth and death registries; environmental data such as sample databases of water conditions and air quality records; census data such as geographic databases with accurate locations of population; and remotely sensed data providing information on land use patterns or air pollution distribution. All of these databases have different temporal and spatial assumptions (for example, different frequencies of collection, different spatial resolution (by state, by county, by zip code, by square kilometer), etc.). In short, the data sets are large, complex, and heterogeneous and contain many errors.[8]

Streaming Data Analysis

Much modern data analysis involves data that streams by so quickly or in such quantity that one has only one shot at it. A similar issue arises when the data is concerned with problems to which a rapid response is essential. This situation is certainly true of much of the data that the intelligence community looks at in trying to assess or detect new terrorist threats. The technical term the TCS community has come to use for the analysis of this type of data is *streaming data analysis*. (See, for example, [49, 97, 234].) Streaming data analysis is widely used to detect trends and sound alarms in telecommunications and finance. For instance, AT&T uses this to detect fraudulent use of credit cards or impending billing defaults, and a group at Columbia University has developed methods for detecting fraudulent behavior in financial systems. (For references on intrusion detection and billing faults, see [14, 38, 47, 118, 225]. For references on applications to financial systems, see [61, 190].) Streaming data analysis uses algorithms based in TCS, but these algorithms have not as yet been applied to disease or bioterrorist event detection.[9]

Among the research issues that face mathematical scientists in the attempt to develop methods of streaming data analysis and adapt them to bioterrorist event detection are the need to modify methods of data collection, transmission, processing, and visualization. The use of decision trees, vector-space methods, and Bayesian and neural nets should be explored. One needs to understand how the results of monitoring systems are best reported and visualized and to understand the extent to which they can lead to fast and safe automated responses. We need to understand also how relevant queries are best expressed, giving the user sufficient power while implicitly restraining him/her from incurring unwanted computational overhead.

Cluster Analysis

Cluster analysis offers the promise of pattern extraction from complex data sets involving different temporal and spatial assumptions, arising from different sources, etc. Clustering problems arise in widely varied applications involving molecular databases, astrophysical studies, geographic information systems, software measurement, worldwide ecological monitoring, medicine, and analysis of telecommunications and credit card data for fraud. Methods of clustering developed in these areas (see, for example, [9, 17, 48, 102, 125, 171]) need to be modified for applications in epidemiology and in particular in bioterrorist event surveillance and detection. Clustering methods aim at finding within a given set of entities subsets, called clusters, that are both homogeneous and well separated, and these concepts

[8] Thanks to Ilya Muchnik and David Ozonoff for some of these ideas.

[9] Thanks to Adam Buchsbaum and S. Muthukrishnan for ideas and references.

are central in applications of clustering to many practical problems of detection, decision making, or pattern recognition.

Application of traditional clustering algorithms to bioterrorism defense in particular and epidemiological problems in general is hindered by the extreme heterogeneity of the data; we need to develop new approaches to deal with such heterogeneities. Among the concepts and algorithms that need modification are those used widely in the clustering literature, such as maximum sum of splits, minimum sum of diameters, etc. (See [72, 107, 109, 110].) Promising algorithmic methodologies for clustering heterogeneous data are in the papers [128, 135, 164]. We also need to build on traditional statistical approaches to heterogeneous epidemiological data (see, e.g., [50, 51, 66, 183]).

Visualization of Data

Huge data sets are sometimes best understood by visualizing them. Visualization provides ways to reduce complex quantitative or qualitative information to an easily understood form, such as a map or a color-coded alert. Sheer data sizes require new visualization regimes, which require suitable external memory data structures to reorganize tabular data in secondary storage so that access, usage, and analysis are facilitated (see, e.g., [1]). The mathematical sciences community is faced with the challenge that developing visualization algorithms becomes harder when data arises from various sources and each source contains only partial information, as is the case, for example, with the cancer monitoring data mentioned above and, more generally, with all kinds of epidemiological data.

Visualization focuses on creating insightful visual representations of complex data. Creating the visual representation usually is done in several phases in a visualization pipeline. (See, for example, [13, 165].) First, the raw data are transformed into a higher-level metaphor (e.g., a graph). Then, the metaphor is transformed into geometry by means of some embedding operation. Finally, the geometry is rendered into an image. Throughout the visualization pipeline, emphasis is given to the clarity and aesthetics of these transformations. Quantitative approaches such as minimizing edge crossings (in graphs), area, etc., do not scale when there are more data elements than there are display pixels. New paradigms emphasize perspective views, hierarchical decompositions, overlays, advanced user-interfaces, and real-time rendering techniques. There has been much work defining models and metaphors but relatively little on formal algorithmic definitions and implementations. Research is required on core algorithmic issues as well as their use in discovery and also aspects of modeling.

Data Cleaning

In detecting diseases or bioterrorist events, a major problem is that the data is very "dirty." The data is dirty due to manual entry, lack of uniform standards for content and formats, data duplication, and measurement errors. The TCS community has been developing data cleaning tools (see, e.g., [94, 95, 136]) and the website www.research.att.com/~tamr/dataquality.html). Among the methods developed so far are methods known as duplicate removal, merge purge, and automated detection. Much more work is needed to develop such methods and adapt them to the detection and surveillance problems important for bioterrorism defense.

Dealing with Reports in "Natural Language"

Still another set of research issues arises from the use of natural language in surveillance and reporting systems to devise effective methods for translating natural language input into formats suitable for statistical analysis (prior work on machine natural language processing and information retrieval is relevant); develop computationally efficient methods to provide automated responses consisting of follow-up questions; and develop semiautomatic systems to generate queries based on dynamically changing data, indicating developing epidemiological trends. Relevant to these questions is work in [235] on interpreting natural language reports based on probabilistic models of context, in [138, 156, 250] on communicating uncertain information and summarizing rough trends, and in [116, 117, 238] on decision-theoretic methods for asking followup questions in natural language processing. Earlier work on electronic surveillance reporting from public health reference laboratories is also very relevant here (e.g., [23, 24, 121, 153]).[10]

Cryptography and Security[11]

A key difficulty in biosurveillance, as well as in intelligence work in general and also in epidemiological research in general, is the need to protect the privacy of individuals. There is a pressing need to devise effective methods for protecting privacy of individuals about whom data is provided to biosurveillance teams, data from emergency department visits, doctor visits, prescriptions, and possibly biosensors in public places. There is also a pressing need to develop ways to share information between databases of different agencies while protecting privacy. For example, how can we make a simultaneous query to two datasets without compromising information in those datasets? (For example, is individual xx included in both sets?) Issues here include ensuring accuracy and reliability of responses, authentication of queries, policies for access control and authorization.

As an example, consider the situation where different agencies need to collaborate in order to determine security-relevant outcomes based on their disjoint data; however, they wish to do so without requiring any agency to reveal all its data to the others. The ability to collaborate without revealing information could be instrumental in fostering interagency collaboration. Technically speaking, *privacy-preserving computation* enables several parties that have private inputs to compute a joint function of their inputs while ensuring that the computation process itself does not reveal any additional information except for the final output. For example, in the case of computing the intersection of several datasets, no party should learn any information about the other parties' data, except for the fact that the elements that appear in the intersection appear in every data set.

The algorithms community has investigated related problems, such as private information retrieval and secure multiparty computation. Secure multiparty computation (see, e.g., [42, 52, 88]) is a very general and well-studied paradigm that allows two or more parties to evaluate a function of their inputs in such a way that no more is revealed to a party or a set of parties about other parties' inputs than what is implied by their own inputs and outputs, even if some of the parties act maliciously. However, this approach has so far not been generally efficient enough to be used in practice. There has been some amount of work in the databases community on privacy aspects of data access. However, these techniques have been designed for a few specific problems—data mining and auditing boolean values—and as such do not address many of the central problems such as finding common elements, finding large join values, merging, etc. New methods are needed for dealing with the set of important data access problems with multiple data sets and the challenge in performing them in a privacy-preserving manner.

[10]Thanks to Mike Fredman and Matthew Stone for ideas and references.

[11]I would like to thank Giovanni Di Crescenzo, Stuart Haber, S. Muthukrishnan, Rafail Ostrovsky, Benny Pinkas, Tomas Sander, and Rebecca Wright for ideas used in this section.

Realistic problems require the development of efficient methods for comparing data sets when the entries could be in different, often incompatible formats and there could be errors in the data. The problem can be described as follows: Suppose we are given two datasets (or two sets of responses from two different datasets) where some of the entries in each response in one dataset correspond to the same entity (for example, the same name, or the same address) in the second dataset and our task is to match an entry in one dataset to most likely data entries in the second dataset. The difficulty arises since the names could be slightly misspelled or addresses could be in a different format. (Otherwise a trivial sorting procedure of both datasets would work.) Nevertheless, we wish to find, in an efficient way, all entries that appear to have a lot in common. The naive solution takes each entry in the first dataset and compares it to every entry in the second dataset, which leads to quadratic running time. A reasonable objective is to be able to achieve this task with far greater efficiency. Some preliminary results, which use ideas from dimension reduction, clustering and searching strategy, suggest that we can do significantly better than the trivial quadratic solution, even for higher-dimensional problems. To go beyond this one needs to study how the solution could be further extended to maintain privacy.

1.4.2 Social Networks

Diseases are often spread through social contact. Contact information is often key to controlling an epidemic, man-made or otherwise. There is already a considerable literature in epidemiology devoted to social networks (see, e.g., [6, 29, 134, 161, 243, 244]), since diseases are often spread through social contact. A major focus is the spread of AIDS and other sexually transmitted diseases, but there is interest in a wide variety of other diseases as well, and in particular some of these, such as smallpox, that are considered serious bioterrorist threats.

Simple Threshold Models of Spread of Disease through a Network

Methods of DM have had a long history in the study of social networks [59] since those networks have a natural graph structure.

In the simplest discrete model of spread of disease through a network, we look at dynamically changing networks with change taking place at discrete times. Here, vertices (individuals) are infected or noninfected. An individual becomes infected at time $t + 1$ if sufficiently many of his or her neighbors are infected at time t. We call this the *threshold model*.[12] Saturation models in economics are analogous, as are models of spread of opinions through social networks; we return to this analogy below.

There are complications and variants on the simplest model. We can consider infection only with a certain probability. We can think of individuals as having degrees of immunity and infection taking place only if sufficiently many neighbors are infected and degree of immunity is sufficiently low. We can add a category for recovered individuals or levels of infection.

In these models, a variety of research issues remain. What sets have the property that their infection guarantees the spread of the disease to $x\%$ of the vertices? What vertices need to be "vaccinated" to make sure a disease does not spread to more than $x\%$ of the vertices? How do the answers to these questions depend upon the network structure? How do they

[12] The reader should not confuse the term "threshold" as used here with the term widely used in theoretical epidemiology when one speaks of a bifurcation threshold where stability is transferred from a disease-free equilibrium to an endemic equilibrium.

depend upon the choice of threshold? Some of these questions have been studied. See [69] for a survey.

Analogies to Models of Spread of Opinion

An old subject in sociology, psychology, and economics involves dynamic models for how individuals change their opinions over time until, hopefully, some consensus is reached. We think of individuals as vertices of a graph or network, a social network. Edges (directed or undirected) represent influence. The edges might have real numbers or weights on them, representing relative influence. Models of the process range from Markov models [60, 92, 202] to dynamic models on graphs closely related to neural nets [103, 191, 69].

In the simplest models, opinions are $+1$ or -1, each vertex has an opinion at any given time, and opinions change at discrete times. Dreyer [69] studied the *k-threshold model* in which the value ($+1$ or -1) at a vertex changes at discrete times if at least k of its neighbors have the opposite value. He also studied the *monotone* variant of this, where change can only go from value -1 to value $+1$. As noted in the previous subsection, this model is useful in studying the spread of infectious disease: an individual gets infected if at least k of its neighbors are infected, and once infected an individual always remains infected. The "voter model" [74] involves variants and generalizations of this model, e.g., where the change from -1 to $+1$ takes place, depending upon some function f of the number of neighboring vertices that are $+1$. A similar threshold model arises in the study of approval voting, where each voter approves of all candidates whose "rating" lies above his threshold. This model is developed in or hinted at in [34] and [157]. It should be worthwhile to investigate for k-threshold processes probabilistic models where infection takes place only with a certain probability; these are similar to those used for approval voting in [65, 82, 200]. In the voter model, the function f can give the probability of change from -1 to $+1$. It should be worthwhile to investigate relationships between k-threshold models and the models of stochastic token theory developed in [64, 79, 80, 81, 83, 199]. Still more complicated models arise when individuals have not only opinions but degrees of confidence in their opinions. In epidemic terms, confidence can be interpreted as degree of immunity and then we are again in the situation discussed in the previous subsection. One model has opinions change if enough neighbors hold the opposite opinion with sufficiently high confidence. If an opinion changes, the confidence in the new opinion goes down and otherwise it goes up, in discrete increments. Similar ideas underly models of Hoffman [114], Latane [140], Latane and Fink [141], Latane and Nowak [142], and Merrill [158]. Finally, other interesting variants of these approaches would involve adding a "recovered" category, levels of infection, etc., again as noted in the previous subsection.

A variant of the threshold process is the *majority process* in which an individual changes opinion (between $+1$ and -1) at discrete times if a majority of its neighbors have the opposite opinion. While this process does not seem relevant to epidemiology in the case that opinions can fluctuate back and forth between $+1$ and -1, it could be relevant if opinions can only change from -1 to $+1$. Thus, it is worth reviewing the relevance to epidemiology of research on majority processes by Mustafa and Pekec [168], Pekec [185], Peleg [187], Flocchini et al. [89, 90], and Luccio [152]. Variants of majority processes have been widely studied under the names "automata networks" and "synchronous neural networks." Majority processes have been applied to the study of social impact [191, 202, 203] through social networks, as have the synchronous neural net models [141]. Automata networks have been widely studied as models in biology, chemistry, and physics and as models of abstract computation. (See, e.g., [28].)

Analogous Models in Distributed Computing

It is also worth reviewing the relevance to epidemics of threshold and majority processes arising in distributed computing. Here, the goal is to eliminate damage caused by failed processors (vertices) or at least to restrict their influence, and this is accomplished for example by maintaining replicated copies of crucial data and, when a fault occurs, letting a processor change "state" if a majority of its neighbors are in a different state. (See [11, 33, 77, 137, 146, 186].) Other applications of similar ideas in distributed computing arise in distributed database management, quorum systems, and fault-local mending.

1.4.3 Evolution[13]

The problem of reconstructing phylogenetic trees for viruses and other pathogens is a central one in mathematical modeling of infectious disease systems and has the potential to help us understand and predict the origin, evolution, and likely future development of such pathogens. Thus, models of evolution might shed light on new strains of infectious agents used by bioterrorists. Phylogeny is now a central tool for studies into the origin and diversity of viruses such as HIV (see, e.g., [194, 196]) and dengue fever virus [195]. These and other investigations have provided new insights, such as identifying the possible pattern of transfer of HIV-type viruses between primate species. Phylogenetic techniques have also proved useful in mapping the evolution of different strains of the human influenza A virus [40, 41], with the goal of predicting which strain is most likely to cause future epidemics, with applications to vaccine development.

Phylogenetic analysis might help in identification of the source of an infectious agent. For example, as noted earlier, DNA sequencing is routinely used to characterize pathogens' strain phylogenies (see, e.g., [167]), a critical step in the identification of potential sources of supply of an agent that might have been used in a bioterrorist attack. Phylogenetic models can help to identify the subtype of an organism and narrow the search for its origin. Moreover, because an organism has been cultured between times it is passed from lab to lab, phylogenetic analysis can help to identify which lab an organism came from. A major research challenge is to build on phylogenetic methods developed by computational biologists to explore ways in which such methods can be applied and developed to shed new light on the evolution of viruses and other pathogens. The phylogeny problem has benefited greatly from the algorithmic approaches of DM and TCS. Of note are information-theoretic bounds on tree reconstruction methods [78, 223, 224], optimal tree refinement methods [32], the disk-covering method [120], and maximum parsimony heuristics, nearest-neighbor-joining method, and hybrid methods combining these two approaches [119]. Additional topics of note are ways to compare two phylogenetic trees using distance-based approaches, to find the linear-cost subtree-transfer distance between two trees [54] or to find the consensus phylogeny among those obtained using alternative reconstruction methods (see, e.g., [122, 139, 246, 247]).

Many of the phylogenetic techniques in use today were originally developed to investigate more traditional and well-behaved evolutionary problems, where historical relationships are typically represented by a binary tree with a small number of species appearing as the leaves (tip vertices). In epidemiology the picture is more complex and this observation underlies the future research challenges. Even if there is a single underlying tree, it may typically have thousands of vertices, and many of these may be of high degree. Furthermore,

[13]Thanks to Eddie Holmes and Mike Steel for many of the ideas in this section.

data may be available not just for the species at the leaves of the tree, but for species distributed at vertices throughout the tree, particularly when the evolution of a virus is studied by serial sampling in patients. This is true for retroviruses which have a very high substitution rate, and whose molecular evolution may be up to 106 times more rapid than eukaryotic or prokaryotic genes [70, 209, 215]. New methods for dealing with these complications should be investigated.

To complicate the picture further, it may well be more appropriate to represent the evolution of a virus by a collection of trees, or by a digraph (or network) to recognize the "quasi species" nature of viruses, such as in the application of split decomposition by Dopazo, Dress, and von Haeseler [67]. Relating population genetics considerations (currently handled by the "coalescent" model) to phylogeny considerations is also potentially useful [208]. However, even here, theory has yet to be developed. For instance, the fact that retroviral evolution occurs within a host means that viral sequences sampled from different hosts must take account of the different dynamics of between-host transmission histories and within-host viral genealogies [207]. This has consequences for the inference of epidemiological parameters based on viral sequences obtained from several hosts, and these should be investigated. Finally, if one wishes to test particular epidemiological hypotheses, it would be helpful to have techniques that avoid having to fix attention on one particular tree. This suggests devising fast methods that would average the quantities of interest over all likely trees, weighted by how well they describe the data—a challenge for modern computational tools.

1.4.4 Decision Making and Policy Analysis

DM and TCS have a close historical connection with mathematical modeling for decision making and policy making. In a decision making context, mathematical models can help us to understand fundamental processes, compare alternative policies and interventions, provide a guide for scenario development, guide risk assessment, aid forensic analysis, and predict future trends. Models can be used to determine policy options and to estimate the consequences of various actions. Ultimately, models are a tool used by decision makers and can be useful for training and educating decision makers in advance of bioterrorist attacks.

Consensus

Many decisions have to be made by groups. DM and TCS are fundamental to the theory of group decision making and consensus. Consensus theory [10, 19, 162, 211, 214] is based on fundamental ideas in the theory of "voting" and "social choice." The key problem is to combine expert judgments (e.g., choices or rankings of alternatives) to make policy.

Consensus methods are coming to be widely used in biology in the field beginning to be called *bioconsensus*. Typically, several alternatives (such as alternative phylogenetic trees or alternative molecular sequences) are produced using different methods or models, and then one needs to find a consensus solution. Day and McMorris [56] surveyed the use of consensus methods in molecular biology, and Day [55] and Day and McMorris [57] gave many references; there are literally hundreds of papers in the field of "alignment and consensus." Kannan [124] surveyed consensus methods for phylogenetic tree reconstruction. Also of interest are the books by Day and McMorris [58] and Janowitz et al. [122]. Typical bioconsensus problems involve finding a common pattern in a library of molecular sequences [96, 241, 242, 159] and finding a consensus phylogeny given alternative phylogenies.

A special challenge for bioconsensus research in the bioterrorism arena, and more generally in epidemiology, is the fact that instead of many "decision makers" (voters) and few "candidates" as is traditional in consensus theory, we often have few decision makers and many candidates, where the candidates are many different parameters to modify. (For some relevant work, see [76].)

There is a developing algorithmic point of view in consensus theory (see, for example, [20, 21, 22]): The goal is to find fast algorithms for finding the consensus policy. Surprisingly enough, there are cases where finding the "best alternative" according to some rule is computationally intractable. The algorithmic point of view will be important in the bioterrorism context, where finding consensus alternatives quickly and efficiently could be significant. Much work is needed to develop good algorithms, efficient approximations, and tractable heuristic methods. (See [76, 154] for some work along these lines.)

Combining partial information to reach decisions involves consensus methods. Yet most known results about computational complexity of consensus functions require assumptions about perfect information of individual preferences, and few similar results are known when there is only partial, distributed, or unreliable information or limits to computation. (For some relevant results, however, see [18, 27, 181].)

Decision Science and Operations Research

Decision science is an old subject fundamentally based on mathematical methods. Modern decision science formalizes utilities and costs and benefits as well as uncertainty and risk. These are ideas that are very critical in decision making about policies to counter bioterrorist attacks. (See, for example, [126].) DM and TCS are very important tools in formalizing and solving optimization problems of the type that decision scientists are concerned with: maximizing utility, minimizing pain, etc. In a bioterrorism context, minimizing number of casualties and minimizing economic cost of an attack are considerations. Yet some of the modern ideas of decision science are not well known among epidemiologists and public health professionals. For example, mathematical and computational methods dealing with partially ordered algebraic systems form the foundations of the modern theory of measurement [133, 151, 201, 228]. They can be used to determine what conclusions using scales of measurement are "meaningful" [205, 206]. Measurement theory (a term which has a different connotation in epidemiology) does not seem to be known to practicing epidemiologists. Of particular significance for epidemiology is the following type of question: What kinds of statistical tests are legitimately applied to data where only order matters? (See [105, 205].) This issue is somewhat recognized by epidemiologists, but its order-theoretic subtleties are usually not. A challenge for research is to find ways to modify existing methods of measurement theory to be applicable to epidemiological problems. For the most part, however, challenges for research in decision science as applied to bioterrorism defense are not so much for the development of new methods but for the application of existing tools, appropriately modified, to new applications.

More generally than decision making, in developing appropriate tools for defense against bioterrorist attacks, we will need to solve the types of optimization problems that are of importance in the field of operations research. Indeed, operations research (OR) was developed during World War II in the context of finding optimal solutions to a wide variety of planning and logistics problems in conflict situations. Relevant OR tools include queueing theory, which might be used to design optimal vaccination strategies, sizes of stockpiles of vaccines, allocation of medications, and analysis of bottlenecks in treatment facilities. Inventory theory can help with managing and controlling and distributing supplies of medications and vaccines. Location theory, another important field in OR, could be used to locate vaccination facilities. Scheduling theory is relevant to the planning of a variety of interventions and assignment theory to quarantine planning and assignment of manpower to treatment facilities. Operations researchers are familiar with the development of stochastic versions of deterministic models and the inclusion of stochastic elements will be important in planning for bioterrorism defense. As with decision science, challenges for research are not so much for the development of new methods but for the application of existing tools to new applications.

Game Theory

Game theory has been widely used in military decision making applications and, as such, is directly relevant to the bioterrorism context. Game theory has been used to model conflicts at various levels. Most widely known in game theory are the two-player games, often presented in matrix form. Here, we consider the rows to correspond to strategies available to one player (the attacker) and the columns to correspond to strategies available to a second player (the defender). The entry in a given row and column corresponds to the outcome if the players use the corresponding strategies. The problem in the bioterrorist context is that many of the entries in the matrix are difficult to calculate. Moreover, we might wish to figure out a probability distribution over strategies used by our opponent, and how we do this is difficult. Finding an optimal strategy as the defender, a key problem of two-person game theory, might be very difficult in this context, and research is needed both in determining the entries of appropriate matrices and in developing new methods for dealing with two-person games under such considerable uncertainty or wide ranges of possible values for the matrix entries.

A less well known branch of game theory involves multiperson games. The theory of multiperson games has a wide variety of applications in situations of conflict and cooperation and bargaining [93, 169]. Of particular interest is the use of multiperson game theory to determine appropriate allocation of scarce resources to different players or components of a comprehensive policy. Game-theoretic solutions such as the Shapley value and the nucleolus have long been used to allocate costs to different users in shared projects. Examples of such applications in the past include allocating runway fees to different users of an airport, highway fees to different size trucks, costs to different universities sharing library facilities, fair allocation of telephone calling charges among users, and distribution of costs in multicast routing/multicast packet transmission. (See [86, 149, 150, 163, 177, 216, 218, 219, 226, 231, 249].) Application of this theory to bioterrorism defense will require new concepts and ideas.

Game-theoretic and decision-theoretic methods that arise in the study of congestion problems on the Internet involve models of games in which players share a common resource with negative congestion effects. Such situations arise when users send information over a network, share a database, access a web server, or dial in to a service provider. Challenging mathematical issues arise in designing scheduling paradigms so that even when users act selfishly, the network achieves its performance goals (see [217]). In a bioterrorism context, one might consider analogous ideas when congestion (such as at vaccine distribution sites) is an issue. Although this is somewhat speculative, it could be worthwhile thinking about game-theoretic methods in connection with vaccine distribution protocols.

Modern game theory has an algorithmic point of view not present in traditional game-theoretical analysis. There is increasingly an emphasis on finding efficient procedures for computing the winner in a game or the appropriate resource allocation corresponding to different game-theoretic solution concepts or simply computing values, optimal strategies, equilibria, and other solution concepts. These concepts can be easy, hard, or even impossible to compute (see [106, 131, 132]). Work is needed to develop this algorithmic point of view in the bioterrorism context.

Cryptographic methods are needed to make sure that in the course of a game, a player's views, utilities, preferences, and strategies are kept private. See [63, 145] for examples of work at the interface between cryptography and game theory. This is especially relevant in the conflict situations arising in bioterrorism defense when we wish to keep our defensive strategies private yet share them with a large number of individuals and agencies.

1.4.5 Order Theory[14]

Many practical epidemiological problems involve the comparison of one or more quantities. Most often the quantities are rates or proportions leading to a measure of effect or association, but they may also involve distances, exposure categories, job titles, etc. Often the actual values in question are not important; only whether one value is smaller than or larger than a second, i.e., their order, is. It could be worthwhile to study how fundamental order-theoretic concepts of TCS and DM such as semiorders, interval orders, general partial orders, and lattices [87, 232] can be used to improve the results of epidemiological investigations. This topic is a bit more esoteric and may not be as immediately useful as some of the other topics discussed in this paper, but it is worth pursuing. It builds upon a large literature in DM and TCS dealing with order relations, computing them, approximating them, visualizing them, and assigning measures to them. However, it does not build upon a large body of work connecting these ideas to epidemiology, mostly upon the view of several active epidemiologists that these ideas are relevant.

For instance, one can give epidemiological concepts a careful definition in the language of partial orders and explore the use of visualization of order-theoretic concepts in epidemiologic studies. The latter will involve issues such as how best to visualize a poset through clever presentation of its Hasse diagram, an issue of great interest in the field of TCS known as graph drawing. One application of these ideas will arise in the problem of determining cutoffs or boundaries so as to categorize numerical attributes (as is done, for instance, in constructing age categories or exposure categories in epidemiology). This can be thought of as the problem of assigning a real number to each attribute (e.g., exposure measurement or age) and then considering the set of all subjects whose assigned numbers exceed a threshold, a common construction in the theory of partially ordered sets.

As another example, point lattices may be regarded as a type of order-theoretic lattice. The point lattice construction has found uses in epidemiology through visualizing the relationships of all possible contingency tables to various statistics, effect measures, and cut-off choices (see, e.g., [180]) and has also been used in statistics (see [170]). Challenges arising in extending these ideas include generalizing the concepts to higher-dimensional tables, where there are additional attributes (ordinal, numerical, or nominal) besides case status and exposure, and applying lattice-theoretic approaches to measurement error. The point lattices formed by 2×2 contingency tables can be represented as n-element strings from a two-letter alphabet $\{x, y\}$. Measurement errors can be thought of as being caused by a transposition from a substring xy to a substring yx. Similar transpositions have been studied from a more general viewpoint by lattice theorists (see [25]) as a special case of a Newman commutativity lattice. Many references can be found in [25], which also mentions connections with weak Bruhat orders of Coxeter groups. It should be worthwhile to examine how higher-dimensional contingency tables relate to what are called multinomial lattices in [25]

[14]Thanks to Mel Janowitz and David Ozonoff for many of the ideas in this section.

and study how combinatorial aspects of the Bruhat orders relate to probabilistic questions in epidemiology. Among other things, one can hope that these considerations will give us guidance on how to decide when observed data tables can be explained by chance alone.

As noted in section 1.4.4, it will also be useful to consider the issue of what kinds of statistical tests are legitimately applied to data where only order matters [155, 205]; this issue is somewhat recognized by epidemiologists, but its order-theoretic subtleties are usually not. Mathematical and computational methods dealing with ordered algebraic systems form the foundations of the modern theory of measurement [133, 151, 204, 228] and can be used to analyze this and important related issues such as what conclusions using scales of measurement are "meaningful" [205, 206]. It should be useful to analyze epidemiological studies from a measurement theory point of view.

1.4.6 Combinatorial Group Testing[15]

Natural or human-induced epidemics might require us to test many samples from large populations at once. One important modern topic in DM and TCS that arose in epidemiology is the theory of group testing, which arose in connection with testing millions of World War II military draftees for syphilis [68]. The idea is to avoid testing each individual and instead to divide them into groups and determine if some individual in the group is positive for the disease, updating the process by further subdividing groups that test positive. The subject is very relevant to schemes for large scale blood testing for viruses such as HIV [99, 233] and has clear relevance to massive testing that might be called for in the event of a bioterrorist attack. The modern theory of group testing [71] is heavily influenced by combinatorial methods, in particular by the methods of combinatorial designs and coding theory, and many modern algorithmic methods of group testing, developed by theoretical computer scientists, are not yet widely known or used in epidemiology.

Group testing also arises in connection with the mapping of genomes (see, e.g., [15, 16, 39, 85, 172]). Here, we have a long list of molecular sequences, form a library of subsequences (clones), and test whether or not a particular sequence (a probe) appears in the library by testing to see which clones it appears in. Because clone libraries can be huge, we do this by pooling the clones into groups.

While the theory of combinatorial group testing has been extensively studied by a small group of researchers, it is not widely known. Moreover, to be readily applicable in an epidemiological context, a variety of new developments are needed. For instance, the "test kit" may have a size constraint, limiting the number of items that can be tested at once. Sometimes, an item can undergo only a limited number of tests. In some applications, we might have certain items as neighbors or lined up in a linear fashion, and then only a set of consecutive items can be tested together. Most importantly, tests have errors and so the theory of group testing under errors needs further development.

1.5 Concluding Comments

So would DM and TCS help with a deliberate outbreak of anthrax? What about a deliberate release of smallpox? Similar approaches, using mathematical models based in DM and TCS, have proven useful in many other fields to make policy, plan operations, analyze risk, compare interventions, and identify the cause of observed events. Why shouldn't these approaches work in the defense against bioterrorism?

[15]Thanks to Frank Hwang for some of the ideas in this section.

Acknowledgments

This research was supported by NSF grants EIA-0205116 and EIA-0209761 to Rutgers University. Parts of this paper were borrowed from grant proposals, and many people contributed pieces to the proposals. In particular, I would like to thank Lora Billings, Sally Blower, Adam Buchsbaum, Mike Fredman, Steve Greenfield, Stuart Haber, Eddie Holmes, Donald Hoover, Frank Hwang, Mel Janowitz, Matt Keeling, Denise Kirschner, Brenda Latka, Simon Levin, Marc Lipsitch, David Madigan, Denis Mollison, Megan Murray, Ilya Muchnik, S. Muthukrishnan, Rafail Ostrovsky, David Ozonoff, Benny Pinkas, Tomas Sander, Mike Steel, Matthew Stone, Ira Schwartz, Dan Wartenberg, Tom Webster, and Rebecca Wright for ideas used in this paper. My apologies to those whose names I may have inadvertently omitted. Some parts of this paper owe a great deal to the DIMACS Working Group Meeting on Mathematical Sciences Methods for the Study of Deliberate Releases of Biological Agents and their Consequences (http://dimacs.rutgers.edu/Workshops/WGDeliberate/), and in particular the report that came out of that meeting: [45]. I would like to thank all of the participants in that meeting for their insights, in particular the discussion group on challenges for computer science. I would like to acknowledge my debt to the organizers and writers of several sections of that report: Irene Eckstrand, John Glasser, Mac Hyman, Ed Kaplan, Jim Koopman, Simon Levin, Ellis McKenzie, Marcello Pagano, and especially my coauthor, Carlos Castillo-Chavez.

Bibliography

[1] J. M. Abello and J. S. Vitter, *External Memory Algorithms*, DIMACS Ser. Discrete Math. Theoret. Comput. Sci. 50, AMS, Providence, RI, 1999.

[2] R. Albert, H. Jeong, and A. L. Barabasi, *Diameter of the World Wide Web*, Nature, 401 (1999), pp. 130–131.

[3] R. M. Anderson, *Transmission dynamics and control of infectious disease agents*, in Population Biology of Infectious Diseases, R. M. Anderson and R. M. May, eds., Springer-Verlag, Berlin, 1982, pp. 149–176.

[4] R. W. Anderson, M. S. Ascher, and H. W. Sheppard,, *Direct HIV cytopathicity cannot account for CD 4 decline in AIDS in the presence of homeostasis: A worst-case dynamical analysis*, J. AIDS and Human Retrov., 17 (1998), pp. 245–252.

[5] R. M. Anderson, C. A. Donnelly, N. M. Ferguson, M. E. Woolhouse, C. J. Watt, H. J. Udy, S. MaWhinney, S. P. Dunstan, T. R. Southwood, J. W. Wilesmith, J. B. Ryan, L. J. Hoinville, J. E. Hillerton, A. R. Austin, and G. A. Wells, *Transmission dynamics and epidemiology of BSE in British cattle*, Nature, 382 (1996), pp. 779–788.

[6] R. M. Anderson and R. M. May, *Infectious Diseases of Humans*, Oxford University Press, Oxford, UK, 1991.

[7] R. Antia, J. C. Koella, and P. Perrott, *Models of the within-host dynamics of persistent mycobacterial infections*, Proc. Roy. Soc. London, 263 (1996), pp. 257–263.

[8] R. Antia, B. Levin, and R. B. May, *Within-host population dynamics and the evolution and maintenance of microparasite virulence*, Amer. Naturalist, 144 (1994), pp. 457–472.

[9] R. Arratia and E. S. Lander, *The distribution of clusters in random graphs*, Adv. in Appl. Math., 11 (1990), pp. 36–48.

[10] K. J. Arrow, *Social Choice and Individual Values*, Wiley, New York, 1951.

[11] H. Attiya and J. Welch, *Distributed Computing: Fundamentals, Simulations and Advanced Topics*, McGraw–Hill, London, 1998.

[12] N. T. J. Bailey, *The Mathematical Theory of Infectious Diseases*, 2nd ed., Hafner, New York, 1975.

[13] C. Bajaj, V. Pascucci, and D. Schikore, *The contour spectrum*, in Proceedings of the IEEE Visualization '97 Conference, ACM Press, New York, 1997, pp. 167–173.

[14] J. S. Balasubramaniyan, J. O. Garcia-Fernandez, D. Isacoff, E. H. Spafford, and D. Zamboni, *An architecture for intrusion detection using autonomous agents*, in Proceedings of the 14th IEEE Conference on Computer Security Applications, 1998.

[15] D. J. Balding and D. C. Torney, *Optimal pooling designs with error detection*, J. Combin. Theory Ser. A, 74 (1996), pp. 131–140.

[16] D. J. Balding and D. C. Torney, *The design of pooling experiments for screening a clone map*, Fungal Genetics and Biology, 21 (1997), pp. 302–307.

[17] A. Baritchi, D. J. Cook, and L. B. Holder, *Discovering structural patterns in telecommunications data*, in Proceedings of the 13th Annual Florida AI Research Symposium, 2000, pp. 82–85.

[18] A. Bar-Noy, X. Deng, J. A. Garay, and T. Kameda, *Optimal amortized distributed consensus*, Inform. and Comput., 120 (1995), pp. 93–100.

[19] J. P. Barthelemy, *Social welfare and aggregation procedures: Combinatorial and algorithmic aspects*, in Applications of Combinatorics and Graph Theory to the Biological and Social Sciences, IMA Vol. Math. Appl. 17, F. S. Roberts, ed., Springer-Verlag, New York, 1989, pp. 39–73.

[20] J. J. Bartholdi, C. A. Tovey, and M. A. Trick, *The computational difficulty of manipulating an election*, Soc. Choice Welf., 6 (1989), pp. 227–241.

[21] J. J. Bartholdi, C. A. Tovey, and M. A. Trick, *Voting schemes for which it can be difficult to tell who won the election*, Soc. Choice Welf., 6 (1989), pp. 157–165.

[22] J. J. Bartholdi, C. A. Tovey, and M. A. Trick, *How hard is it to control an election*, Math. Comput. Modelling, 16 (1992), pp. 27–40.

[23] N. H. Bean and M. Martin, *Implementing a network for electronic surveillance reporting from public health reference laboratories: An international perspective*, Emerging Infectious Diseases, 7 (2001), pp. 773–779.

[24] N. H. BEAN, M. MARTIN, AND H. BRADFORD, *PHLIS: An electronic system for reporting public health data from remote sites*, Amer. J. Public Health, 82 (1992), pp. 1273–1276.

[25] M. K. BENNETT AND G. BIRKHOFF, *Two families of Newman lattices*, Algebra Universalis, 32 (1994), pp. 115–144.

[26] B. BERGER AND M. SINGH, *An iterative method for improved protein structural motif recogntion*, J. Comput. Biol., 4 (1997), pp. 261–273.

[27] P. BERMAN AND J. A. GARAY, *Cloture votes: $n/4$-resilient distributed consensus in $t+1$ rounds*, Math. Systems Theory, 26 (1993), pp. 3–20.

[28] E. BIENENSTOCK, F. FOGELMAN-SOULIE, AND G. WEISBUCH, EDS., *Disordered Systems and Biological Organization*, NATO ASI Series F: Computing and System Sciences, Springer-Verlag, Berlin, 1986.

[29] P. BLANCHARD, G. F. BOLZ, AND T. KRUGER, *Mathematical modelling on random graphs of the spread of sexually transmitted diseases with emphasis on HIV infection*, in Dynamics and Stochastic Processes, R. Lima, L. Streit, and R. Vilela Mendes, eds., Springer-Verlag, Berlin, 1990, pp. 55–75.

[30] S. M. BLOWER, A. R. MCLEAN, T. C. PORCO, P. M. SMALL, P. C. HOPEWELL, M. A. SANCHEZ, AND A. R. MOSS, *The intrinsic transmission dynamics of tuberculosis epidemics*, Nature Med., 1 (1995), pp. 815–821.

[31] S. M. BLOWER, P. M. SMALL, AND P. C. HOPEWELL, *Control strategies for tuberculosis epidemics: New models for old problems*, Science, 273 (1996), pp. 497–500.

[32] M. BONET, M. STEEL, T. WARNOW, AND S. YOOSEPH, *Faster algorithms for solving parsimony and compatibility*, J. Comput. Biol., 5 (1998), pp. 409–422.

[33] G. BRACHA, *An $O(\log n)$ expected rounds randomized Byzantine generals protocol*, J. ACM, 34 (1987), pp. 910–920.

[34] S. J. BRAMS AND P. C. FISHBURN, *Approval Voting*, Birkhäuser, Boston, 1983.

[35] F. BRAUER AND C. CASTILLO-CHAVEZ, *Mathematical Models in Population Biology and Epidemiology*, Springer-Verlag, New York, 2001.

[36] S. E. BROSSETTE, W. T. JONES, A. P. SPRAGUE, J. M. HARDIN, AND S. A. MOSER, *DMSS: A knowledge discovery system for epidemiologic surveillance*, Internat. J. Knowledge Discovery and Data Mining, (1999).

[37] S. E. BROSSETTE, A. P. SPRAGUE, J. M. HARDIN, K. B. WAITES, W. T. JONES, AND S. A. MOSER, *Association rules and data mining in hospital infection control and public health surveillance*, J. Amer. Med. Informatics Assoc., 5 (1998), pp. 373–381.

[38] J. N. BRUNKEN, R. MAGER, AND R. A. PUTZKE, *NEMOS—the network management system for the AT&T long distance network*, in Proceedings of the 1989 IEEE International Conference on Communications, Vol. 3, 1989, pp. 1193–1197.

[39] W. J. BRUNO, D. J. BALDING, E. H. KNILL, D. BRUCE, C. WHITTAKER, N. DOGGETT, R. STALLINGS, AND D. C. TORNEY, *Design of efficient pooling experiments*, Genomics, 26 (1995), pp. 21–30.

[40] R. M. BUSH, C. A. BENDER, K. SUBBARAO, N. J. COX, AND W. M. FITCH, *Predicting the evolution of human influenza A*, Science, 286 (1999), pp. 1921–1925.

[41] R. M. BUSH, W. M. FITCH, C. A. BENDER, AND N. J. COX, *Positive selection on the $H3$ hemagglutinin gene of human influenza virus A*, Molecular Biol. Evol., 16 (1999), pp. 1457–1465.

[42] R. CANETTI AND R. GENNARO, *Incoercible multiparty computation*, in 37th IEEE Symposium on Foundations of Computer Science (FOCS '96), 1996.

[43] T. E. CARPENTER, *Methods to investigate spatial and temporal clustering in veterinary epidemiology*, Preventive Veterinary Med., 48 (2001), pp. 303–320.

[44] C. CASTILLO-CHAVEZ AND Z. FENG, *To treat or not to treat: The case of tuberculosis*, J. Math. Biol., 35 (1997), pp. 629–659.

[45] C. CASTILLO-CHAVEZ AND F. S. ROBERTS, *Report on DIMACS Working Group Meeting: Mathematical Sciences Methods for the Study of Deliberate Releases of Biological Agents and Their Consequences*, DIMACS Center, Rutgers University, Piscataway, NJ, 2002; available online from http://dimacs.rutgers.edu/Workshops/WGDeliberate/FinalReport5-20-02.doc.

[46] Y. C. CHAGNON, I. B. BORECKI, L. PRUSSE, S. ROY, M. LACAILLE, M. CHAGNON, M. A. HO-KIM, T. RICE, M. A. PROVINCE, D. C. RAO, AND C. BOUCHARD, *Genome-wide search for genes related to the fat-free body mass in the Quebec family study*, Metabolism, 49 (2000), pp. 203–207.

[47] P. CHAN, W. FAN, A. PRODROMIDIS, AND S. STOLFO, *Distributed data mining in credit card fraud detection*, IEEE Intell. Systems, 14 (1999), pp. 67–74.

[48] K. CHURCH, *Massive data sets and graph algorithms in telecommunications systems*,in Session on Mathematical, Statistical, and Algorithmic Problems of Very Large Data Sets, American Mathematical Society Meeting, San Diego, CA, 1997.

[49] C. CORTES, K. FISHER, D. PREGIBON, A. ROGERS, AND F. SMITH, *Hancock: A language for extracting signatures from data streams*, in KDD 2000, 2000.

[50] L. H. COX AND W. W. PIEGORSCH, *Combining environmental information* I: *Environmental monitoring, measurement and assessment*, Environmetrics, 7 (1996), pp. 299–308.

[51] L. H. COX AND W. W. PIEGORSCH, *Combining environmental information* II: *Environmental epidemiology and toxicology*, Environmetrics, 7 (1996), pp. 309–324.

[52] R. CRAMER, I. DAMGÅRD, AND U. MAURER, *Efficient general secure multi-party computation from any linear secret-sharing scheme*, in Proceedings of EUROCRYPT '00, Brugge, Belgium, Lecture Notes in Comput. Sci. 1807, Springer-Verlag, Berlin, 2000, pp. 316–334.

[53] A. CURTIS, *Using a spatial filter and a geographic information system to improve rabies surveillance data*, Emerging Infectious Diseases, 5 (1999), pp. 603–606.

[54] B. DASGUPTA, X. HE, T. JIANG, M. LI, AND J. TROMP, *On the linear-cost subtree-transfer distance between phylogenetic trees*, Algorithmica, 25 (1999), pp. 176–195.

[55] W. H. E. DAY, *The Sequence Analysis and Comparison Bibliography*, 1995; available online from http://edfu.lis.uiuc.edu/~class/sequence/.

[56] W. H. E. DAY AND F. R. MCMORRIS, *Critical comparison of consensus methods for molecular sequences*, Nucleic Acids Res., 20 (1992), pp. 1093–1099.

[57] W. H. E. DAY AND F. R. MCMORRIS, *Discovering consensus molecular sequences*, in Proceedings of 16 Jahrestagung, Gesellschaft fuer Klassifikation e.V., Dortmund, Germany, 1–3 April, 1992, Stud. Classification Data Anal. Knowledge Organization, Springer-Verlag, Berlin, 1992.

[58] W. H. E. DAY AND F. R. MCMORRIS, *Axiomatic Consensus Theory in Group Choice and Biomathematics*, SIAM, Philadelphia, PA, 2003.

[59] A. DEGENNE AND M. FORSE, *Introducing Social Networks*, Sage Publications, London, 1999.

[60] M. H. DEGROOT, *Reaching a consensus*, J. Amer. Statist. Assoc., 69 (1974), pp. 167–182.

[61] V. DHAR, *Data mining in finance: Using counterfactuals to generate knowledge from organizational information systems*, Inform. Systems, 23 (1998), pp. 423–437.

[62] O. DIEKMANN AND J. A. P. HEESTERBEEK, *Mathematical Epidemiology of Infectious Diseases*, Wiley, New York, 2000.

[63] Y. DODIS, S. HALEVI, AND T. RABIN, *A cryptographic solution to a game theoretic problem*, in Advances in Cryptology—Crypto2000 Proceedings, Lecture Notes in Comput. Sci. 80, Springer-Verlag, Berlin, 2000, pp. 112–130.

[64] J.-P. DOIGNON AND J.-C. FALMAGNE, *Well graded families of relations*, Discrete Math., 173 (1997), pp. 35–44.

[65] J.-P. DOIGNON AND M. REGENWETTER, *An approval-voting polytope for linear orders*, J. Math. Psych., 41 (1997), pp. 171–188.

[66] F. DOMINICI, G. PARMIGIANI, R. WOLPERT, AND V. HASSELBLAD, *Meta-analysis of migraine headache treatments: Combining information from heterogeneous designs*, J. Amer. Statist. Assoc., 94 (2000), pp. 16–28.

[67] J. DOPAZO, A. DRESS, AND A. VON HAESELER, *Split decomposition: A technique to analyse viral evolution*, Proc. Natl. Acad. Sci. USA, 90 (1993), pp. 10320–10324.

[68] R. DORFMAN, *The detection of defective members of a large population*, Ann. Math. Statist., 14 (1943), pp. 436–440.

[69] P. A. DREYER, *Applications and Variations of Domination in Graphs*, Ph.D. thesis, Department of Mathematics, Rutgers University, New Brunswick, NJ, 2000.

[70] A. DRUMMOND AND A. G. RODRIGO, *Reconstructing genealogies of serial samples under the assumption of a molecular clock using serial-sample UPGMA (sUPGMA)*, Molecular Biol. and Evolution, 17 (2000), pp. 1807–1815.

[71] D.-Z. DU AND F. K. HWANG, *Combinatorial Group Testing and Its Applications*, 2nd ed., World Scientific, Singapore, 2000.

[72] O. Du Merle, P. Hansen, B. Jaumard, and M. Mladenovic, *An interior point algorithm for minimum sum of squares clustering*, SIAM J. Sci. Comput., 21 (2000), pp. 1485–1505.

[73] W. DuMouchel, *Bayesian data mining in large frequency tables, with an application to the FDA spontaneous reporting system, with discussion*, Amer. Statist., 53 (1999), pp. 177–202.

[74] R. Durrett, *Ten lectures on particle systems*, in Lectures on Probability Theory, École d'Été de St. Flour XXIII, P. Bernard, ed., Springer-Verlag, New York, Berlin, 1993, pp. 97–201.

[75] R. Durrett, *Stochastic spatial models*, SIAM Rev., 41 (1999), pp. 677–718.

[76] C. Dwork, R. Kumar, M. Naor, and D. Sivakumar, *Rank Aggregation, Spam Resistance, and Social Choice*, preprint, IBM Almaden, San Jose, CA, 2000.

[77] C. Dwork, D. Peleg, N. Pippenger, and E. Upfal, *Fault tolerance in networks of bounded degree*, SIAM J. Comput., 17 (1988), pp. 975–988.

[78] P. L. Erdos, M. A. Steel, L. A. Székely, and T. J. Warnow, *A few logs suffice to build (almost) all trees* I, Random Structures Algorithms, 14 (1999), pp. 153–184.

[79] J. Cl. Falmagne, *A stochastic theory for the emergence and the evolution of preference relations*, Math. Social Sci., 31 (1996), pp. 63–84.

[80] J. Cl. Falmagne, *Stochastic token theory*, J. Math. Psych., 41 (1997), pp. 152–159.

[81] J. C. Falmagne and J.-P. Doignon, *Stochastic evolution of rationality*, Theory and Decision, 43 (1997), pp. 107–138.

[82] J. C. Falmagne and M. Regenwetter, *Random utility models for approval voting*, J. Math. Psych., 40 (1996), pp. 152–159.

[83] J. C. Falmagne, M. Regenwetter, and B. Grofman, *A stochastic model for the evolution of preferences*, in Choice, Decision and Measurement: Essays in the Honor of R. Duncan Luce, A. A. J. Marley, ed., Erlbaum, Mahwah, NJ, 1997.

[84] M. Faloutsos, P. Faloutsos, and C. Faloutsos, *On power-law relationships of the internet topology*, in Proceedings of SIGCOMM '99, 1999, pp. 251–262.

[85] M. Farach, S. Kannan, E. Knill, and S. Muthukrishnan, *Group testing problems with sequences in experimental molecular biology*, in Proceedings of of Sequences '97 (Conference on Compression and Complexity of Sequences, Positano, Salerno, Italy), 1997.

[86] J. Feigenbaum, C. H. Papadimitriou, and S. Shenker, *Sharing the cost of multicast transmissions*, J. Comput. System Sci., to appear; preliminary version in Proceedings of the 32nd ACM Symposium on Theory of Computing, 2000.

[87] P. C. Fishburn, *Interval Orders and Interval Graphs*, Wiley, New York, 1985.

[88] N. Fitzi, J. A. Garay, U. Maurer, and R. Ostrovsky, *Minimal complete primitives for secure multi-party computation*, in Advances in Cryptology—CRYPTO '01, Lecture Notes in Comput. Sci. 2139, J. Kilian, ed., Springer-Verlag, Berlin, 2001, pp. 80–100.

[89] P. FLOCCHINI, E. LODI, F. LUCCIO, L. PAGLI, AND N. SANTORO, *Irreversible dynamos in tori*, in Proceedings of Europar, 1998, pp. 554–562.

[90] P. FLOCCHINI, E. LODI, F. LUCCIO, L. PAGLI, AND N. SANTORO, *Monotone dynamos in tori*, in Proceedings of the 6th Colloquium on Structural Information and Communication Complexity, 1999.

[91] G. A. FORGIONNE, A. GANGOPADHYAY, AND M. ADYA, *Cancer surveillance using data warehousing, data mining, and decision support systems*, Topics Health Inform. Management, 21 (2000), pp. 21–34.

[92] J. R. P. FRENCH, JR., *A formal theory of social power*, Psych. Rev., 63 (1956), pp. 181–194.

[93] D. FUDENBERG AND J. TIROLE, *Game Theory*, MIT Press, Cambridge, MA, 1991.

[94] H. GALHARDAS, D. FLORESCU, D. SHASHA, AND E. SIMON, *An extensible framework for data cleaning*, in Proceedings of the 16th International Conference on Data Engineering (ICDE), 2000, p. 312.

[95] H. GALHARDAS, D. FLORESCU, D. SHASHA, AND E. SIMON, *AJAX: An extensible data cleaning tool*, in SIGMOID Conference, 2000, p. 590.

[96] D. J. GALAS, M. EGGERT, AND M. S. WATERMAN, *Rigorous pattern recognition methods for DNA sequences: Analysis of promoter sequences from E. coli*, J. Molecular Biol., 186 (1985), pp. 117–128.

[97] V. GANTI, J. GEHRKE, AND R. RAMAKRISHNAN, *Mining very large databases*, IEEE Comput. J., 32 (1999), pp. 38–45.

[98] G. P. GARNETT AND H. C. WADDELL, *Public health paradoxes and the epidemiological impact of an HPV vaccine*, J. Clinical Virology, 19 (2000), pp. 101–111.

[99] J. GASTWIRTH AND W. JOHNSON, *Screening with cost-effective quality control, potential application to HIV and drug testing*, J. Amer. Statist. Assoc., 89 (1994), pp. 972–993.

[100] W. GILKS, S. RICHARDSON, AND D. SPIEGELHALTER, EDS., *Markov Chain Monte Carlo in Practice*, Chapman and Hall, London, 1996.

[101] G. E. GLASS, B. S. SCHWARTZ, J. M. MORGAN III, D. T. JOHNSON, P. M. NOY, AND E. ISRAEL, *Environmental risk factors for Lyme disease identification with geographic information systems*, Amer. J. Public Health, 85 (1995), pp. 944–948.

[102] E. GODEHART, *Graphs as Structural Models*, 2nd. ed., Vieweg, Braunschweig, Germany, 1990.

[103] E. GOLES AND S. MARTINEZ, *Statistical Physics, Automata Networks, and Dynamical Systems*, Kluwer Academic Publishers, Dordrecht, The Netherlands, 1992.

[104] D. M. GORDON AND M. A. RILEY, *A theoretical and experimental analysis of bacterial growth in the bladder*, Molecular Microbiol., 6 (1992), pp. 555–562.

[105] B. T. GRENFELL AND B. M. BOLKER, *Cities and villages: Infection hierarchies in a measles metapopulation*, Ecol. Lett., 1 (1998), pp. 63–70.

[106] M. D. Grigoriadis and L. G. Khachiyan, *A sublinear-time randomized approximation algorithm for matrix games*, Oper. Res. Lett., 18 (1995), pp. 53–58.

[107] A. Guenoche, P. Hansen, and B. Jaumard, *Efficient algorithms for divisive hierarchical clustering with the diameter criterion*, J. Classification, 8 (1991), pp. 5–30.

[108] K. P. Hadeler, *Spatial epidemic spread by correlated random walk, with slow infectives*, in Ordinary and Partial Differential Equations, Pitman Research Notes in Mathematics Ser. 370, D. Smith and R. J. Jarvis, eds., Longman, Harlow, UK, 1997, pp. 18–32.

[109] P. Hansen, O. Frank, and B. Jaumard, *Maximum sum of splits clustering*, J. Classification, 6 (1989), pp. 177–193.

[110] P. Hansen and B. Jaumard, *Minimum sum of diameters clustering*, J. Classification, 4 (1987), pp. 215–226.

[111] S. H. Heisterkamp, G. Doornbos, and N. J. Nagelkerke, *Assessing health impact of environmental pollution sources using space-time models*, Statist. Med., 19 (2000), pp. 2569–2578.

[112] H. W. Hethcote, *Simulations of pertussis epidemiology in the United States: Effects of adult booster doses*, Math. Biosci., 158 (1999), pp. 47–73.

[113] H. W. Hethcote and J. A. Yorke, *Gonorrhea Transmission Dynamics and Control*, Springer-Verlag, Berlin, 1984.

[114] F. Hoffman, *Social impact models*, in Southeastern Conference on Combinatorics, Graph Theory and Computing, Boca Raton, FL, 1995.

[115] J. H. Holmes, D. R. Durbin, and F. K. Winston, *Discovery of predictive models in an injury surveillance database: An application of data mining in clinical research*, in Proceedings of the AMIA Symposium, 2000, pp. 359–363.

[116] E. Horvitz and T. Paek, *A computational architecture for conversation*, in User Modeling Conference, 1999, pp. 201–210.

[117] E. Horvitz and T. Paek, *Harnessing models of users' goals to mediate clarification dialog in spoken language systems*, in User Modeling Conference, 2001.

[118] A. Hume, S. Daniels, and A. MacLellen, *Gecko: Tracking a very large billing system*, in USENIX, 1999.

[119] D. Huson, S. Nettles, K. Rice, T. Warnow, and S. Yooseph, *The hybrid tree reconstruction method*, J. Experimental Algorithmics, 4 (1999), pp. 178–189.

[120] D. Huson, S. Nettles, and T. Warnow, *Disk-covering, a fast converging method for phylogenetic tree reconstruction*, J. Comput. Biol., 6 (1999), pp. 369–386.

[121] L. C. Hutwagner, E. K. Maloney, N. H. Bean, L. Slutsker, and S. M. Martin, *Using laboratory-based surveillance data for prevention: An algorithm for detecting salmonella outbreaks*, Emerging Infectious Diseases, 3 (1997), pp. 395–400.

[122] M. Janowitz, F.-J. LaPointe, F. R. McMorris, B. Mirkin, and F. S. Roberts, eds., *Bioconsensus*, DIMACS Ser. Discrete Math. Theoret. Comput. Sci. 61, AMS, Providence, RI, 2003.

[123] D. L. Johnson, K. McDade, and D. Griffith, *Seasonal variation in paediatric blood lead levels in Syracuse, NY, USA*, Environmental Geochemistry and Health, 17 (1995), pp. 81–88.

[124] S. Kannan, *A survey of tree consensus criteria and methods*, in Proceedings of Phylogeny Workshop, DIMACS Technical Report 95-48, S. Tavare, ed., DIMACS Center, Rutgers University, Piscataway, NJ, 1995.

[125] D. Karger, *Information retrieval: Challenges in interactive-time manipulation of massive text collections*, in Session on Mathematical, Statistical, and Algorithmic Problems of Very Large Data Sets, American Mathematical Society Meeting, San Diego, CA, 1997.

[126] A. F. Kaufman, M. I. Meltzer, and G. P. Schmid, *The economic impact of a bioterrorist attack: Are prevention and postattack intervention programs justifiable?*, Emerging Infectious Diseases, 3 (1997), pp. 83–94.

[127] M. Keeling, *Multiplicative moments and measures of persistence in ecology*, J. Theoret. Biol., 205 (2000), pp. 269–281.

[128] Y. Kempner, B. Mirkin, and I. Muchnik, *Monotone linkage clustering and quasi-concave set functions*, Appl. Math. Lett., 4 (1997), pp. 19–24.

[129] U. Kitron, H. Pener, C. Costin, L. Orshan, Z. Greenberg, and U. Shalom, *Geographic information system in malaria surveillance: Mosquito breeding and imported cases in Israel*, Amer. J. Tropical Med., 50 (1994), pp. 550–556.

[130] J. Koch-Weser, E. M. Sellers, and R. Zacest, *The ambiguity of adverse drug reactions*, European J. Clinical Pharmacology, 11 (1977), pp. 75–78.

[131] D. Koller and N. Megiddo, *The complexity of two-person zero-sum games in extensive form*, Games and Economic Behavior, 4 (1992), pp. 528–552.

[132] D. Koller and N. Megiddo, *Finding mixed strategies with small supports in extensive form games*, Internat. J. Game Theory, 25 (1996), pp. 73–92.

[133] D. H. Krantz, R. D. Luce, P. Suppes, and A. Tversky, *Foundations of Measurement*, Vol. I, Academic Press, New York, 1971.

[134] M. Kretzschmar and M. Morris, *Measures of concurrency in networks and the spread of infectious disease*, Math. Biosci., 133 (1996), pp. 165–195.

[135] E. Kuznetsov and I. Muchnik, *Analysis of the distribution of functions in an organization*, Autom. Remote Control, 43 (1982), pp. 1325–1331.

[136] D. Lambert, J. Pinheiro, and D. Sun, *Reducing Transaction Databases, without Lagging behind the Data or Losing Information*, Technical Report, Statistics and Data Mining Research Department, Bell Labs, Lucent Technologies, Murray Hill, NJ, 1999.

[137] L. Lamport, R. Shostak, and M. Pease, *The Byzantine generals problem*, ACM Trans. Programming Languages Systems, 4 (1982), pp. 382–401.

[138] C. P. Langlotz, E. H. Shortliffe, and L. M. Fagan, *A Methodology for Computer-Based Explanation of Decision Analysis*, Report KSL-86-57, Stanford University, 1986.

[139] F.-J. Lapointe and G. Cucumel, *How many consensus trees?*, in Bioconsensus, DIMACS Ser. Discrete Math. Theoret. Comput. Sci. 61, M. Janowitz, F.-J. LaPointe, F. R. McMorris, B. Mirkin, and F. S. Roberts, eds., AMS, Providence, RI, 2003.

[140] B. Latane, *The emergence of clustering and correlation from social interaction*, in Models of Social Dynamics: Order, Chaos, and Complexity, R. Hegelmann and H. O. Peitgen, eds., Holder, Picker, Tempsky, Vienna, 1996, pp. 79–104.

[141] B. Latane and E. Fink, *Symposium: Dynamical social impact theory and communication*, J. Communication, 46 (1996), pp. 4–77.

[142] B. Latane and A. Nowak, *Attitudes as catastrophes: From dimensions to categories with increasing involvement*, in Dynamical Systems in Social Psychology, R. R. Vallacher and A. Nowak, eds., Academic Press, New York, 1994, pp. 219–249.

[143] S. A. Levin and R. Durrett, *From individuals to epidemics*, Philos. Trans. Roy. Soc. London Ser. B, 351 (1997), pp. 1615–1621.

[144] S. A. Levin, B. Grenfell, A. Hastings, and A. S. Perelson, *Mathematical and computational challenges in population biology and ecosystems science*, Science, 17 (1997), pp. 334–343.

[145] N. Linial, *Games computers play: Game-theoretic aspects of computing*, in Handbook of Game Theory with Economic Applications, Vol. II, R. J. Aumann and S. Hart, eds., North-Holland, Amsterdam, 1994, Chapter 38, pp. 1340–1395.

[146] N. Linial, D. Peleg, Y. Rabinovitch, and M. Saks, *Sphere packing and local majorities in graphs*, in Proceedings of the 2nd ISTCS, 1993, pp. 141–149.

[147] M. Lipsitch and B. R. Levin, *The population dynamics of antimicrobial chemotherapy*, Antimicrobial Agents and Chemotherapy, 41 (1997), pp. 363–370.

[148] M. Lipsitch and B. R. Levin, *Population dynamics of tuberculosis treatment*, Internat. J. Tuberculosis and Lung Disease, 2 (1998), pp. 187–199.

[149] E. Loehman and A. Whinston, *An axiomatic approach to cost allocation for public investment*, Public Finance Quart., 2 (1974), pp. 236–251.

[150] W. F. Lucas, *Applications of cooperative games to equitable allocations*, in Game Theory and Its Applications, Proc. Sympos. Appl. Math. 24, AMS, Providence, RI, 1981, pp. 19–36.

[151] R. D. Luce, D. H. Krantz, P. Suppes, and A. Tversky, *Foundations of Measurement*, Vol. III, Academic Press, New York, 1990.

[152] F. Luccio, *Almost exact minimum feedback vertex sets in meshes and butterflies*, Inform. Process. Lett., 66 (1998), pp. 59–64.

[153] B. E. Mahon, D. D. Rohn, S. R. Pack, and R. V. Tauxe, *Electronic communication facilitates investigation of a highly dispersed foodborne outbreak: Salmonella on the superhighway*, Emerging Infectious Diseases, 1 (1995), pp. 94–95.

[154] P. Marcotorchino and P. Michaud, *Heuristic approach of the similarity aggregation problem*, Methods Oper. Res., 43 (1981), pp. 395–404.

[155] H. Marcus-Roberts and F. S. Roberts, *Meaningless statistics*, J. Educ. Statist., 12 (1987), pp. 383–394.

[156] R. McConachy, K. B. Korb, and I. Zuckerman, *Deciding what not to say: An attentional probabilistic approach to argument presentation*, in Proceedings of the 20th Annual Conference of the Cognitive Science Society (CogSci 98), 1998.

[157] S. Merrill III, *Making Multicandidate Elections More Democratic*, Princeton University Press, Princeton, NJ, 1988.

[158] S. Merrill III, *A Unified Theory of Voting: Directional and Proximity Spatial Models*, Cambridge University Press, New York, 1999.

[159] B. Mirkin and F. S. Roberts, *Consensus functions and patterns in molecular sequences*, Bull. Math. Biol., 55 (1993), pp. 695–713.

[160] D. Mollison, ed., *Epidemic Models: Their Structure and Relation to Data*, Cambridge University Press, Cambridge, UK, 1995.

[161] M. Morris and L. Dean, *Effects of sexual behavior change on long-term human immuno-deficiency virus prevalence among homosexual men*, Amer. J. Epidemiology, 140 (1994), pp. 217–231.

[162] H. Moulin, *Axioms of Cooperative Decision Making*, Cambridge University Press, Cambridge, UK, 1988.

[163] H. Moulin, *Incremental cost sharing: characterization by coalition strategy-proofness*, Soc. Choice Welf., 16 (1999), pp. 279–320.

[164] I. Muchnik and L. Shvartser, *Maximization of generalized characteristics of functions of monotone systems*, Autom. Remote Control, 51 (1990), pp. 1562–1572.

[165] T. Munzner, *Exploring large graphs in 3D hyperbolic space*, IEEE Comput. Graphics Appl., 18 (1998), pp. 18–23.

[166] J. D. Murray, *Mathematical Biology*, Springer-Verlag, Berlin, 1980.

[167] M. Murray, *Determinants of cluster distribution in the molecular epidemiology of tuberculosis*, Proc. Natl. Acad. Sci. USA, 99 (2002), pp. 153–154.

[168] M. Mustafa and A. Pekec, *Majority consensus and the local majority rule*, in Proceedings of the 28th International Colloquium on Automata, Languages and Programming—ICALP '01, Lecture Notes in Comput. Sci. 2076, Springer-Verlag, Vienna, 2001, pp. 530–542.

[169] R. Myerson, *Game Theory*, Harvard University Press, Cambridge, MA, 1991.

[170] T. Narayana, *Lattice Path Combinatorics with Statistical Applications*, Mathematical Expositions 23, University of Toronto Press, Toronto, 1979.

[171] M. D. Neil, *Multivariate assessment of software products*, J. Software Testing, 1 (1992), pp.17–37.

[172] H. Q. Ngo and D.-Z. Du, *A survey on combinatorial group testing algorithms with applications to DNA library screening*, in Discrete Mathematical Problems with Medical Applications, DIMACS Ser. Discrete Math. Theoret. Comput. Sci. 55, D.-Z. Du, P. M. Pardalos, and J. Wang, eds., AMS, Providence, RI, 2000.

[173] M. A. Nowak, D. C. Krakauer, A. Klug, and R. M. May, *Prion infection dynamics*, Integrative Biol., 1 (1998), pp. 3–15.

[174] M. A. Nowak and R. M. May, *Virus Dynamics*, Oxford University Press, Oxford, UK, 2000.

[175] M. A. Nowak, R. M. May, and R. M. Anderson, *The evolutionary dynamics of HIV-1 quasispecies and the development of immunodeficiency disease*, AIDS, 4 (1990), pp. 1095–1103.

[176] J. R. Nuckols, D. Ellington, and M. Faidi, *Addressing the non-point source implications of conjunctive water use with a geographic information system (GIS)*, in Application of Geographic Information Systems in Hydrology and Water Resource Management (HydroGIS 96), Vienna, 1996, pp. 341–348.

[177] N. Okada, T. Hashimoto, and P. Young, *Cost allocation in water resources development*, Water Resources Res., 18 (1982), pp. 361–373.

[178] P. D. O'Neill, D. J. Balding, N. G. Becker, M. Eerola, and D. Mollison, *Analyses of infectious disease data from household outbreaks by Markov chain Monte Carlo methods*, J. Roy. Statist. Soc. Ser. C, 49 (2000), pp. 517–542.

[179] S. Openshaw, I. Turton, and J. MacGill, *Using the geographic analysis machine to analyze limiting long-term illness census data*, Geographic and Environmental Model., 3 (1999), pp. 83–99.

[180] D. Ozonoff and T. Webster, *The lattice diagram and* 2×2 *tables*, in Hazardous Waste: Impacts on Human and Ecological Health (Proceedings of the 2nd International Congress on Hazardous Waste), B. L. Johnson, C. Xintaras, and J. S. Andrews, eds., Princeton Scientific, Princeton, NJ, 1997, pp. 441–458.

[181] C. H. Papadimitriou and M. Yannakakis, *On the value of information in distributed decision-making*, in PODC 91, 1991.

[182] P. M. Pardalos, D. Shalloway, and G. Xue, eds., *Global Minimization of Nonconvex Energy Functions: Molecular Conformation and Protein Folding*, DIMACS Ser. Discrete Math. Theoret. Comput. Sci. 23, AMS, Providence, RI, 1996.

[183] G. P. Patil, *Encountered data, statistical ecology, environmental statistics, and weighted distribution methods*, Environmetrics, 2 (1991), pp. 377–423.

[184] C. Paulu and D. Ozonoff, *Exploring associations between residential history and breast cancer risk in a case-control study*, in International Society for Environmental Epidemiology/International Society of Exposure Analysis, Annual Conference, Boston, MA, August 15–18, 1998.

[185] A. Pekec, *Monochromatic Outcomes in Majority Processes*, preprint, Department of Mathematics, Rutgers University, New Brunswick, NJ, 1995.

[186] D. Peleg, *Local majority voting, small coalitions and controlling monopolies in graphs: A review*, in Proceedings of the 3rd Colloquium on Structural Information and Communication Complexity, 1996, pp. 170–179.

[187] D. Peleg, *Size bounds for dynamic monopolies*, Discrete Appl. Math., 86 (1998), pp. 263–273.

[188] P. C. Pendharkar, J. A. Rodger, G. J. Yaverbaum, H. Herman, and M. Benner, *Association, statistical, mathematical and neural approaches for mining breast cancer patients*, Expert Systems with Appl., 17 (1999), pp. 223–232.

[189] A. S. Perelson, A. U. Neumann, M. Markowitz, J. M. Leonard, and D. D. Ho, *HIV-*1 *dynamics in vivo: Virion clearance rate infected cell life span and viral generation time* Science, 271 (1996), pp. 1582–1586.

[190] T. Podding and G. Huber, *Data mining for the detection of turning points in financial time series*, in Proceedings of Advances in Intelligent Data Analysis, Lecture Notes in Comput. Sci. 1642, Springer-Verlag, Berlin, 1999, pp. 427–436.

[191] S. Poljak and M. Sura, *On periodical behavior in society with symmetric influences*, Combinatorica, 3 (1983), pp. 119–121.

[192] S. P. Proctor, T. Heeren, R. F. White, J. Wolfe, M. S. Borgos, J. D. Davis, L. Pepper, R. Clapp, P. B. Sutker, J. J. Vasterling, and D. Ozonoff, *Health status of Persian Gulf War veterans: Self-reported symptoms, environmental exposures and the effect of stress*, Internat. J. Epidemiology, 27 (1998), pp. 1000–1010.

[193] M. A. Province and A. Single, *Sequential, genome-wide test to identify simultaneously all promising areas in a linkage scan*, Genetic Epidemiology, 19 (2000), pp. 301–322.

[194] O. Pybus, A. Rambaut, and P. H. Harvey, *An integrated framework for the inference of viral population history from reconstructed genealogies*, Genetics, 155 (2000), pp. 1429–1437.

[195] A. Rambaut, *Estimating the rate of molecular evolution: Incorporating noncontemporaneous sequences into maximum likelihood phylogenies*, Bioinformatics, 16 (2000), pp. 395–399.

[196] A. Rambaut, D. L. Robertson, O. G. Pybus, M. Peeters, and E. Holmes, *Phylogeny and the origin of HIV-*1, Nature, 410 (2001), pp. 1047–1048.

[197] D. A. Rand, M. Keeling, and H. B. Wilson, *Invasion, stability and evolution to criticality in spatially extended, artificial host-pathogen ecologies*, Proc. Roy. Soc. Lond., 259 (1995), pp. 55–63.

[198] M. D. Rawlins, *Pharmacovigilance: Paradise lost, regained or postponed?*, The William Withering Lecture, J. Roy. Coll. Physicians London, 29 (1995), pp. 41–49.

[199] M. Regenwetter, J.-C. Falmagne, and B. Grofman, *A stochastic model of preference change and its application to 1992 presidential election panel data*, Psychological Rev., 2 (1999), pp. 362–384.

[200] M. Regenwetter and B. Grofman, *Approval voting, Borda winners and Condorcet winners: evidence from seven elections*, Management Sci., 44 (1998), pp. 520–533.

[201] G. Richards, V. J. Rayward-Smith, P. H. Sonksen, S. Carey, and C. Weng, *Data mining for indicators of early mortality in a database of clinical records*, Artif. Intell. Med., 22 (2001), pp. 215–231.

[202] F. S. Roberts, *Discrete Mathematical Models, with Applications to Social, Biological, and Environmental Problems*, Prentice–Hall, Englewood Cliffs, NJ, 1976.

[203] F. S. Roberts, *Graph theory and the social sciences*, in Applications of Graph Theory, R. Wilson and L. Beineke, eds., Academic Press, New York, 1979, pp. 255–291.

[204] F. S. Roberts, *Measurement Theory, with Applications to Decisionmaking, Utility, and the Social Sciences*, Addison–Wesley, Reading, MA, 1979.

[205] F. S. Roberts, *Limitations on conclusions using scales of measurement*, in Operations Research and the Public Sector, A. Barnett, S. M. Pollock, and M. H. Rothkopf, eds., Elsevier, Amsterdam, 1994, pp. 621–671.

[206] F. S. Roberts, *Meaningless statements*, in Contemporary Trends in Discrete Mathematics, DIMACS Ser. Discrete Math. Theoret. Comput. Sci. 49, R. L. Graham, J. Kratochvil, J. Nesetril, and F. S. Roberts, eds., AMS, Providence, RI, 1999, pp. 257–274.

[207] A. G. Rodrigo, *HIV evolutionary genetics* [*Commentary*], Proc. Natl. Acad. Sci. USA, 96 (1999), pp. 10559–10561.

[208] A. G. Rodrigo and J. Felsenstein, *Coalescent approaches to HIV-*1 *Population Genetics*, in Molecular Evolution of HIV, K. A. Crandall, ed., Johns Hopkins University Press, Baltimore, MD, 1999.

[209] A. G. Rodrigo, E. G. Shpaer, E. L. Delwart, A. K. N. Iversen, M. V. Gallo, J. Brojatsch, M. S. Hirsch, B. D. Walker, and J. I. Mullins, *Coalescent estimates of HIV-*1 *generation time in vivo*, Proc. Natl. Acad. Sci. USA, 96 (1999), pp. 2187–2191.

[210] P. Rohani, D. J. D. Earn, and B. T. Grenfell, *Impact of immunisation on pertussis transmission in England and Wales*, Lancet, 355 (2000), pp. 285–286.

[211] D. G. Saari, *Chaotic Elections! A Mathematician Looks at Voting*, AMS, Providence, RI, 2001.

[212] N. L. Saccone, J. M. Kwon, J. Corbett, A. Goate, N. Rochberg, H. J. Edenberg, T. Foroud, T. Li, H. Begleiter, T. Reich, J. P. and Rice, *A genome screen of maximum number of drinks as an alcoholism phenotype*, Amer. J. Med. Genetics, 96 (2000), pp. 632–637.

[213] M. C. Schuette and W. H. Hethcote, *Modeling the effects of varicella vaccination programs on the incidence of chickenpox and shingles*, Bull. Math. Biol., 61 (1999), pp. 1031–1064.

[214] A. K. Sen, *Collective Choice and Social Welfare*, Holden–Day, San Francisco, 1970.

[215] R. Shankarappa, R. B. Margolick, S. J. Gange, A. G. Rodrigo, D. Upchurch, H. Farzadegan, P. Gupta, C. R. Rinaldo, G. H. Learn, X. He, X.-L. Huang, and J. I. Mullins, *Consistent viral evolutionary changes associated with progression of HIV-1 infection*, J. Virology, 73 (1999), pp. 10489–10502.

[216] W. Sharkey, *Network models in economics*, in Network Routing, Handbooks in Operations Research and Management Science 8, M. O. Ball, T. L. Magnenti, C. L. Monma, and G. L. Nemhauser, eds., North–Holland, Amsterdam, 1995, pp. 713–765.

[217] S. Shenker, *Making greed work in networks*, ACM/IEEE Trans. Networking, 3 (1995), pp. 819–831.

[218] S. Shenker and H. Moulin, *Strategyproof sharing of submodular costs: Budget balance versus efficiency*, Economic Theory, 18 (2001), pp. 511–533.

[219] M. Shubik, *Incentives, decentralized controls, the assignment of joint costs and internal pricing*, Management Sci., 8 (1962), pp. 325–343.

[220] M. Singh, B. Berger, and P. S. Kim, *LearnCoil-VMF: Computational evidence for coiled-coil-like motifs in many viral membrane-fusion proteins*, J. Molecular Biol., 290 (1999), pp. 1031–1041.

[221] M. Singh, B. Berger, P. S. Kim, J. M. Berger, and A. G. Cochran, *Computational learning reveals coiled coil-like motifs in histidine kinase linker domains*, Proc. Natl. Acad. Sci. USA, 95 (1998), pp. 2738–2743.

[222] D. Stauffer and A. Aharony, *Introduction to Percolation Theory*, 2nd ed., Taylor and Francis, London, 1992.

[223] M. A. Steel and L. A. Székely, *Inverting random functions*, Ann. Combin., 3 (1999), pp. 103–113.

[224] M. A. Steel and L. A. Székely, *Inverting Random Functions* II: *Explicit Bounds for Parametric and Non-Parametric MLE, with Applications*, preprint, 2001.

[225] S. Stolfo, W. Fan, W. Lee, A. Prodromidis, and P. Chan, *Cost-based modeling for fraud and intrusion detection: Results from the JAM project*, in Proceedings of the DARPA Information Survivability Conference and Exposition, Vol. 2, IEEE Computer Press, Los Alamitos, CA, 2000, pp. 130–144.

[226] P. D. Straffin and J. P. Heaney, *Game theory and the Tennessee Valley Authority*, Internat. J. Game Theory, 10 (1981), pp. 35–43.

[227] B. L. Strom and P. Tugwell, *Pharmacoepidemiology: Current status, prospects, and problems*, Ann. Internal Med., 113 (1990), pp. 179–181.

[228] P. Suppes, D. H. Krantz, R. D. Luce, and A. Tversky, *Foundations of Measurement*, Vol. II, Academic Press, New York, 1989.

[229] M. Szklo and F. J. Nieto, *Epidemiology, Beyond the Basics*, Aspen, Gaithersburg, MD, 2000.

[230] T. O. Talbot, S. P. Forand, and V. B. Haley, *Geographic analysis of childhood lead exposure in New York State*, in 3rd National GIS in Public Health Conference, 1998.

[231] A. Thomas, *A Behavioral Analysis of Joint Cost Allocation and Transfer Pricing*, Arthur Andersen and Company Lecture Series, Stipes Publishing Company, Champaign, IL, 1980.

[232] W. T. Trotter, *Combinatorics and Partially Ordered Sets: Dimension Theory*, Johns Hopkins University Press, Baltimore, MD, 1992.

[233] X. Tu, E. Litvak, and M. Pagano, *Screening tests: Can we get more by doing less*, Studies of Aids and HIV Surveillance, Statist. Med., 13 (1994), pp. 1905–1919.

[234] J. D. Ullman, *The MIDAS data-mining project at Stanford*, in Proceedings of the 1999 IEEE Symposium on Database Engineering and Applications, 1999, pp. 460–464.

[235] K. VanLehn and Z. Niu, *Bayesian student modeling, user interfaces and feedback: A sensitivity analysis*, Internat. J. Artif. Intell. Education, 12 (2001), pp. 154–184.

[236] V. Vieira, T. Webster, A. Aschengrau, and D. Ozonoff, *A method for spatial analysis of risk in a population-based case-control study*, Internat. J. Hygiene and Environmental Health, 205 (2002), pp. 115–120.

[237] M. Vine, D. Degnan, and C. Hanchette, *Geographic information systems: Their use in environmental epidemiologic research*, Environmental Health Perspectives, 105 (1997), pp. 598–605.

[238] M. A. Walker, *An application of reinforcement learning to dialogue strategy selection in a spoken dialogue system for email*, J. Artif. Intell. Res., 12 (2000), pp. 387–416.

[239] D. Wartenberg, *Use of geographic information systems for risk screening and epidemiology*, in Hazardous Waste and Public Health: International Congress on Health Effects of Hazardous Waste, J. S. Andrews, Jr., H. Frumkin, B. L. Johnson, M. A. Mehlman, C. Xintaras, and J. A. Bucsela, eds., Princeton Scientific, Princeton, NJ, 1994, pp. 853–859.

[240] D. Wartenberg, M. Greenberg, R. Lathrop, R. Manning, and S. Brown, *Using a geographic information system to identify populations living near high voltage electric power transmission lines in New York State*, in World Computer Graphics Foundation, 1997, p. 20.

[241] M. S. Waterman, *Consensus patterns in sequences*, in Mathematical Methods for DNA Sequences, M. S. Waterman, ed., CRC Press, Boca Raton, FL, 1989, pp. 93–115.

[242] M. S. Waterman, D. Galas, and R. Arratia, *Pattern recognition in several sequences: Consensus and alignment*, Bull. Math. Biol., 46 (1984), pp. 515–527.

[243] D. J. Watts, *Small Worlds: The Dynamics of Networks between Order and Randomness*, Princeton University Press, Princeton, NJ, 1999.

[244] D. J. Watts and S. H. Strogatz, *Collective dynamics of "small-world" networks*, Nature, 393 (1998), pp. 440–442.

[245] J. E. Wigginton and D. E. Kirschner, *A model to predict cell-mediated immune regulatory mechanisms during human infection with mycobacterium tuberculosis*, J. Immunology, 166 (2001), pp. 1951–1967.

[246] M. Wilkinson, *Majority-rule reduced consensus methods and their use in bootstrapping*, Molecular Biol. and Evolution, 13 (1996), pp. 437–444.

[247] M. Wilkinson and J. Thorley, *Reduced consensus and supertree methods*, in Bioconsensus, DIMACS Ser. Discrete Math. Theoret. Comput. Sci. 61, M. Janowitz, F.-J. LaPointe, F. R. McMorris, B. Mirkin, and F. S. Roberts, eds., AMS, Providence, RI, 2003.

[248] S. S. Yoon, *Geographical information systems: A new tool in the fight against schistosomiasis*, in The Added Value of Geographical Information Systems in Public and Environmental Health, M. J. C. de Lepper, H. J. Scholten, and R. M. Stern, eds., Kluwer Academic Publishers, Boston, 1995, pp. 201–213.

[249] H. P. Young, *Cost Allocation: Methods, Principles, Applications*, North-Holland, Amsterdam, 1985.

[250] I. Zuckerman, R. McConachy, and K. B. Korb, *Bayesian reasoning in an abductive mechanism for argument generation and analysis*, in Proceedings of the 15th National Conference on Artificial Intelligence (AAAI 98), 1998.

Chapter 2

Worst-Case Scenarios and Epidemics

Gerardo Chowell and Carlos Castillo-Chavez†*

2.1 Introduction

The potential deliberate release of biological agents such as smallpox, influenza, and foot and mouth disease (FMD) is a source of continuous concern. It is believed that "worst" epidemics (single outbreaks) are most likely to occur in populations where individuals mix randomly (proportionate mixing). Therefore, the definition and identification of landscapes or topologies that support worst-case scenarios is critical. Here, SIR (susceptible-infective-recovered) epidemics that result from the introduction of single or multiple infectious sources are studied on various topologies including small-world and scale-free networks.

Most of the models studied in classical mathematical epidemiology fall in the class of *compartmental models* because the population under consideration is divided into classes or compartments defined by epidemiological status [1]. The simplest version assumes that individuals mix uniformly (*homogeneous mixing*) within each compartment. The study of the transmission dynamics of communicable diseases in human populations via mathematical (compartmental) models can be traced back to the work of Kermack and McKendrick (1927). Their simple SIR (susceptible-infective-removed) epidemic model (Figure 2.1) not only was capable of generating realistic single-epidemic outbreaks but also provided important theoretical epidemiological insights.

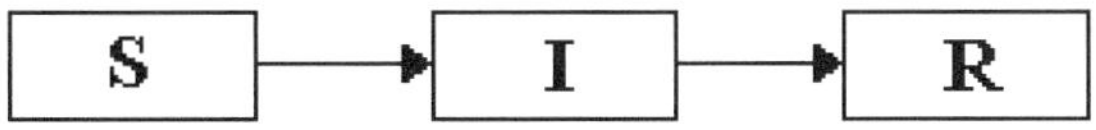

Figure 2.1. *Simple SIR epidemic model.*

*Center for Nonlinear Studies, Los Alamos National Laboratory, Los Alamos, NM 87545.
†Department of Biological Statistics and Computational Biology, Cornell University, Ithaca, NY 14853-7801, and Center for Nonlinear Studies, Los Alamos National Laboratory, Los Alamos, NM 87545.

The Kermack and McKendrick (K-M) model in essence captures the theoretical underpinnings associated with the framework that it is currently used to define (and model) worst-case epidemic outbreaks. The K-M model is given by the following system of nonlinear differential equations:

$$\begin{aligned} \dot{S} &= -\lambda(N)SI, \\ \dot{I} &= \lambda(N)SI - \gamma I, \\ \dot{R} &= \gamma I, \end{aligned} \tag{2.1}$$

where $S(t)$ denotes susceptible individuals at time t; $I(t)$ infective (assumed infectious) individuals at time t; $R(t)$ recovered (assumed permanently immune) individuals at time t; $\lambda(N)$ the transmission rate when the total population is $N(N = S + I + R)$; and γ the recovery rate. In the case of a fatal disease, $N = S + I$ as $R(t)$ would denote those removed by death and γ the per-capita death rate.

The K-M threshold theorem establishes (quantitatively) the conditions required for successful disease invasion. This threshold theorem says that a disease will invade whenever its basic reproductive number

$$R_o = \frac{\lambda(N(0))N(0)}{\gamma} > 1.$$

R_o is interpreted as the number of secondary infectious individuals generated by a "typical" infectious individual when introduced into a fully susceptible population [2, 3]. An alternative interpretation is that, in a randomly mixing population, a disease will invade if there are enough susceptibles, that is, if

$$N(0) > \frac{\gamma}{\lambda(N(0))}.$$

Typically, it has been assumed that either $\lambda(N) = \frac{\beta_o}{N}$ or $\lambda(N) = \beta_o$, and consequently, the interpretation of R_o depends on the definition of $\lambda(N)$ [5, 1, 2].

Nold [6] introduced the concept of *proportionate mixing* as a way of modeling a simple form of heterogeneous mixing. She divided the population into K groups, each with population size $N_i(t) = S_i(t) + I_i(t) + R_i(t)$ $(i = 1, \ldots, K)$ and modeled an SIR epidemic that considered interactions of various intensities among individuals. In order to describe her framework we introduce the mixing matrix $P(t) = (P_{ij}(t))$, where $P_{ij}(t)$ denotes the proportion of contacts of individuals in group i with individuals in group j *given* that i-individual had a contact with a member of the population at time t. Nold's proportionate mixing corresponds to the case where P_{ij} is independent of i, that is,

$$P_{ij} \equiv \bar{P}_j = \frac{C_j N_j}{\sum_{l=1}^{k} C_l N_l}, \qquad (\Pi) \tag{2.2}$$

where C_l denotes the average activity level (contact rate) of individuals in group $l = 1, \ldots, k$. Other forms of mixing can be found in [2, 7, 8, 9] and references therein.

A slightly modified version of Nold's generalization of the K-M model (assumes that

$\lambda_j \equiv \frac{\beta_j}{N_j}$, $j = 1, \ldots, K$) is given by

$$\begin{aligned}
\dot{S}_i &= -S_i(t)\sum_{j=1}^{k}\beta_j\bar{P}_j\frac{I_j}{N_j},\\
\dot{I}_i &= S_i(t)\sum_{j=1}^{k}\beta_j\bar{P}_j\frac{I_j}{N_j} - \gamma_i I_i,\\
\dot{R}_i &= \gamma_i I_i, \quad i = 1, \ldots, K.
\end{aligned} \tag{2.3}$$

Examples of mathematical studies for compartmental models of this type can be found in [5, 1, 2, 6]. Extensions of these models to (local) populations interconnected via migrating individuals (*metapopulation models*) have been carried out in some situations (see [9] and references therein). In the K-M model and Nold's models, individuals mix at random (see Figure 2.2), an *uncommon* and extreme situation. Recent analyses of worst-case scenarios for the deliberate release of biological agents (smallpox in particular) have been carried out under the assumption that random mixing supports the worst epidemic outbreaks [10]. The focus of this paper (as suggested to us by Ed Kaplan) consists of the preliminary examination of the validity of this assumption.

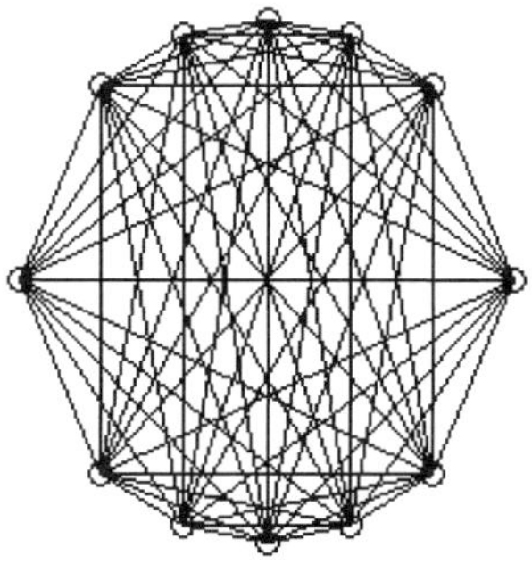

Figure 2.2. *Fully mixed transmission network.*

The identification of worst-case scenarios would require an approach that considers "all" measures of mixing [6]. The study of the meaning of worst-case scenarios using "mean" field models like the K-M or Nold's model is problematic (mixing assumes a predetermined number of groups or types). Hence, we look at this question in the context of simple individual-based models where mixing is embedded in a preselected (fixed) topology. This approach also has its limitations as the nodes (individuals) of the network have no dynamics. However, we feel that this approach is useful if the goal is to study the "strength" of a single epidemic outbreak, that is, in the study of situations where transient disease dynamics are critical.

2.2 Individual-Based Models

Population structures are often represented by networks (graphs) composed of nodes (individuals) and edges (representing predefined relationships between nodes). Examples include family trees, traffic networks that describe street intersections by nodes and traffic direction

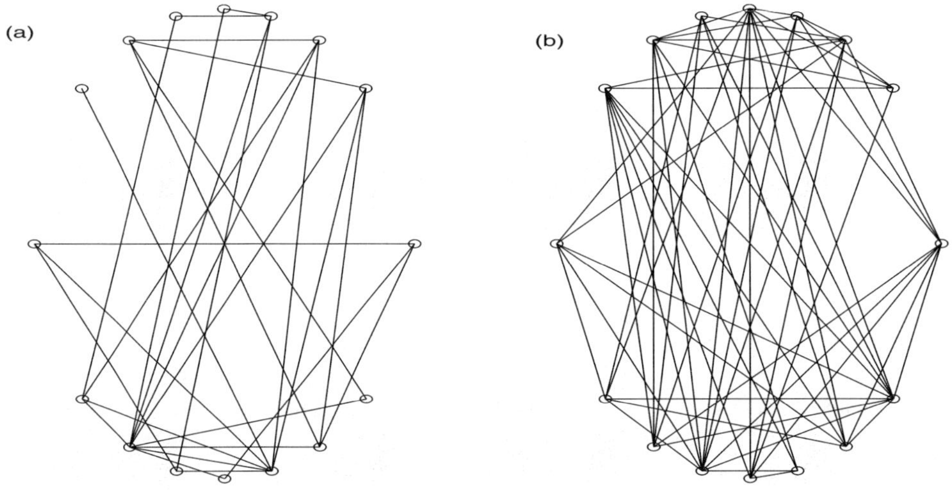

Figure 2.3. *The Erdős and Rényi random graph with* $N = 16$, (a) $p_{\text{ER}} = 0.25$, (b) $p_{\text{ER}} = 0.5$.

by arrows (edges), and airline traffic networks. Graphs (networks) can be represented by the adjacency matrix T, where $T(i, j) = 1$ implies that vertex i is connected to j. If the network is undirected (edges have no direction), then T is symmetric; that is, $T = T^t$ (transpose of T). An *adjacency* list, a compilation of the vertices of a graph and the vertices adjacent to such a vertex, is also used to represent graphs. The analysis of network models can be traced back to the work of Erdős and Rényi in the 1960s. These researchers introduced the following simple algorithm for the construction of *random* networks [11]: start with a fixed number of disconnected nodes N and connect each pair of nodes independently by an edge with probability p_{ER}. Hence, $p_{\text{ER}} = 0$ corresponds to the case where no node is connected to any of the other $N - 1$ nodes, while $p_{\text{ER}} = 1$ corresponds to the case where every node is connected to all other nodes in the network (complete graph). Examples are depicted in Figure 2.3. The total number of edges when $p_{\text{ER}} = 1$ is $\binom{N}{2}$; the average number of edges is $\frac{N(N-1)p_{\text{ER}}}{2}$; and the average degree of a node (number of edges incident from a node) is $z = (N - 1)p_{\text{ER}} \simeq Np_{\text{ER}}$ (for large N).

Erdős and Rényi [11] showed that for large systems (large N) the probability that a node has k edges follows the Poisson distribution $P(k) = \frac{\exp(-z)z^k}{k!}$ $(k = 0, 1, \ldots, N)$. Furthermore, they showed that there is a critical value of z (z_c) such that whenever $z > z_c$, a *connected component* (a subset of vertices in the graph each of which is reachable from the others by some path through the network) forms. Such a component is often referred to as a *spanning cluster* [14]. The Erdős and Rényi random graph provides a null-model for the "comparative" study of the disease transmission on other network topologies. The case $p_{\text{ER}} = 1$ (totally connected network) is naturally believed to be the generator of the landscape most conducive to disease spread and, in some sense, "corresponds" to Nold's version of the K-M model. Hence, here we address, via simulations, whether or not populations modeled by graphs *comparable* to those of the Erdős and Rényi model support the worst possible epidemics. It is not clear, however, whether or not diseases spread at a faster rate in highly clustered networks, that is, in networks where there is a higher probability that neighbors of a particular node are also neighbors of each other. The importance of this possibility

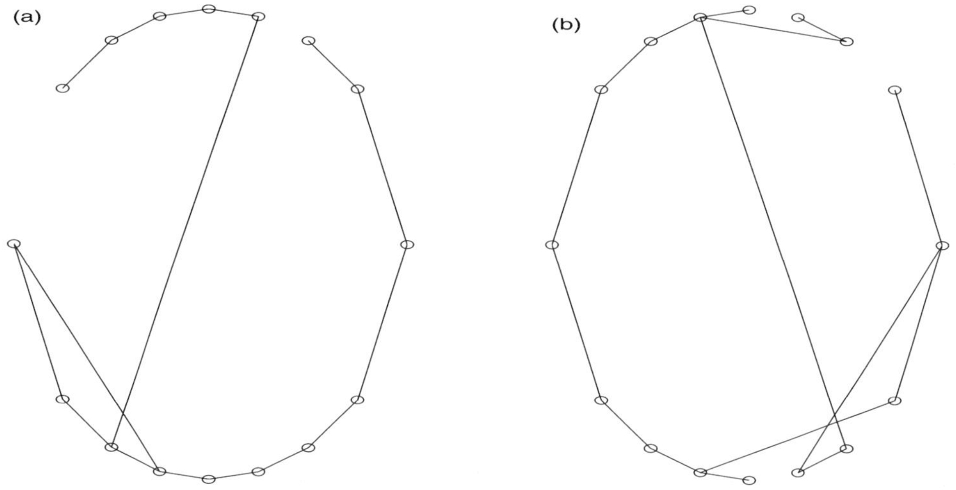

Figure 2.4. *Small-world networks with $N = 16$, $K = 1$, (a) $p = 0.1$, (b) $p = 0.3$.*

comes in part from the fact that Watts and Strogatz [15] showed that networks with high degree of aggregations (clustering), a characteristic absent in random networks (see Erdős and Rényi [11]), are not uncommon. In order to explain our simulations of epidemics in various network setups, we need to describe their construction.

Watts and Strogatz's [15] algorithm for the construction of networks is as follows: a one-dimensional ring lattice of N nodes connected to its $2K$ nearest neighbors (K is known as the coordination number) and periodic boundary conditions are preselected. The algorithm goes through each of the edges in turn and, independently with probability p_{WS}, "rewires" it to a randomly selected node. That is, the Watts–Strogatz algorithm shifts one end of the edge to a new randomly chosen node from the whole lattice (except that no two nodes can have more than one edge running between them, and no node can be connected by an edge to itself (see Figure 2.4). Watts and Strogatz [15] classified networks by their level of randomness, as measured by their own disorder parameter p_{WS} (from "regular" $p_{\mathrm{WS}} = 0$ to completely random, $p_{\mathrm{WS}} = 1$). Whenever each node in a network is just connected to its nearest two neighbors one on its right and on the other on its left, the network is *regular* [15]. A completely random network has $p_{\mathrm{WS}} = 1$; that is, all nodes are randomly connected to each other. Watts and Strogatz showed that the introduction of a small number of random connections ($p_{\mathrm{WS}} \simeq 0.01$) significantly reduces the average distance between any two nodes (characteristic path length), a property that facilitates disease spread. In fact, Watts and Strogatz showed that such average distance grows like $O(\log(N))$ and not as $O(N)$. Networks constructed via the Watts–Strogatz algorithm also support high levels of clustering. The small-world effect (short average distance between nodes and high levels of clustering) has been detected in networks that include a network of actors in Hollywood, the power generator network in the western US, and the neural network of *Caenorhabditis elegans* [15]. This "small-world effect" had already been documented by the psychologist Stanley Milgram, using data from the letter-passing experiments that he conducted in the 1960s [16]. Newman and Watts [17] studied a slight variation of the Watts–Strogatz model. They added shortcut edges with probability ϕ per edge in the underlying ring latice instead of "rewiring" the existing edges. The Newman and Watts model turned out to be easier to

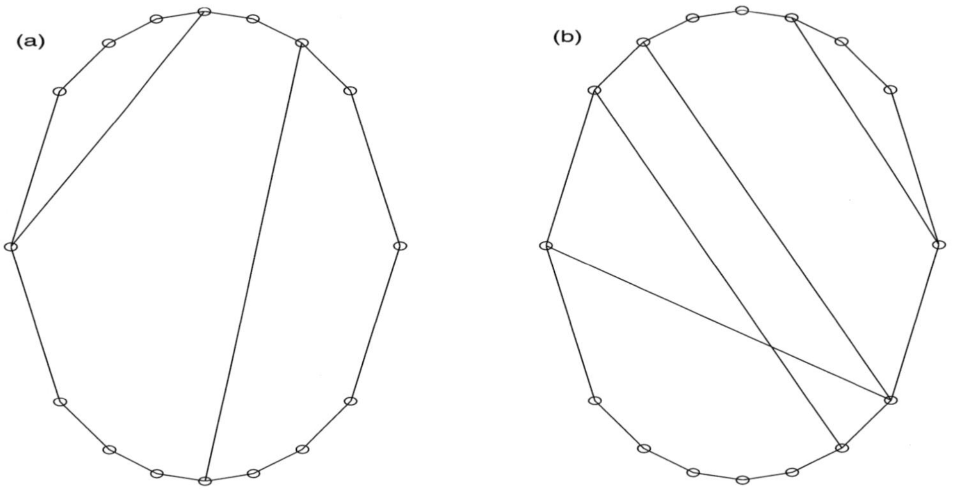

Figure 2.5. *Small-world networks with* $N = 16$, $K = 1$, (a) $\phi = 0.1$, (b) $\phi = 0.3$.

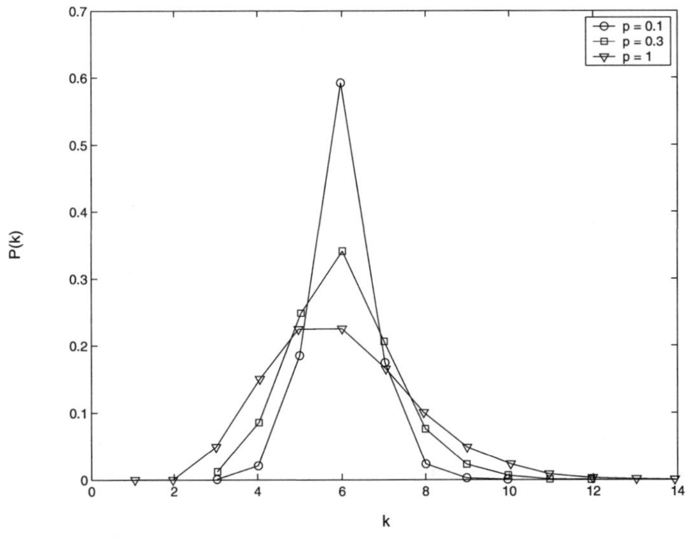

Figure 2.6. *Connectivity distributions* $P(k)$ *of the small-world network model with three different disorder parameters* $p = 0.1$, $p = 0.3$, *and* $p = 1$ *with networks of size* 10^4 *and* $K = 3$.

analyze since the network cannot become disconnected after rewiring. Figure 2.5 shows a small-world network with $N = 16$, $K = 1$, and $\phi = 0.1$. The degree or connectivity distribution $P(k)$ (k is the degree or number of connections per node) of the small-world network model depends on the disorder parameter p_{WS}. That is, $p_{WS} = 0$ implies that the connectivity distribution is given by the delta function $\delta(k-2K)$, where K is the coordination number in the network. As p_{WS} approaches 1, the connectivity distribution converges to that

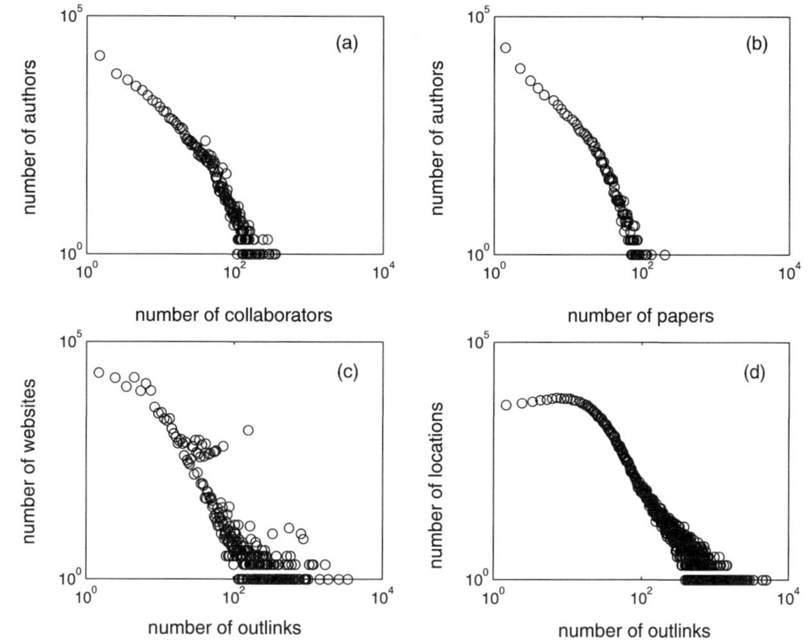

Figure 2.7. *The power-law distributions observed in* (a), (b) *scientific collaboration networks (The Los Alamos e-Print Archive)* [12]; (c) *the World Wide Web (nd.edu domain)* [13]; *and* (d) *the location-based network of the city of Portland* [25].

obtained from the Erdős and Rényi model. Figure 2.6 shows the degree distribution for a small-world model as a function of the disorder parameter p_{WS}, when $N = 10^4$ and $K = 3$. "Unfortunately," small-world networks do not exhaust all possibilities. The bell-shaped node degree distributions observed in the Erdős–Rényi, Watts–Strogatz, and Newman–Watts models are in contrast with the power-law degree distributions observed in a number of biological [18], social [19, 20, 21, 22, 23, 24, 25], and technological [19, 20, 26, 27] networks (see Figure 2.7). (Power-law degree distributions—also known as Pareto distributions—are given by the parametric family

$$P(k) = Ck^{-\gamma},$$

where γ is typically between 2 and 3 (infinite variance) and C is a normalization constant (makes $P(k)$ a probability density function)). Networks that fit power-law degree distributions (free of characteristic scale) well have a small number of highly connected nodes; that is, most nodes have a small number of connections. Barabási and Albert [19] dubbed this type of structures *scale-free* networks.

The number of sexual partners in the 1996 Swedish survey of sexual behavior [24] fits a power-law distribution and the number of sexual partners of Cornell University undergraduates from the *1990 Cornell Undergraduate Social and Sexual Patterns (CUSSP) Survey* [28] can also be fitted well by such distribution (see Figure 2.8). These observations, for example, support the view that sex-education campaigns must target individuals with the highest number of partners (core group) [29]. The location-based network of the city of Portland, Oregon also exhibits a scale-free structure (Chowell et al. [25]). Here, nodes represent locations while directed connections between locations represent the average movement

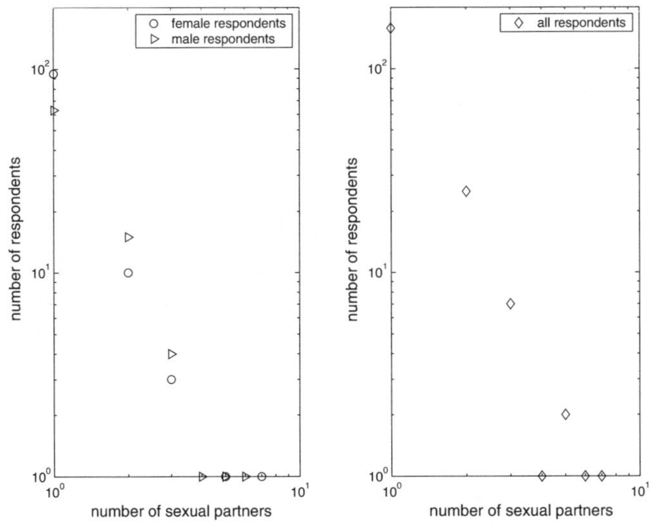

Figure 2.8. *Number of sexual partners of Cornell University undergraduates from the* 1990 Cornell Undergraduate Social and Sexual Patterns (CUSSP) Survey [28]. *The power-law exponent for females is* 2.86, *for males is* 2.90, *and for the overall distribution is* 2.78.

of individuals in the city. The scale-free (Figure 2.7(d)) topology implies the existence of a high a number of locations with a low number of connections (i.e., households) and a small number of highly connected *hubs* (i.e., schools, hospitals, etc.). Barabási and Albert [19, 20] introduced a simple theoretical model that generates networks with a power-law degree distribution (see Figure 2.9). The Barabási–Albert (BA) algorithm starts with a small number of nodes (m_o), and at each time step, a new node connects (with m links), with higher probability, to nodes that have already accumulated a higher number of connections. The resulting network has a power-law exponent of 3 and a mean connectivity of $2m$. Thus, the BA model captures features that seem characteristic of real-world networks, namely, *growth and preferential attachment*. Figure 2.10 shows a network generated using the BA model.

Several modifications of the BA model have been studied, including edge rewiring [30], edge removal [31], growth constraints [32, 33], and edge competition [34]. Klemm and Eguíluz [35] developed an alternative algorithm for the generation of scale-free networks. These researchers incorporated memory as part of a node's ability to acquire additional links. The Klemm–Eguíluz model produces scale-free networks with high clustering coefficients, a property not generated by the BA model.

The capacity of networks to maintain essential properties when nodes are removed is a measure of their robustness. In scale-free networks, most of the nodes have low degree; hence their removal "typically" does not impact the connectivity of the remaining vertices. However, the removal of nodes with the highest degree (pressure points) of connectivity can have dramatic consequences (see Figure 2.11). This effect was first demonstrated independently and numerically by Albert [36] and Broder et al. [37], using subsets of data of the World Wide Web. The practical relevance of network robustness was highlighted by the February 2000 service denial of highly connected web servers, including Yahoo, CNN, Amazon, eBay, Excite, and E∗TRADE, following a network attack.

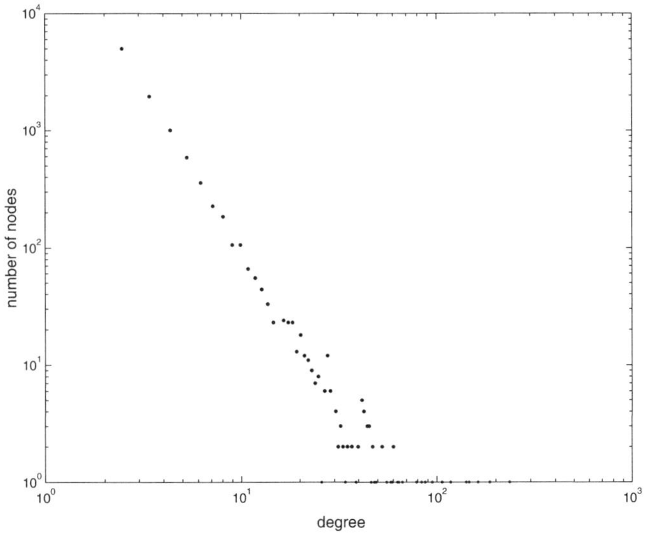

Figure 2.9. *Connectivity distribution of the BA model decays as a power law with* $\gamma = 3$. *Here,* $N = 10000$, $m_o = 3$, *and* $m = 2$.

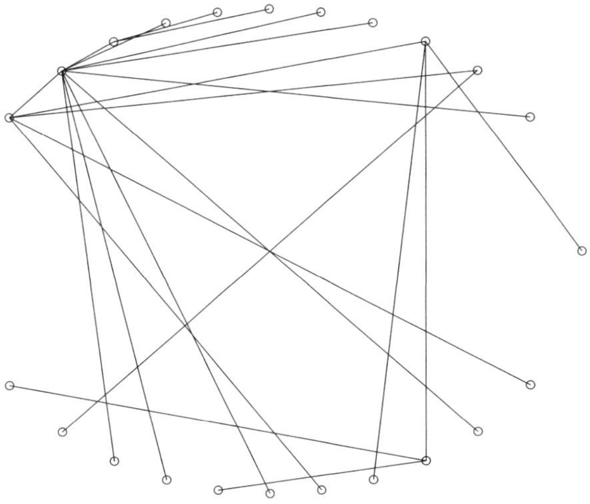

Figure 2.10. *Network of size* $N = 23$ *generated using the BA model described in the text with* $m_o = 2$ *and* $m = 1$.

Important work on the use of worst-case scenarios in the development of response policy has been carried out by Kaplan [38, 39, 10] in the context of HIV and smallpox. However, the nature of his approach does not allow for the incorporation of population structures such as those identified in [19, 20, 21, 22, 23, 25]. The importance and frequency of these networks is the main motivation behind our efforts to look at the rate of growth of epidemic outbreaks on these graphs. The focus of this paper (instigated by Ed Kaplan) is driven by these questions:

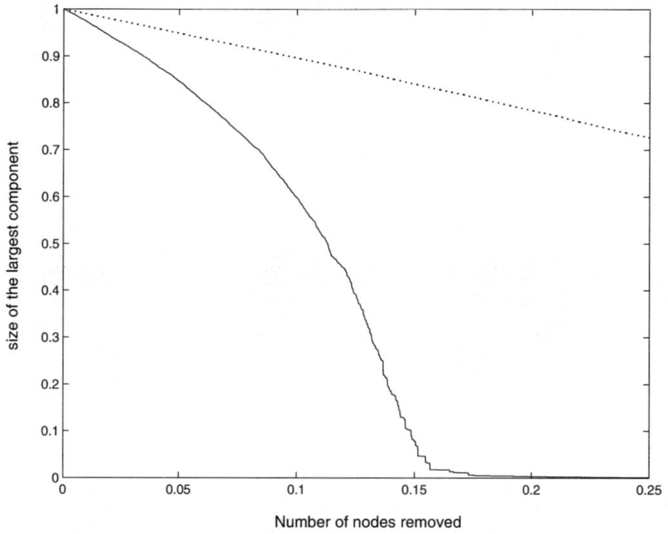

Figure 2.11. *The size of the largest component in a scale-free networks as a function of the number of nodes removed randomly (dashed line) or in decreasing order of degree; that is, hubs are removed first (solid line). The network was constructed using the BA model with* $N = 10000$, $m_o = 3$, *and* $m = 2$.

- How is the initial rate of epidemic growth affected by a population's structure?
- What are the roles of "social" topologies and the number of initial infectious sources on the rate of growth of an epidemic?

The following sections represent an initial attempt to address these questions in the context of small-world [40, 15] and scale-free networks [19]. The organization of the rest of this paper is as follows: Section 2.3 corroborates Kaplan's view of mixing in worst-case epidemics when the transmission topology is given by small-world networks; section 2.4 focuses on the study of epidemics in scale-free networks where a natural "node" hierarchy often emerges. (This structure, in some sense, "equivalent" to the concept of core group developed by Hethcote and Yorke [29], seems to provide a fluid landscape for disease spread); section 2.5 collects our conclusions, caveats, and views on implications of these results in the study of the impact of deliberate releases of biological agents.

2.3 Epidemics on Small-World Networks

Simple epidemic models such as the susceptible-infected-recovered (SIR) model have been studied on small-world networks. Moore and Newman [41] studied SIR epidemics on small world networks via site and bond percolation. In site percolation, nodes (sites) are occupied (by spins) or not and any two spins occupying nearest neighbor sites are connected by an open bond. In bond percolation, the relevant entities are bonds or edges. Bonds are sequentially visited and set open with probability p or closed with probability $1 - p$ (independently). The percolation threshold is the smallest probability p at which an infinite cluster of sites emerges when sites or bonds (depending on the type of percolation) are occupied with that probability.

SIR epidemic processes are built on the assumption that nodes are occupied by individuals who can be infected by neighbors connected by edges or bonds (Grassberger [42]). When the total probability of transmission from one individual to another is greater than this threshold, the disease explodes; that is, a "giant" component (whose size is the size of the epidemic) appears [14, 43]. In an epidemic that starts with a single infectious source and spreads as a bond percolation process, the subset of nodes (individuals) that can be reached from the initial infective individuals by traversing only open bonds is the size of the epidemic outbreak. Newman, Jensen, and Ziff [44] studied SIR epidemics on a two-dimensional small-world network via bond percolation. His investigation was motivated by the study of disease transmission in plants coming from two sources: from nearest neighbor (plants) and long-distance contacts (vectors). Epidemics on small-world networks can exhibit phase transition behavior; that is, there is a critical value of the disorder parameter (p_c) such that for values of $p_{\mathrm{WS}} > p_c$ self-sustained oscillations in the number of infected individuals in susceptible-infected-susceptible (SIS) epidemics are possible (Kuperman and Abramson [45]).

In order to study the role of the disorder parameter p_{WS} (small-world networks) on the initial rate of growth of disease spread, the following algorithm is used to compute the initial (empirical) rate of growth. The number of infected individuals is computed as a function of time $I(t)$ (time is discrete) for a small time range ($t < t_c$). The value of t_c is selected so that $I(t)$ is still in its exponentially growing phase. The algorithm follows three steps:

1. *Computation of* t_c. The time t_c at which $I(t)$ changes cancavity, that is, the value of t at which the second derivative of $I(t)$ changes from positive to negative (see Figure 2.12).

2. *Rescaling of* $I(t)$. $\hat{I}(t) \equiv \log_e[I(t)]$ for $t < t_c$, where t_c is the value computed in Step 1.

3. *Regression on* $\hat{I}(t)$. Compute the average slope r of the best fitting line to $\hat{I}(t)$. r is the average "r" that results from 50 realizations. The average initial rate of growth r is computed as a function of the disorder parameter (p_{WS}) of small-world networks ($p_{\mathrm{WS}} \in [0, 1]$ is changed in increments of 0.01).

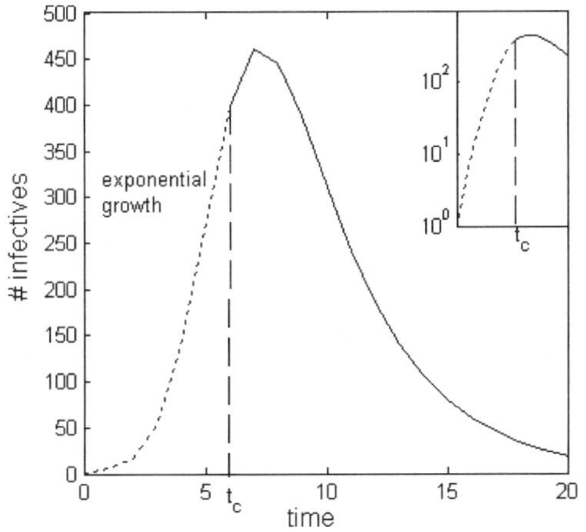

Figure 2.12. *Computation of the empirical rate of growth of epidemics on networks.*

2.3.1 Epidemiological Model

We consider an stochastic SIR epidemiological model. Hence, individuals can be in one of three epidemiological states: susceptible (S), infected (I), or recovered (R). A susceptible individual in contact with i infectious individuals may become infected in a short period of time δt with a probability given by $\hat{\beta} i \delta t$, where $\hat{\beta}$ is the constant risk of infection per unit of time and $\delta t = 1$ in this discrete time model. Similarly, infected recover with a probability given by $\hat{\gamma}\delta t$, where $\frac{1}{\hat{\gamma}}$ is the mean period of infectivity. After recovery, individuals get full immunity to the disease.

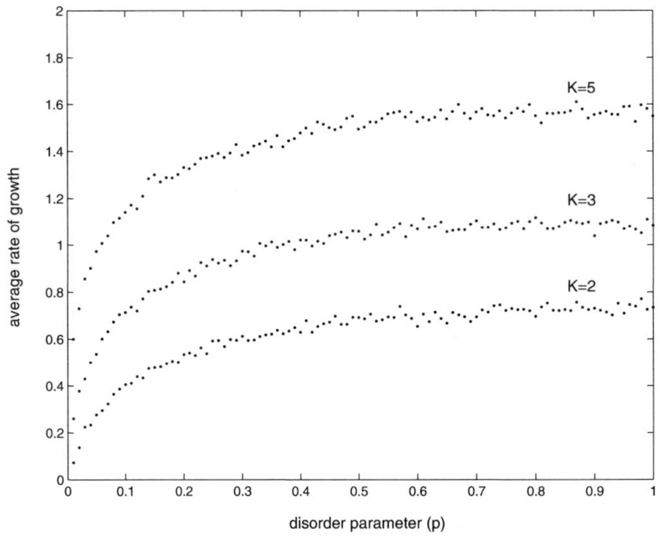

Figure 2.13. *Rate of growth of epidemics in small-world networks of size* $N = 10^4$ *with* $K = 2$, $K = 3$, *or* $K = 5$ *and as a function of the disorder parameter* $p \in [0, 1]$. *Averages are taken from* 50 *realizations.* p *is incremented by* 0.01. *Disease parameters are* $\hat{\beta} = \frac{4}{7}$, $\gamma = \frac{2}{7}$, *and* $I(0) = 1$.

Epidemics were simulated on small-world networks of size $N = 10^3$ with $K = 2$, $K = 3$, and $K = 5$ (K is the coordination number of small-world networks). Empirical results on the average rate of growth were obtained from the mean of 50 realizations with disease parameters $\hat{\beta} = \frac{4}{7}$ and $\hat{\gamma} = \frac{2}{7}$. Simulations were started by placing a single infectious source randomly at a node. The average rate of growth increases in a nonlinear fashion as the disorder in the network grows. It saturates when it is close to 1 (totally random networks) with $r_{\text{random}} \approx 0.7481$ and $K = 2$. Figure 2.13 shows the average (from 50 realizations) rate of growth of epidemics in small-world networks of size $N = 10^3$ (with $K = 2$, $K = 3$, and $K = 5$) as a function of the disorder parameter p_{WS}. The rate of growth in small-world networks also increases as the coordination number (K) increases.

The initial rate of growth of epidemics depends on the network topology. The simulations start with an initial (small) group of infective individuals chosen from those with highest connectivity. The resulting growth rate is computed and compared to that resulting from epidemics where the initial infectious sources are chosen (uniformly) at random. Naturally, epidemics that started at the most connected nodes exhibited a higher average rate of growth (see Figure 2.14). Higher rates of growth are observed as the number of initial infec-

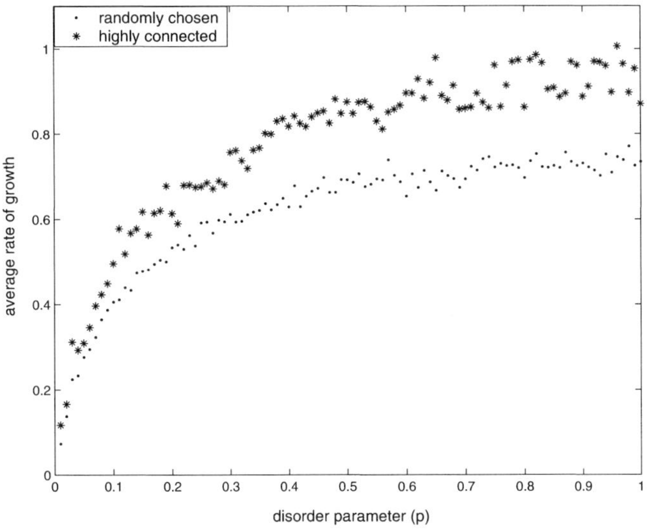

Figure 2.14. *Rate of growth is higher when the epidemics start at the individuals (nodes) with the highest connectivity* (∗) *rather than chosen uniformly at random* (.) *with* $N = 10^4$ *and* $K = 2$. *Averages are taken from* 50 *realizations. Disease parameters are* $\hat{\beta} = \frac{4}{7}$, $\gamma = \frac{2}{7}$, *and* $I(0) = 1$.

tious sources (always small compared to the size of the network) in the network increases (see Figure 2.15).

2.4 Epidemics on Scale-Free Networks

Pastor-Satorras and Vespignani [46] studied an SIS epidemic model on scale-free networks (generated using the BA model) and found that the disease may persist independently of its transmissibility. That is, the *basic reproductive number* R_0, routinely computed in classical mathematical epidemiology sometimes loses its meaning in their setting. The small number of nodes with a high connectivity (hubs) observed in scale-free networks are responsible for "zero" threshold behavior. This observation gives rise to the following question: Can a control strategy be implemented that restores a positive epidemic threshold?

This question was studied by Pastor-Satorras and Vespignani [47] and independently by Dezso and Barabási [48]. Both groups concluded that targeted immunization campaigns towards the most connected nodes or hubs increase the probability of recovering finite epidemic threshold behavior. A similar result has been observed on the spread of foot and mouth disease (FMD) in three distinct regions of Uruguay (Rivas et al. [49]). A contrasting result has been established on alternative highly clustered scale-free networks [35]. Here, a finite epidemic threshold has been observed on (SIS) epidemics (Eguíluz and Klemm [50]).

An extensive number of simulations have been carried out of SIR epidemics on scale-free networks. We compute the average rate of growth from the mean of 50 realizations of the epidemic process with two sets of initial conditions: We place the infective source at a randomly selected node or at the most connected node (highest degree). Simulations are carried out on small-world and scale-free networks of the same size ($N = 10^4$) and

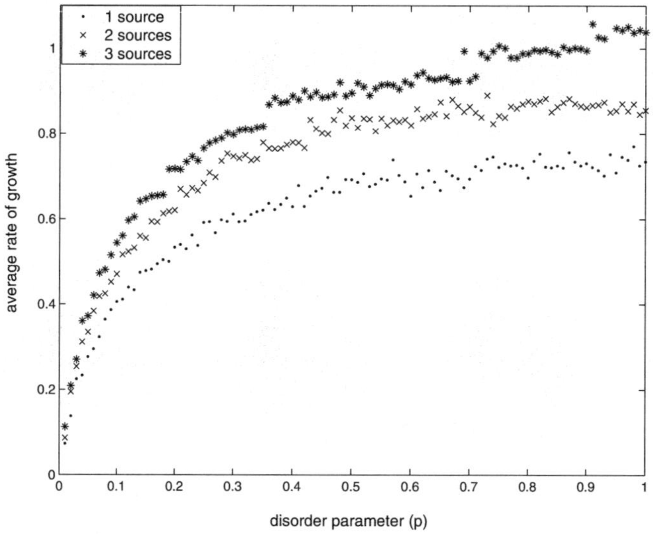

Figure 2.15. *Average rate of growth of multiple source epidemics: one source* $[I(0) = 1]$ (.), *two sources* $[I(0) = 2]$ $(\times)$, *and three sources* $[I(0) = 3]$ $(*)$ *with* $N = 10^4$ *and* $K = 2$. *Averages are taken from* 50 *realizations. Disease parameters are* $\hat{\beta} = \frac{4}{7}$, $\gamma = \frac{2}{7}$.

average connectivity ($\bar{k} = 4$). Significantly higher rates of growth are observed in scale-free networks (see Figure 2.16). Hence, scale-free topologies seem to provide ideal scenarios for the study of worst-case epidemics. The existence of highly connected nodes or hubs in scale-free networks plays a central role on the rate at which viruses (or information) spreads. Or, in other words, the concept of core group [29] is still critical to disease spread on topologically defined networks.

2.5 Conclusions and Caveats

The development and implementation of policies that deal with the deliberate release of biological agents must consider worst possible situations and such scenarios are highly dependent on the network of individual interactions (social topology). Hence, gaining some understanding of the nature of social structures that facilitate disease spread is critical. Kaplan et al. [10] assume that random mixing corresponds to a worst-case scenario and (using such a setup) conclude that, in the case of a bioterrorist smallpox attack, mass vaccination is a better policy than ring vaccination. Halloran et al. [51], using a stochastic model with a structured community of 2000 people, conclude that targeted vaccination outperforms mass vaccination. The disagreement in results may be directly related to the preassumed population structure and mixing topology (network of interactions).

Here, we have tried to identify under what conditions random mixing can be used to model worst-case scenarios. We have found that on small-world networks, random mixing indeed supports *epidemics* with the highest average rate of growth. However, this is not necessarily the case on scale-free networks. The nature of the mixing between individuals (the connectivity hierarchy in scale-free networks) plays a key role on the initial average

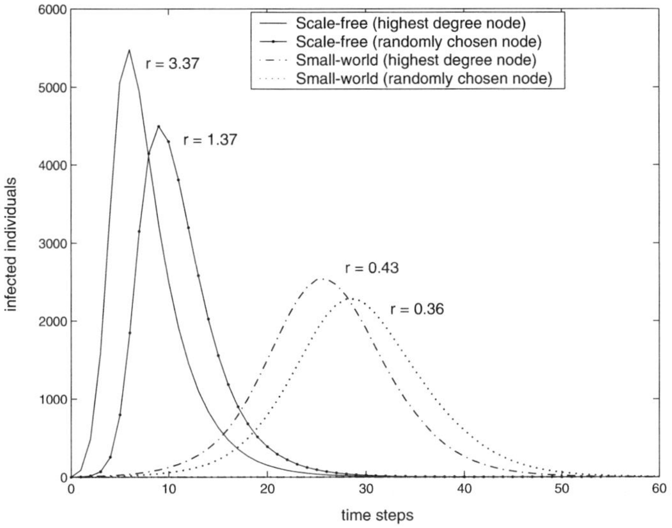

Figure 2.16. *Average number of infected individuals from* 50 *realizations over time in small-world* ($p = 0.1$) *and scale-free networks of the same size* ($N = 10^4$) *and average connectivity* ($\bar{k} = 4$). *Two different initial conditions are considered: The initial infectious source is placed in a randomly selected or in the most connected node (highest degree). The rates of growth of the epidemics are higher in scale-free networks. Disease parameters are* $\hat{\beta} = \frac{4}{7}$, $\gamma = \frac{2}{7}$, *and* $I(0) = 1$.

rate of growth of an initial epidemic outbreak. The inherent connectivity hierarchy of scale-free networks and the "sensitivity" (lack of robustness) to the removal of key nodes (most connected individuals), in some sense, corresponds to the concept of core group [29]. Highly connected nodes are pressure points in the network, and consequently, their identification and management must be considered in the development and implementation of a logistic plan of response to the threat of a bioterrorist attack.

The location-based network of the simulated city of Portland possesses a scale-free nature (Chowell et al. [25]). Hence, the initial rate of growth of epidemics in this city can be expected to be (on average) significantly higher whenever a source is placed at a hub (see Figure 2.16).

While the use of classical epidemiological approaches has been and will continue to be central in the study of disease spread [5, 1, 4], it is clear that the study of the potential initial impact of the deliberate release of pathogens on unsuspecting populations must be addressed on various setups. The reasons for such an approach are multiple, the interest is no longer on long-term disease dynamics but on the elaboration of policies that result on timely responses (see Rivas et al. [49]). Such a degree of urgency requires the identification of the most sensitive points of release (pressure points) and the use of models that account for multiple releases. The approach used here has its own drawbacks. The most important comes from the fact that temporal dynamics are not explicitly considered. The incorporation of temporal dynamics on networks poses challenges and opportunities for serious theoretical work. Although we have fallen short in addressing the challenge posed by Ed Kaplan, we hope that our results have at least helped clarify it because the development of contingency plans in worst-case scenarios is fundamental.

Acknowledgments

We thank Shu-Fang Hsu Schmitz, Albert László Barabási, and Mark Newman for providing the CUSSP (Cornell Undergraduate Social and Sexual Patterns Survey), World Wide Web (nd.edu domain), and scientific collaboration (Los Alamos Archive) datasets, respectively.

Bibliography

[1] F. Brauer and C. Castillo-Chavez, *Mathematical Models in Population Biology and Epidemiology*, Springer-Verlag, New York, 2000.

[2] O. Diekmann and J. Heesterbeek, *Mathematical Epidemiology of Infectious Diseases: Model Building, Analysis and Interpretation*, Wiley, New York, 2000.

[3] C. Castillo-Chavez, Z. Feng, and W. Huang, *On the computation of R_0 and its role on global stability*, in Mathematical Approaches for Emerging and Reemerging Infectious Diseases: An Introduction, IMA Vol. Math. Appl. 125, C. Castillo-Chavez, S. Blower, P. van den Driessche, D. Kirschner, and A. A. Yakubu, eds., Springer-Verlag, Berlin, Heidelberg, New York, 2002, pp. 229–250.

[4] C. Castillo-Chavez, S. Blower, P. van den Driessche, D. Kirschner, and A.-A. Yakubu, *Mathematical Approaches for Emerging and Reemerging Infectious Diseases Part* I: *An Introduction to Models, Methods, and Theory; Part* II: *Models, Methods, and Theory*, IMA Vol. Math. Appl. 125, Springer-Verlag, Berlin, Heidelberg, New York, 2002.

[5] R. M. Anderson and R. M. May, *Infectious Diseases of Humans*, Oxford University Press, Oxford, UK, 1991.

[6] A. Nold, *Heterogenity in disease-transmission modeling*, Math. Biosci., 52 (1980), pp. 227–240.

[7] S. Busenberg and C. Castillo-Chavez, *Interaction, pair formation and force of interaction terms in sexually transmitted diseases*, in Mathematical and Statistical Approaches to AIDS Epidemiology, C. Castillo-Chavez, ed., Springer-Verlag, Berlin, 1989, pp. 289–300.

[8] C. Castillo-Chavez, J. X. Velasco-Hernandez, and S. Fridman, *Modeling contact structures in biology*, in Frontiers of Theoretical Biology, Lecture Notes in Biomathematics 100, S. A. Levin, ed., Springer-Verlag, New York, 1994, pp. 454–491.

[9] C. Castillo-Chavez and A. Yakubu, *Intra-specific competition, dispersal and disease dynamics in discrete-time patchy environments*, in Mathematical Approaches for Emerging and Reemerging Infectious Diseases: An Introduction, IMA Vol. Math. Appl. 125, C. Castillo-Chavez, S. Blower, P. van den Driessche, D. Kirschner, and A. A. Yakubu, eds., Springer-Verlag, Berlin, Heidelberg, New York, 2002, pp. 165–181.

[10] E. H. Kaplan, D. Craft, and L. Wein, *Emergency response to a smallpox attack: The case of mass vaccination*, Proc. Natl. Acad. Sci. USA, 99 (2002), pp. 10935–10940.

[11] B. Bollobás, *Random Graphs*, Academic Press, London, 1985.

[12] http://www.nd.edu/~networks/database/index.html.

[13] http://www.santafe.edu/~mark/collaboration/.

[14] D. STAUFFER AND A. AHARONY, *Introduction to Percolation Theory*, revised 2nd ed., Taylor and Francis, London, 2002.

[15] D. J. WATTS AND S. H. STROGATZ, *Collective dynamics of "small-world" networks*, Nature, 383 (1998), pp. 440–442.

[16] S. MILGRAM, *The small world problem*, Psychology Today, 2 (1967), pp. 60–67.

[17] M. E. J. NEWMAN AND D. J. WATTS, *Renormalization group analysis of the small-world network model*, Phys. Lett. A, 263 (1999), pp. 341–346.

[18] H. JEONG, B. TOMBOR, R. ALBERT, Z. N. OLTVAI, AND A.-L. BARABÁSI, *The large-scale organization of metabolic networks*, Nature, 407 (2000), pp. 651–654.

[19] A.-L. BARABÁSI AND R. ALBERT, *Emergence of scaling in random networks*, Science, 286 (1999), pp. 509–512.

[20] A.-L. BARABÁSI, R. ALBERT, AND H. JEONG, *Mean-field theory for scale-free random networks*, Phys. A, 272 (1999), pp. 173–187.

[21] A.-L. BARABÁSI, H. JEONG, R. RAVASZ, Z. NÉDA, T. VICSEK, AND A. SCHUBERT, *Evolution of the social network of scientific collaboration*, Phys. A, 311 (2002), pp. 590–614.

[22] M. E. J. NEWMAN, *The structure of scientific collaboration networks*, Proc. Natl. Acad. Sci. USA, 98 (2001), pp. 404–409.

[23] M. E. J. NEWMAN, *Who is the best connected scientist? A study of scientific coauthorship networks*, Phys. Rev. E, 64 (2001), 016131; Phys. Rev. E, 64 (2001), 016132.

[24] F. LILJEROS, C. R. EDLING, L. A. N. AMARAL, H. E. STANLEY, AND Y. ABERG, *The web of human sexual contacts*, Nature, 411 (2001), pp. 907–908.

[25] G. CHOWELL, J. M. HYMAN, S. EUBANK, AND C. CASTILLO-CHAVEZ, *Analysis of a Real-World Network: The City of Portland*, Los Alamos Unclassified Report LA-UR-02-6658, BSCB Department Report BU-1604-M, Cornell University, Ithaca, NY, 2002.

[26] R. KUMAR, P. RAGHAVAN, S. RAJAGOPALAN, D. SIVAKUMAR, A. S. TOMKINS, E. UPFAL, *The web as a graph*, in Proceedings of the 19th ACM SIGACT-SIGMOD-AIGART Symposium on Principles of Database Systems, ACM Press, New York, 2000, pp. 1–10.

[27] M. FALOUTSOS, P. FALOUTSOS, AND C. FALOUTSOS, *On power-law relationships of the Internet topology*, in Proceedings of SIGCOMM '99, ACM Press, New York, 1999, pp. 251–262.

[28] C. M. CRAWFORD, S. J. SCHWAGER, AND C. CASTILLO-CHAVEZ, *Research design for the Cornell undergraduate social and sexual patterns survey*, Technical Report HSS-90-CC1/BU-1083-M, Human Service Studies, Cornell University, Ithaca, NY, 1990.

[29] H. W. HETHCOTE AND J. A. YORKE, *Gonorrhea Transmission Dynamics and Control*, Lecture Notes in Biomathematics 56, Springer-Verlag, Berlin, 1984.

[30] R. ALBERT AND A.-L. BARABÁSI, *Topology of evolving networks: Local events and universality*, Phys. Rev. Lett., 85 (2000), pp. 5234–5237.

[31] S. N. DOROGOVTSEV AND J. F. F. MENDES, *Language as an evolving word web*, Proc. Roy. Soc. London Ser. B, 268 (2001), pp. 2603–2606.

[32] L. A. N. AMARAL, A. SCALA, M. BARTÉLÉMY, AND H. E. STANLEY, *Classes of small-world networks*, Proc. Natl. Acad. Sci. USA, 97 (2000), pp. 11149–11152.

[33] S. N. DOROGOVTSEV AND J. F. F. MENDES, *Scaling properties of scale-free evolving networks: Continuous approach*, Phys. Rev. E, 63 (2001), pp. 1–19.

[34] G. BIANCONI AND A.-L. BARABÁSI, *Competition and multiscaling in evolving networks*, Europhys. Lett., 54 (2001), pp. 436–442.

[35] K. KLEMM AND V. M. EGUÍLUZ, *Highly clustered scale-free networks*, Phys. Rev. E, 65 (2002), 036123.

[36] R. ALBERT, H. JEONG, AND A.-L. BARABÁSI, *Diameter of the World Wide Web*, Nature, 401 (1999), pp. 130–131.

[37] A. BRODER, R. KUMAR, F. MAGHOUL, P. RAGHAVAN, S. RAJAGOPALAN, R. STATA, A. TOMKINS, AND J. WIENER, *Graph structure in the web*, in Proceedings of the 9th International World Wide Web Conference, 2000, pp. 309–320.

[38] E. H. KAPLAN, *Asymptotic worst-case mixing in simple demographic models of HIV/AIDS*, Math. Biosci., 108 (1992), pp. 141–156.

[39] E. H. KAPLAN, *Mean-max bounds for worst-case endemic mixing models*, Math. Biosci., 105 (1991), pp. 97–109.

[40] S. H. STROGATZ, *Exploring complex networks*, Nature, 410 (2001), pp. 268–276.

[41] C. MOORE AND M. E. J. NEWMAN, *Epidemics and percolation in small-world networks*, Phys. Rev. E, 61 (2000), pp. 5678–5682.

[42] P. GRASSBERGER, *On the critical behavior of the general epidemic process and dynamical percolation*, Math. Biosci., 63 (1983), pp. 157–172.

[43] R. DURRETT, *Ten Lectures on Particle Systems*, Lecture Notes in Math. 1608, Springer-Verlag, New York, 1995, pp. 97–201.

[44] M. E. J. NEWMAN, I. JENSEN, AND R. M. ZIFF, *Percolation and epidemics in a two-dimensional small world*, Phys. Rev. E, 65 (2002), 021904.

[45] M. KUPERMAN AND G. ABRAMSON, *Small-world effect in an epidemiological model*, Phys. Rev. Lett., 86 (2001), pp. 2909–2912.

[46] R. PASTOR-SATORRAS AND A. VESPIGNANI, *Epidemic spreading in scale-free networks*, Phys. Rev. Lett., 86 (2001), pp. 3200–3203.

[47] R. PASTOR-SATORRAS AND A. VESPIGNANI, *Immunization of complex networks*, Phys. Rev. E, 65 (2002), 036104.

[48] Z. DEZSO AND A.-L. BARABÁSI, *Halting viruses in scale-free networks*, Phys. Rev. E. 65 (2002), 055103.

[49] A. L. Rivas, S. E. Tennenbaum, J. P. Aparicio, A. L. Hoogesteyn, R. W. Blake, and C. Castillo-Chavez, *Critical response time (time available to implement effective measures for epidemic control): Model building and evaluation*, Canadian J. Veterinary Res., to appear.

[50] V. M. Eguíluz and K. Klemm, *Epidemic threshold in structured scale-free networks*, Phys. Rev. Lett., 89 (2002), 108701.

[51] M. E. Halloran, I. M. Longini, Jr., A. Nizam, and Y. Yang, *Containing bioterrorist smallpox*, Science, 298 (2002), pp. 1428–1432.

Chapter 3

Chemical and Biological Sensing: Modeling and Analysis from the Real World

Ira B. Schwartz,[*] Lora Billings,[†] David Holt,[‡] Anne W. Kusterbeck,[‡] and Ioana Triandaf[*]

3.1 Introduction

Under the threat of biological and chemical attacks, new technologies are needed to detect minute amounts of chemical and biological substances prior to their dissemination on a large scale. However, current biochemical sensors fail to detect low levels of environmental changes which come in the form of biochemical toxins, electrical stimuli, or even physical stress. Our goal is to examine a few cases from the real sensing world and introduce a few techniques which lead to an understanding of the biochemical information. The goal of these techniques is to be able to extract information changes due to changes in environment using analytic, geometric, and mathematical modeling tools.

Even prior to September 11, the field of biological and chemical sensors has flourished. The term "biosensor" itself has become synonymous with various laboratory instruments and has resulted in a wealth of literature on the subject covering both theory and experiment, much of it related to the commercial instruments. (See [20, 19, 18].) With all of the development of these sensors, many of them are still not rugged enough to take into the field and be of practical use. An important application in the field is that of TNT molecule detection. As a result, a real workhorse called the continuous flow displacement immunosensor (CFI) has filled the need for use in practice. The CFI sensors have exploded on the scene and are in use in detecting trace amounts of explosives in soil and water and steroids in fluids, as well as illicit drugs.

In all of the above situations, the CFI sensors operate in a constant flow regime. However, we will show that sensitivity can be improved by changing the steady flow to a pulsatile flow in a reaction-convection model. The results predict an increase in signal

[*]Nonlinear Dynamics Section, U.S. Naval Research Laboratory, Code 6792, Washington, DC 20375.
[†]Department of Mathematical Sciences, Montclair State University, Upper Montclair, NJ 07043.
[‡]Center for Bio-Molecular Science and Engineering, U.S. Naval Research Laboratory, Code 6900, Washington, DC 20375.

sensitivity by a factor of two, which clearly improves detection discrimination.

In another application, we consider a situation of a cellular sensor where a good model fails to exist but where there are good experimental data to analyze. Spiking phenomena in cells have long been considered a transmitter of information. Recently, analysis has focused on a dynamic cellular response to a signal that is controlled externally by a parameter [33, 3]. A great deal of research has been performed to decode this signal from spikes, or action potentials, in voltage time series. Difficulty arises in short data sets with relatively few spikes, common in rapid detection applications. Therefore, we describe two new biosensor methods for short data sets that use area measurements to detect subtle changes in the signal, namely, spike area distribution and local expansion rates derived from the dynamic reconstruction of a time series.

Many results describing spiking onset in neuron cells have been studied theoretically as well as experimentally [30, 26, 9, 17, 31, 13, 5, 14]. One interpretation is that a signal is encoded by the time intervals between spikes, known as the interspike intervals (ISI) [33, 24, 28]. ISI studies have complemented linear statistical methods currently used to interpret coding and parameter changes in spiking systems [29]. Using information based on the ISI generated by neurons, numerous studies have used neurons as biological detectors of external agents, such as ionic channel blockers and synaptic modulators [40, 41, 39]. Most of this analysis is linearly based and examines the effect of toxins on spike rate and amplitude change. In general, these methods may require sufficiently long runs, consisting of thousands of spikes, in order to detect a change due to the presence of a foreign substance [40, 41]. Other nonlinear approaches require long data runs as well [8]. In our experiments, data sets typically consisted of 100–400 spikes, far fewer than required for these methods.

In this chapter, we have made a comparison of the dynamics between normal (control) spiking cells and cells treated with small concentrations of a channel blocker. Here, we use two approaches to detect the presence of very small amounts of a foreign substance, or toxin, when operating with relatively few spikes. (By few spikes, we mean an order of magnitude fewer than in most previous studies.) One approach is a nonlinear dynamics measurement of the distribution of local expansion rates, or how groups of points expand/contract locally in time when examined in dynamically reconstructed space (see below for details). The other uses spike area distribution, which is a measure of the statistical spread in computed areas under measured spikes in the data. The main hypothesis which was tested in each of two methods was whether or not a significant statistical change could be detected between the control and treated samples of cell data. Since measurable differences between control and treated samples are observed, it suggests that the methods introduced here might lead to improved techniques for rapid cell drug and toxin detection.

Our final topic of discussion is how to analyze data from a high-dimensional theoretical model. Although the cellular section is based on time series analysis, there exists no current model of the cells used experimentally that captures the observed behavior. Therefore, to illustrate a comparison between modeling and time series analysis, we consider the problem of nonstationary parameter detection in a well defined physical problem [37]. We illustrate how the time series embedding procedure compares with an exact computation when applied to discovering external and internal changes in a continuum mechanical system.

3.2 Continuous Flow Sensors: Predicting an Increase in Performance

When a biochemical assay is technologically packaged into a "biosensor," it is customary to deliver the reagents to and remove products from the reactive zone(s) using fluid streams

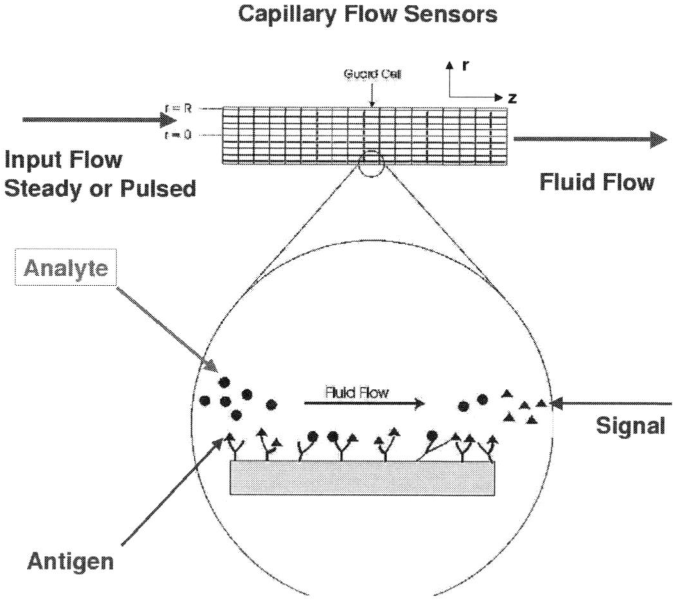

Figure 3.1. *Schematic representation of a displacement assay and a computational grid used to model the bead, membrane, and capillary solid supports. Antibodies are immobilized on a substrate and then saturated with labeled antigen (△). A sample containing the analyte (•) is injected over the support and causes a displacement of the labeled species, which is detected downstream.*

driven by pressure or electro-osmotic forces. Except for transients at the initiation and stoppage of flow, biosensors are typically operated under steady flow conditions. The continuous flow immunosensor (CFI) developed at the Naval Research Laboratory is a biosensor of this nature. (See Figure 3.1.)

Antibodies against target analytes are immobilized on the surface of a solid support and saturated with a fluorescent tagged antigen analogue. A buffer stream flowing over the antibodies establishes a slowly decreasing fluorescent baseline as natural dissociation of the antigen analogue from the immobilized antibodies occurs. Upon injection of a sample, the target species competes with the analogue for binding sites, which results in a spike in the level of the fluorescent compound detected downstream. Again, this "displacement" assay procedure was developed and is utilized at steady flow rates. It is likely that, as with most dynamic systems, more interesting and potentially improved behavior can be obtained when the forcing function (the flow rate) itself is time dependent. It may be possible to bring the fluid convection into a quasi-resonance condition with the reaction and/or diffusion processes occurring in the biosensor and enhance its performance in terms of sensitivity. Numerical simulations of an idealized displacement CFI biosensor system operating under pulsed-flow conditions (sinusoidal ramping) have been used to investigate this hypothesis. This idealized biosensor is taken to be a cylindrical capillary in which the biomolecule components are immobilized on the inner wall, as depicted in Figure 3.1. The capillary and the corresponding continuity equation are spatially discretized in the axial and radial directions to account for convective and diffusive mass transfer.

3.2.1 The Reaction-Convection Model

In the sensing zone, the following reactions are handled through appropriate boundary conditions:

$$Ab + Ag \leftrightarrow AbAg, \tag{3.1}$$

$$Ab + Ag^* \leftrightarrow AbAg^*. \tag{3.2}$$

The sum of (3.1) and (3.2) is the displacement reaction:

$$AbAg^* + Ag \leftrightarrow AbAg + Ag^*. \tag{3.3}$$

The capillary consists of two other sections in addition to the biosensing area: the upstream sample loading zone and the downstream signal detection zone. A detailed description of the model can be found in [20]. Here we just give a sketch of the modeling equations and some results.

Most capillary systems may be modeled as a generic two-dimensional grid representing the plane which bisects a cylinder along its axis, as depicted in Figure 3.1. The grid is subdivided into an injection of sample loading region, a reaction region, and a detection region. The sensor is characterized by antigen-antibody interactions at the boundaries (walls), while the systems based on porous media (beads and membranes) have binding events occurring in volume of the reactive region. The fate of a species j is governed by mass conservation. The continuity equation is given in

$$\frac{\partial c_j}{\partial t} = -\nabla \cdot (c_j \widehat{u}) - \nabla \cdot \widehat{J_j} + R_j, \tag{3.4}$$

where $\widehat{u}$ is the velocity of the fluid, $\widehat{J_j}$ is the diffusive flux, and R_j is the change in concentration c_j due to reactions. To model the sensor, several assumptions were made to reduce the complexity of (3.4):

1. Convection occurs only in the axial (z) direction.
2. Diffusion is significant only in the radial direction.
3. Diffusivities are spatially invariant (i.e., independent of concentration).

Based on the above assumptions, we have that

$$\frac{\partial c_j}{\partial t} = -\frac{\partial}{\partial z}(c_j u(r)) + D_j \left[r^{-1} \frac{\partial}{\partial r} \left(r \frac{\partial c_j}{\partial r} \right) \right] + R_j. \tag{3.5}$$

Assuming that k_1 and k_2 are forward and reverse rates for (3.1), while k_3 and k_4 are those for (3.2), the combined rate laws for the three immobilized species are

$$\begin{aligned} \frac{\partial \sigma_{AbAg}}{\partial t} &= k_1 \sigma_{Ab}[Ag] - k_2 \sigma\, AbAg, \\ \frac{\partial \sigma_{AbAg^*}}{\partial t} &= k_3 \sigma_{Ab}[Ag^*] - k_4 \sigma_{AbAg^*}, \\ \frac{\partial \sigma_{Ab}}{\partial t} &= -\left(\frac{\partial \sigma_{AbAg^*}}{\partial t} + \frac{\partial \sigma_{AbAg}}{\partial t} \right). \end{aligned} \tag{3.6}$$

In (3.6), σ_j represents surface densities, and the square brackets $[\cdot]$ denote concentrations. In a well-mixed system, the rate of change in the concentration of a solution species due to a surface reaction is related to the rate of that surface reaction through the surface to volume ratio. Thus, letting ϕ denote the surface area–void volume ratio leads to

$$\begin{aligned} \frac{\partial [Ag]}{\partial t} &= -\phi \frac{\partial \sigma_{AbAg}}{\partial t}, \\ \frac{\partial [Ag^*]}{\partial t} &= -\phi \frac{\partial \sigma_{AbAg^*}}{\partial t}. \end{aligned} \tag{3.7}$$

At the boundary, we have

$$\begin{aligned} D_{Ag} \frac{\partial [Ag]}{\partial r} \Big|_{r=R} &= -\frac{\partial \sigma_{AbAg}}{\partial t}, \\ D_{Ag^*} \frac{\partial [Ag^*]}{\partial r} \Big|_{r=R} &= -\frac{\partial \sigma_{AbAg^*}}{\partial t}. \end{aligned} \tag{3.8}$$

3.2.2 Numerical Results

For simplicity, laminar fluid velocity profiles were imposed for both steady and pulsed-flow simulations. Pulse frequencies were investigated in the 0.0001–10 Hz range at a constant pulse amplitude. All stages of the assay were simulated. These include a label saturation step, a washdown step to establish a baseline, and finally the sample injection phase. The results of the simulations indicate that CFI biosensor sensitivity can indeed be modulated via pulsing the flow rate about some average and that the pulse frequency is an optimizing parameter.

Figure 3.2 depicts the simulated raw signal traces for the steady-state (0 Hz) flow condition (solid line) and the pulsed-flow frequency which gave the most signal enhancement. The pulsed-flow signal is significantly stronger than that from the steady flow simulation.

Figure 3.3 reflects the pulsed biosensor performance relative to that of the steady flow condition as a function of pulse frequency. At high ($>$ 0.1 Hz) and low ($<$ 0.001 Hz) frequencies, the signal peak levels become similar to that from the steady flow simulation. At the intermediate frequencies, the system is much more responsive to the pulse frequency. The parameter in the biosensor model to which flow pulse frequencies in the range studied can couple is the AbAg* dissociation rate. The largest signal in the simulations resulted when the pulse frequency to dissociation rate ratio was ~ 1. In Figure 3.4, a direct comparison of the spatiotemporal response of the system is made. The left panel denotes the signal concentration of the steady flow mode, while the panel on the right denotes the pulsed configuration. The practical implication of these results is that even moderate sensitivity (2x) improvements from using a pulsatile flow "tuned" to the optimum frequency can be extremely useful since the detection limits of steady flow CFIs are typically at threshold levels listed in regulatory guidelines. Thus the ability to reliably detect at some level below these thresholds would allow these sensor systems to truly function in the monitoring roles for which they were designed. In addition, real-world samples often contain components which mask or quench signals (often called a "matrix" effect) when samples contain only trace levels of the target compound. If the quenching species does not act through changes in the dissociation rate of the AbAg* complex, then the pulsed-flow technique could possibly be used to generate a signal large enough to overcome the masking phenomenon.

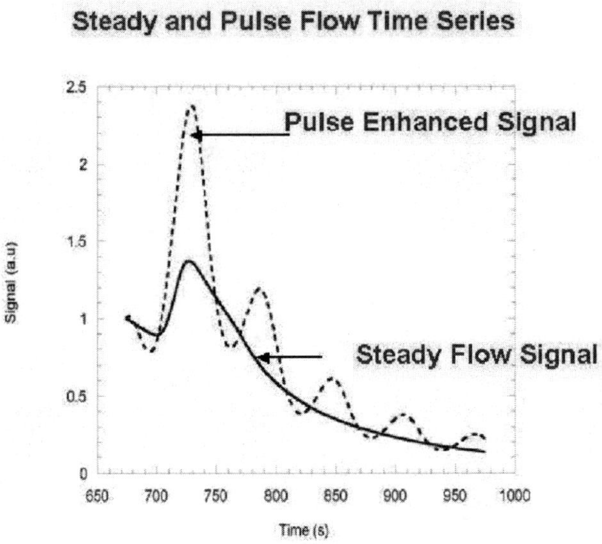

Figure 3.2. *Simulated raw signal traces for the steady state (0 Hz) flow condition (solid line) and the pulsed-flow frequency which gave the best signal enhancement.*

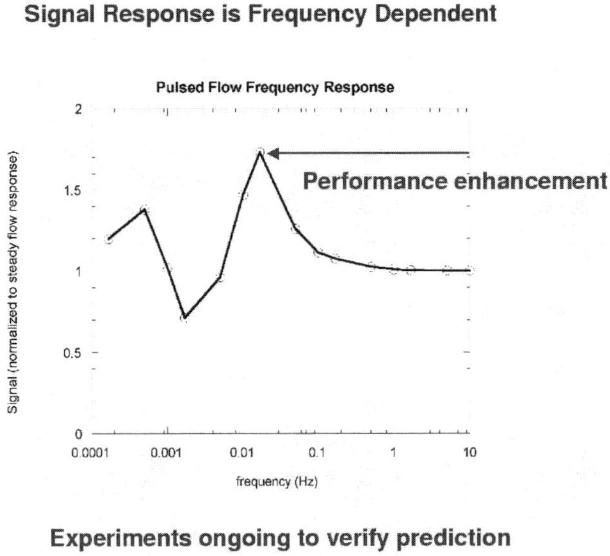

Figure 3.3. *A comparison of the sensor performance in a pulsed mode to steady flow mode.*

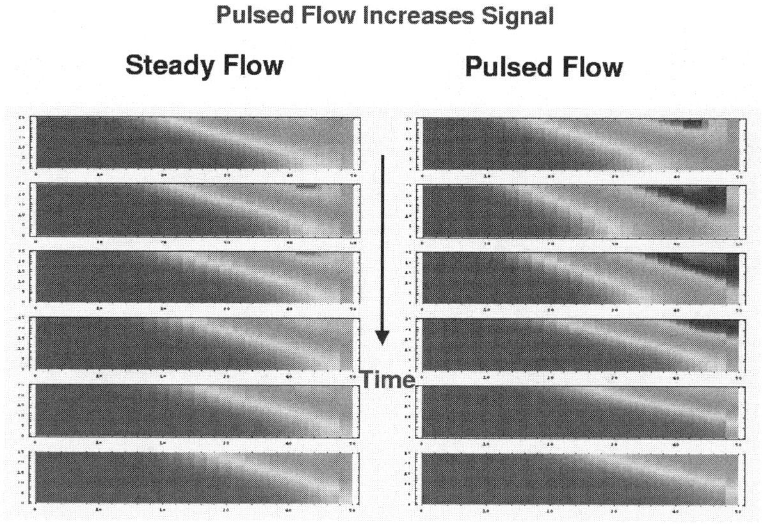

Figure 3.4. *A comparison of the space-time results of solutions to model. Detected signal concentration is measured in the darkest shades in the upper right corners. The left panel is the steady flow model, while the panel on the right is the pulsed flow model.*

3.3 Cell-Based Biosensors[1]

In this section, we examine the change in the dynamic behavior of mammalian neuronal cells subject to neurotoxins. Neuronal cells are highly sensitive to biological warfare agents, as pictured in Figure 3.5. It is a controlled experiment which reflects how nerve agents affect humans. Nerve agents, such as VX (O-ethyl S-(2-diisopropylaminoethyl) methylphosphonothioate) and GD (Pinacolyl methylphosphonofluoridate), are compounds not found in nature and are highly toxic to humans. Symptoms to overexposure are many, but in severe cases, they progress to convulsions and respiratory failure. Therefore, it is important to detect the presence of small concentrations of such nerve agents as soon as possible. This is true not just for the battlefield but also for places where the agents are destroyed on a regular basis.

In measuring the cellular response to nerve agents such as VX or GD, one can measure the action potential, or voltage, across the cell membrane. Throughout the cell membrane, there exist pores which act as ionic pumps. Nerve cells maintain a balance between several ions that travel through the pumps, such as Na^+, K^+, Cl^-, and Ca^{2+}. Flow of such ions through the pores generates a current through the cell membrane and, when measured through a resistor, becomes a measure of voltage across the membrane. The effect of VX and GD on the cell membrane is to block the ionic flow through the pores, causing a cessation of firing in the action potential.

Notice in Figure 3.5 that in the VX application to the cell, the membrane action potential ceases and then recovers. In the case when GD is added to the cell, there is no recovery. In both cases, the concentration are on the order of 10 micromolar (μM), which is quite small. Since mammalian cells are known to react in the presence of smaller concentrations, it is important to determine what kind of limits exist for such cells.

We study data from differentiated NG108-15 neuroblastoma x glioma cells. These

[1] Portions of this section have appeared previously in [36], and an early version has appeared in [4].

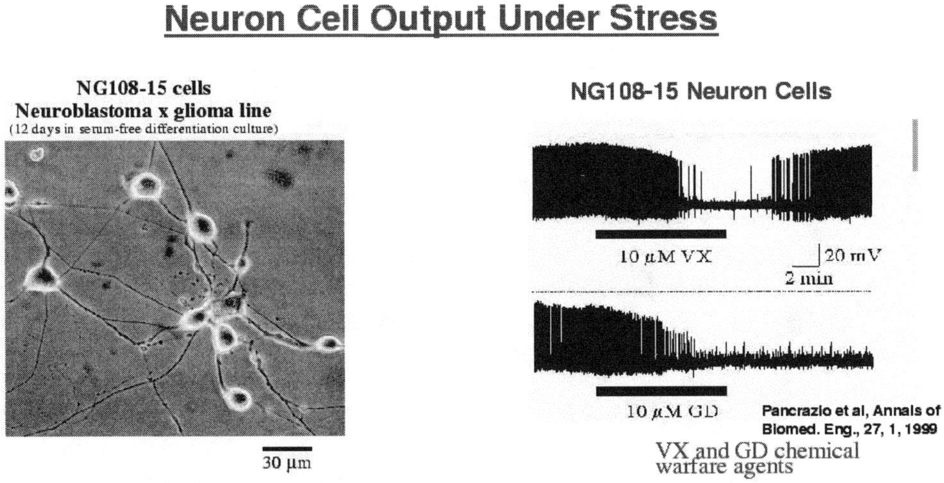

Figure 3.5. *An example picture of the type of cells used in the biosensor experiments is in the left panel. Notice that the scale is on the order of* 30 *microns, which is quite smaller than those of lower life forms, such as crustaceans. The right panel shows the membrane action potential of the cells before, during, and after treatment with chemical toxins VX and GD.*

NG108-15 cells provide a neuronal isolated cell model capable of action potentials [25], and firing repeatability is possible in both spontaneous and driven cell preparations. The experiment records the membrane action potential of the cells before, during, and after treatment with very small amounts of $CdCl_2$. See Figure 3.6 for a comparison of a spontaneously bursting cell with and without $CdCl_2$ added. We use data from 16 separate experiments with differing amounts of this toxin. (See [36, 4] for experimental details of the patch clamp experiment.)

The short time series data provided in patch clamp experiments have a contrasting situation where there is heavy oversampling but relatively few spikes to analyze. Therefore, many standard tests to tell the difference between regular behavior from the effect of the added toxin prove inconclusive. The first observation is that there is no significant change in the mean spike height. Similarly, the firing rate variations are no larger than the variations in the control time series itself. Because of low spike numbers, the firing rate data is misleading. For example, in row 3 in Table 3.1, the firing rate of the control data has an average of 1.6904 spikes/second. But if that same sequence is split into two subsequences, the first half has an average of 1.8741 spikes/second and the second half has 1.5067 spikes/second. The average of the first half is very similar to the average for the associated $CdCl_2$ data, 1.8868 spikes/second. Another standard test is to find the probability density function (PDF) of the ISI. The PDF is a histogram of ISI time lengths for a given data set. As in the firing rate, the small changes are usually not larger than control fluctuations.

Low spike numbers also imply very large error bars, making hypothesis testing using a T-test difficult at best. In conclusion, we think the problem lies in the length of the data sample, not in the methods themselves since longer data samples will yield more accurate results by generating better statistics.

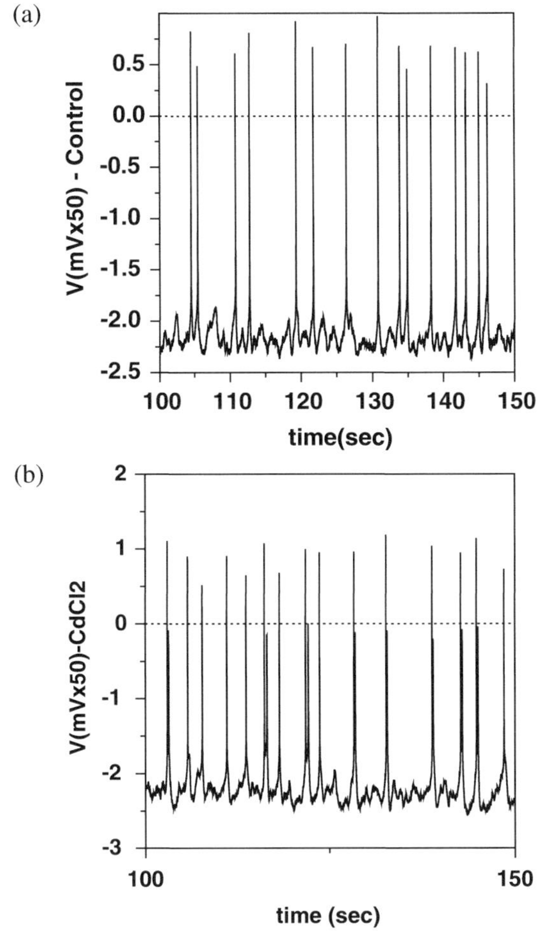

Figure 3.6. *Measured action potentials of an autonomous spiking* $NG108$-15 *cell:* (a) *control,* (b) *with* $CdCl_2$. *The concentration of the added* $CdCl_2$ *solution was at* $1\mu M$. *The dotted line marks the potential at* 0 *mV. The sampling rate was* 1 *ms.*

3.3.1 Local Expansion Rates

Because longer time series data is not available from these experiments, we take a dynamical systems approach to the problem. Our hypothesis is that small concentrations of toxins can affect the local statistics of the dynamics, and by local we mean those time intervals which are sufficiently short such that determinism dominates any stochastic behavior that is present. That is, it is assumed that there exists a function such that if I_n denotes the nth ISI, then $I_{n+1} = G(I_n)$. We choose to analyze the entire time series and make use of the fact that although noise is present and may affect the onset and timing variability of the spikes [13, 16], it may also be critical in producing reliability in spike timing [35].

We examine the time series in a space that may be thought of as a dynamical phase space approximated by a nonlinear change of variables. We take the measured time series, $s(t)$, and create d_E-dimensional vectors by using the well-known time delay method [7, 6, 42, 34].

Table 3.1. *Firing rates of the $CdCl_2$ data. Note that in the spiking type, if the driving force is not noted, it is a spontaneously bursting cell.*

Test #	$CdCl_2$ Amount	Spiking Type	Spikes/sec Control	Spikes/sec $CdCl_2$
1.	300 nM	spon.	0.4807	0.3336
2.	300 nM	0.5Hz	1.3553	1.1957
3.	1 μM	spon.	1.6904	1.8868
4.	1 μM	0.5Hz	2.4246	2.8570
5.	1 μM	spon.	0.2118	0.3009
6.	1 μM	0.5Hz	1.1842	1.4316
7.	1 μM	spon.	0.4528	0.6169
8.	1 μM	0.5Hz	1.0296	2.5215
9.	1 μM	1.0Hz	0.8748	1.3586
10.	3 μM	spon.	0.1057	0.2519
11.	3 μM	0.5Hz	1.1350	0.6540
12.	3 μM	0.5Hz	0.2401	0.0551
13.	3 μM	0.5Hz	0.4204	0.4748
14.	3 μM	spon.	0.1308	0.0150
15.	3 μM	0.5Hz	0.4700	0.5303
16.	3 μM	0.5Hz	0.3248	0.6234

Theory states that by doing this construction for a deterministic process, the resulting picture is equivalent (topologically) to measuring all independent variables, such as state space variables and their time derivatives. In practice, we find that a delay of roughly 20% of the decorrelation time of the measured time series produces a picture form the embedding that is representative of the original phase space.

For a given time series measurement from a cell, we assume a fixed delay τ, which is typically between 10% and 20% of the average decorrelation time of the membrane potential. As a function of the time delay chosen in this range, we do not see much variation in the results of the statistics. This agrees with the general theory about time delay embedding which states that for almost *any* delay, the dynamical picture is equivalent to the original phase space.

In order to preserve the predictability of the short time dynamics, the embedding dimension must be chosen such that the trajectories in the embedded space do not cross. (The Whitney embedding theorem states that any smooth dynamical system of dimension D may be faithfully reconstructed in a space of $2D+1$.) To measure the fraction of self-crossings as a function of embedding dimension, we use the concept of false nearest neighbors [1]. A false nearest neighbor test was performed on the all data and converged to a dimension of 5. For the local expansion rates method, we have included results from dimensions 4 and 5 to illustrate the effect of choosing a dimension that is too low.

We now discuss the embedding procedure. The measured action potential sampled at delay time τ is denoted by $s_i = s(i\tau)$. We choose an embedding dimension in d_E and define vectors $v_i = (s_i, s_{i+1}, \dots, s_{i+d_E-1})$. We assume the dynamics may be represented by $v_{n+1} = F(v_n) + \eta_n$, where η is a d_E-dimensional noise vector and F is an unknown function. We vary the embedding dimension from 4 to 7. Local predictability is quantified

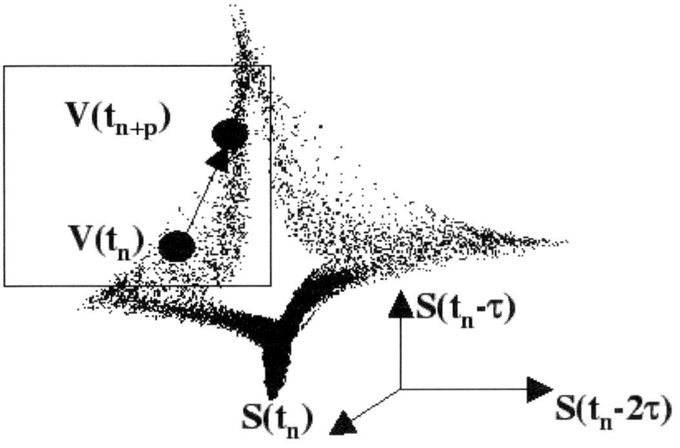

Figure 3.7. *A three-dimensional projection of an embedding of a typical time series membrane potential. For a given delay* τ*, the discrete time series* $s(t_n)$ *is embedded in three-dimensional vectors as* $V(t_n) = [s(t_n), s(t_n - \tau), s(t_n - 2\tau)]$*. The set of vectors is then plotted for each n. Given a reference vector* $V(t_n)$*, and its image in the future,* $V(t_{n+p})$*, points in near neighborhoods are used to construct a local linear (affine) map which takes* $V(t_n)$ *to* $V(t_{n+p})$*.*

by computing the local maps in a chosen embedded dimension. That is, we examine the statistics of local maps along the entire time series from one point in the embedded phase to another. We describe how the dynamics map a reference point v_i and its neighbors, v_j, to $v_{(i+1)}$ by constructing a local linear approximation given by $M_i(v_j - v_i) = v_{j+1} - v_{i+1}$ [7, 6, 32]. Figure 3.7 shows a three-dimensional embedding of an autonomous neuron. Notice that there is a great deal of structure revealing a high degree of correlation in the embedded space, although the actual dimension needed for local predictability is at least 5.

If the number of points in a neighborhood of a reference point is greater than or equal to embedding dimension, d_E, we can compute the matrix M_i using linear least squares methods. Given a local linear map, we use the linear matrix to determine the maximal local expansion, γ_i. The maximal (finite time) d_E Lyapunov exponent λ_i is just the average of the logarithm of γ_i over the number of chosen reference points. But this number is of no real utility since noise eventually will obscure the determinism on a sufficiently long time scale. That is, noise will cause the underlying deterministic process to look like diffusion in the long time limit.

Therefore, we look locally in time and for each map, M_i, and compute a largest expanding eigenvalue of the collection of local maps. Given a set of reference points which generate a collection of reference maps, the set of local expansion rates will generate a histogram. The results of local expansion rates are presented in Figure 3.8 as that fraction contained in an expansion rate interval for an autonomous spiking cell in the control and treated cases. The plot shows the results for the full domain of expansions. (The sum of the fraction is equal to one, but the expansion rates are unnormalized, so the area under the curve of the histogram is not unity.) Notice that around the neutral expanding value of unity, there is a dominant peak. (The inset highlights its asymmetry.) In a neighborhood of this peak, there is a noticeable change in the distribution of local expansion rates between the control and treated cells. We will exploit this observation in our analysis.

The set of reference points in the embedded space is crucial for the analysis. That

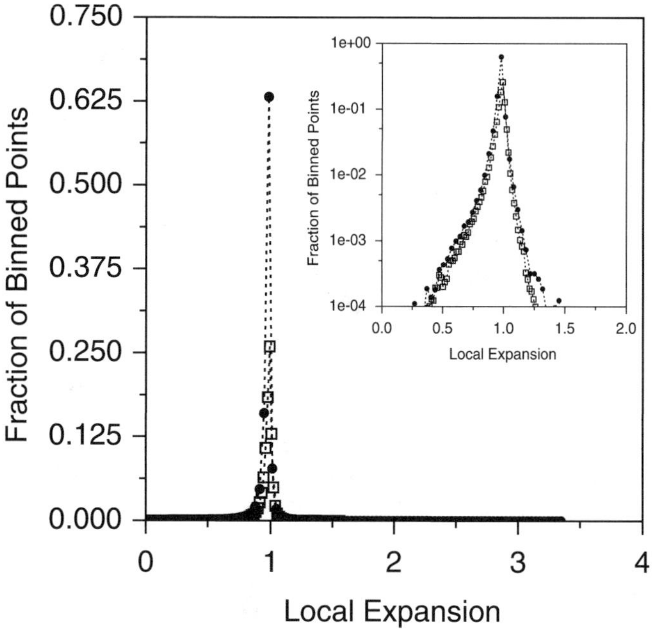

Figure 3.8. *The fraction of points with local expansion rates for the control and $3\mu M$ $CdCl_2$ over the entire range of expansions. Notice that the only differences occur in a neighborhood of unit expansion. The blowup (inset) highlights the asymmetry between the control and treated cases around unit expansion rate. The cell is spiking autonomously.*

is, given a reference point chosen at random, and its associated image point, we say the reference point is mappable if there is a sufficient number of points to carry out a successful least squares solution to find the map M. Since the volume of the embedded space increases with embedding dimension, it is to be expected that the number of mappable reference points decreases with dimension. Moreover, if the data set is sampled at a lower frequency, the same number of mappable points also decreases. To see this, we have taken the data set at 3μM $CdCl_2$ concentration, and computed the number of mappable points as a function of decimated data sets. See section 3.3.3 for results. It should be noted, however, that extrapolation of the mappable points for dimensions of 8 or greater will result in a only a few hundred points.

Next, we define a metric to the difference of control and perturbed cases. We use the area under part of the curve of the fraction of local expansion rates but above a threshold of 0.01. We use a Riemann sum–type calculation, which is an approximation using a sum of rectangles. For each cell, the areas are computed for both control and treated cases and then differenced. The absolute values of the differences were then analyzed for mean and standard deviation for the control groups cells. The results are plotted in Figure 3.9 for the 3μM $CdCl_2$ case, showing the difference between areas for spontaneous and periodically stimulated cells as a function of embedding dimension. The standard deviation of the control groups for spontaneous and driven cells is also plotted. Notice that in the driven cells, all but one case was below the control variation, implying that the dynamic approach to detecting $CdCl_2$ fails. In contrast, the dynamic approach succeeds in the spontaneous case when the

(a)

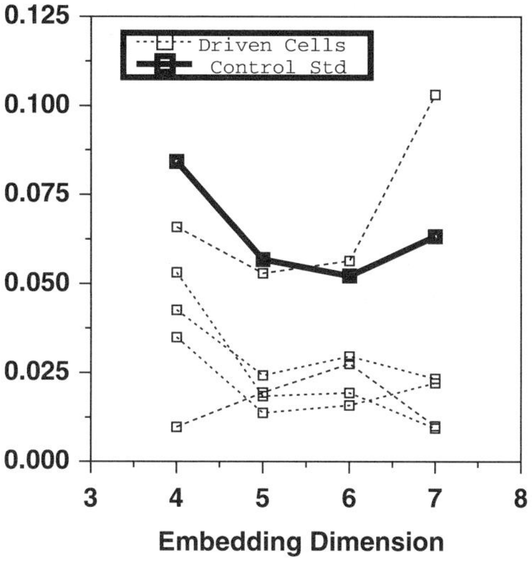

(b)

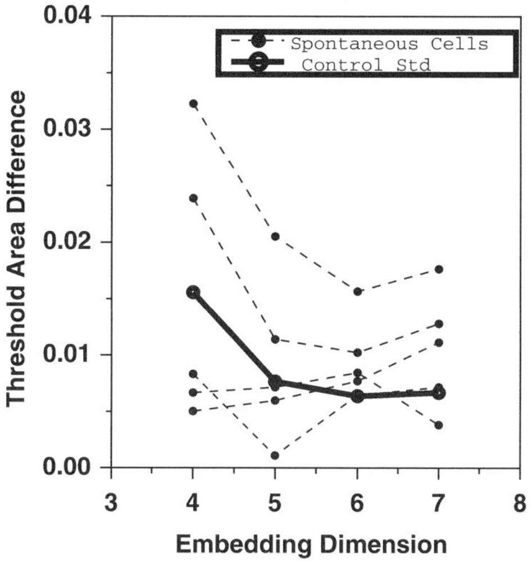

Figure 3.9. *The absolute value difference of threshold areas for local expansion rate PDFs between control and* $3\mu M\ CdCl_2$ *cases. The results are shown as a function of the embedding dimension. (See text for details on the computation.) Both* (a) *spontaneous and* (b) *stimulated cases are shown. Standard deviations of the controlled cells are also shown for the spontaneous (bold open circle) and stimulated cases (bold open squares). Notice that the local expansion method is not sensitive in the driven cell case, but it does quite well in the spontaneous case when compared to the standard deviation of the data.*

dimension is 6 or 7 in all but one case.

To check the embedding dimension, we also test the data using dimensions 4 and 5. Recall that the false nearest neighbor measurements applied to the time series imply that a minimum dimension of 6 must be used to achieve local predictability statistically. As expected, in dimensions 4 and 5, detection fails. In these dimensions, the local maps may not be accurate representations of the dynamics, generating erroneous statistics of local

Table 3.2. *Mean statistic of intercell control group.*

d_E	$\mu_{\text{spontaneous}}$	μ_{driven}
4	0.0206	0.0064
5	0.0098	0.0136
6	0.0058	0.0122
7	0.0024	0.0145

expansion rates.

To explain the failure of detection in the driven cases, we reexamine our metric. Ideally, we should be able to assume that the local expansion rate areas of all control cells, when differenced against each other, should have mean zero. Indeed, in Table 3.2, the control spontaneous case has a small mean for embedding dimensions 6 and 7, but the control driven cases do not. From an experimental viewpoint, it may be possible that the injected current biases the data in the driven cells, resulting in a nonzero mean. The implication may be that there is a lack of local determinism in the driven cases, or noise is dominating the dynamics in between spikes. Further research is required here to clarify the nonzero mean in driven control cells.

A sensitivity statistic to detect the presence of $CdCl_2$ is also computed using a cutoff value based on a significance level of 10% of a normal distribution with mean and standard deviation computed from the control fluctuations [23] in both the spontaneous and driven cases. That is, we make the null hypothesis that a measured cell is within the control distribution, $P(X)$, assumed to be normal having measured mean and standard deviation. We then solve the probability equation $P(X > c) = 1 - P(X \leq c) = \alpha$ for cutoff c for a given α typically set at 0.1. If the ratio of the area difference to the cutoff is larger than unity, we say $CdCl_2$ is present. That is, the hypothesis that the cell is not affected by an external agent is rejected. The probability that $CdCl_2$ is detected but not really present is then 10% or less. The differences in the threshold area under the local expansion rate curves are then normalized with respect to this cutoff, and are presented in Table 3.3. Notice that if there is positive detection in a spontaneous cell, the sensitivity correlates with $CdCl_2$ concentration. That is, as concentration of $CdCl_2$ increases, so does the sensitivity of the normalized local expansion rate statistic. Thus, given the current data, statistical analysis of local expansion rates will detect and scale against small amounts of $CdCl_2$ if the cell is spontaneously bursting, but not if the cell is driven.

3.3.2 Area under the Spikes

There is some evidence that the spike changes shape as a toxin is introduced, so an alternative idea is to monitor the areas under the action potential, or spike. Since the data is heavily oversampled, this type of area measurement is sensitive enough to reveal changes between a controlled cell and a perturbed cell. There is much debate over the exact beginning and end of a spike, so we use a threshold idea. Given a threshold amplitude, we record each area associated with a continuous sequence above the threshold and find the area PDF of that sequence. We then sweep through a range of thresholds and record the associated PDFs. It is easiest to visualize the change in area PDFs as a surface shown in Figure 3.10, where the PDF is plotted as a function of threshold and area. If the same number of bins is used for

Table 3.3. *Spike area and local expansion (LE) area $CdCl_2$ detection results. Note that in the spiking type, if the driving force is not noted, it is a spontaneously bursting cell.*

Test #	$CdCl_2$ Amount	Spiking Type	Spike Area Change Ratio	$CdCl_2$ Detect	LE Area Diff (d_E=6) Ratio	$CdCl_2$ Detect	LE Area Diff (d_E=7) Ratio	$CdCl_2$ Detect
1.	300 nM	spon.	2.42	yes	0.03	no	0.85	possible
2.	300 nM	0.5Hz	1.05	yes	0.23	no	0.02	no
3.	1 μM	spon.	0.723	no	1.09	yes	0.83	possible
4.	1 μM	0.5Hz	1.36	yes	0.16	no	0.03	no
5.	1 μM	spon.	1.87	yes	0.87	possible	0.99	yes
6.	1 μM	0.5Hz	1.24	yes	0.02	no	0.03	no
7.	1 μM	spon.	1.14	yes	0.82	possible	0.78	no
8.	1 μM	0.5Hz	1.45	yes	0.06	no	0.07	no
9.	1 μM	1.0Hz	1.02	yes	0.002	no	0.03	no
10.	3 μM	spon.	2.31	yes	1.92	yes	1.5	yes
11.	3 μM	0.5Hz	3.00	yes	0.42	no	0.33	no
12.	3 μM	0.5Hz	1.85	yes	0.39	no	0.31	no
13.	3 μM	0.5Hz	2.11	yes	0.27	no	0.14	no
14.	3 μM	spon.	2.11	yes	3.91	yes	2.5	yes
15.	3 μM	0.5Hz	1.42	yes	0.22	no	0.12	no
16.	3 μM	0.5Hz	2.72	yes	0.8	possible	1.47	yes

each PDF, the surface heights can be recorded in matrix form, which we define as the *area PDF matrix*.

The detection of the toxin depends on the differences in the area PDF matrix for the control time series and the area PDF matrix for the $CdCl_2$, but significant changes are difficult to quantify. Therefore, we created the following metric. For each area bin, or column of the area PDF matrix, take the absolute value of the maximum change over *all* thresholds. Call this list the net area change. We use the standard deviation of each net area change. For each set of controlled experiments, a cutoff was computed from a measured mean and standard deviation. Then, as in the local expansion rate test, the data was normalized with respect to this cutoff, where a number greater than unity means that the cell had a 90% probability of being modified with the $CdCl_2$ agent. The general algorithm is as follows:

1. Normalize the time series to [0, 1].

2. Given a threshold (horizontal line), find the area under each part of the time series above the threshold.

3. Form a sequence of these values and find the PDF.

4. Repeat this process for several threshold values in (0, 1), using the same histogram bins, and record the results in a matrix (area PDF matrix).

5. Repeat steps 1–4 for both the control and $CdCl_2$.

6. Find the absolute value of the difference between the two area PDF matrices.

7. Find the maximum value for each area bin column (net area change).

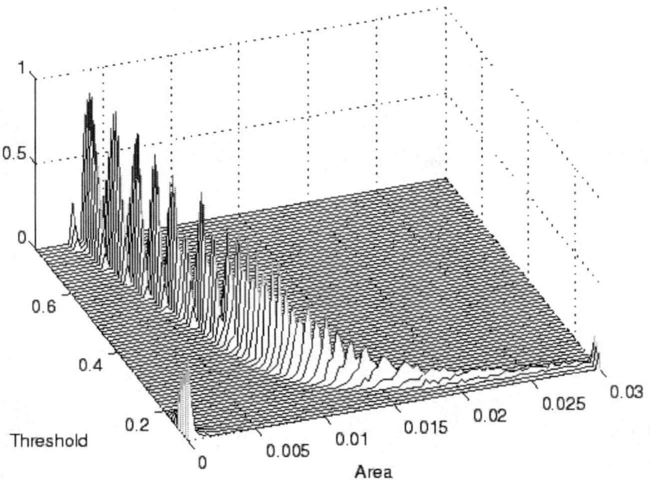

Figure 3.10. *For the sample control time series (#1 in Table 3.3), sequences of areas under the spikes are generated for several thresholds. This graph shows the PDFs of those area sequences as a function of the threshold. This is called the area PDF matrix. (See text for computational details.)*

The natural fluctuations in the system are also computed by performing the same procedure on several subseries from the control as a self-referencing test. In 15 of the 16 experiments, the (normalized) standard deviation of the net change test is clearly distinguishable from the control self test, as shown in Table 3.3. In contrast, dynamic local expansion detected $CdCl_2$ in all spontaneous cases above 3 nanomolar concentrations but failed to detect any concentration in the driven cases.

3.3.3 Decimated Data Results

In this section, we check the sensitivity of the local expansion method and the spike area method to the data sample size by decimating the data. Both the tests require a certain amount of data to detect the presence of the toxin. In the analysis of the embedding method, we saw that the local expansion rate method relied crucially on the number of successful mappable points constructed from the time series in an embedded space. We now quantify these lower limits.

Our decimation algorithm creates successive levels by taking the current data set and then removing every other point, as if the sampling time was doubled. Define the decimation factor as the resulting length time series divided by the length of the original time series. In this way, the structure of the original time series is preserved.

The results for the number of mappable points as a function of the decimating factor for two experiments are shown in Figure 3.11. As the data was decimated by a factor of two, the number of mappable points generated for successful statistical analysis fell sharply for embedding dimensions of 4 and 7. In both cases, the general trend is evident, demonstrating the rapid decrease in the number of mappable points as a function of decimation factor. For decimation factors sufficiently small, it is clear that there will be eventually too few points in the embedded phase space to compute the statistics of the local embedding dimension.

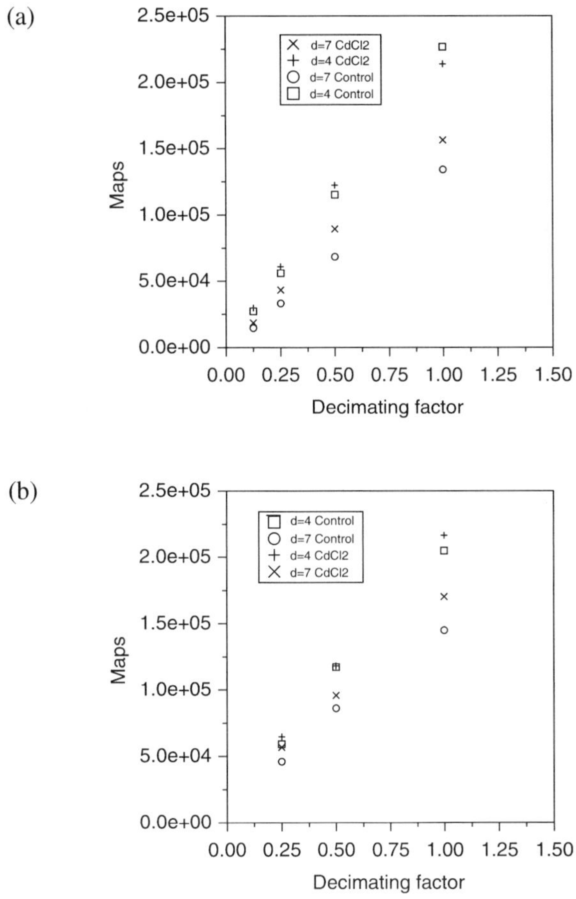

Figure 3.11. *The number of mappable points determined from the linear least squares solution as a function of decimation factor for two data sets. Both control and $CdCl_2$ cases are shown for embedding dimensions 4 and 7. In addition to the decreasing trend with respect to the decimating factor, the higher dimension typically has fewer mappable points.*

Therefore, we will restrict our analysis to decimation factors to greater than or equal to one eighth of the original time series.

In the local expansion rate case, we have examined decimated data sets for the 3μM spontaneous cases over three decimating factors (three orders of magnitude in base two). In Figure 3.12(a), we have shown the result for normalized detection as a function of decimating factor. (Notice that the decimating factor of unity is the original starting series.) In particular, only embedding dimensions of 4 and 5 are shown here for three of the 3μM studies. Notice that when the data is decimated by one half, there is a clear drop in sensitivity. However, there appears to be an increase in sensitivity for two other decimations. In Figure 3.12(b), similar results hold when the embedding dimensions are 6 and 7. Here, as in our earlier analysis, the results are more sensitive due to the fact that the embedding dimensions are large enough to generate adequate statistics. However, there is still a precipitous drop in

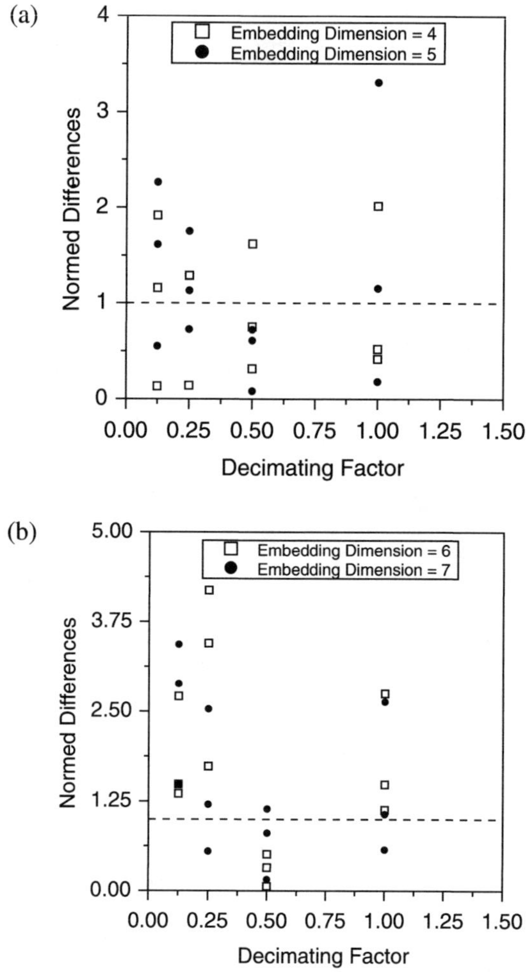

Figure 3.12. *Detection results for normalized local expansion as a function of decimation factor. Only* 3μM *concentrations of spontaneous cells are considered here, since the driven cell analysis had negative results. The embedding dimensions shown:* (a) 4 *and* 5, (b) 6 *and* 7. *A cutoff threshold for rejection of the control hypothesis at* 10%. *Differences above unity are distinguishable from the control statistics, while those below are not. The embedding dimensions* 4 *and* 5 *are too low for the analysis, resulting in poor detection. The sensitivity is much improved in dimensions* 6 *and* 7.

sensitivity when the data is decimated by one half. Further decimations are not affected and yield good sensitivity.

Although the local expansion rate method appears to work over a range of decimating factors, the spike area is even more sensitive. The results, normalized with respect to a cutoff of 0.05, show what appears to be a remarkable detectability over all ranges. The spontaneous cell results are depicted in Figure 3.13(a), and the evoked cell responses are in (b). In both cases, it is clear that detection takes place successfully over all decimation scales for the 3μM cases.

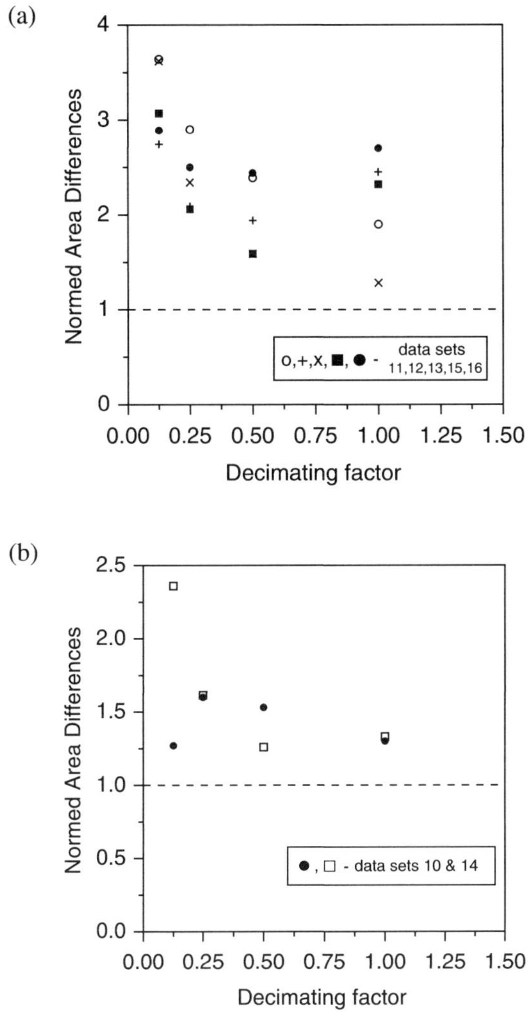

Figure 3.13. *Analysis of the spike area method for* $3\mu M$ (a) *spontaneous cells and* (b) *driven cells. Here the cutoff used to normalize the data was at 5%. Differences above unity are detectable. Notice that the differences increase with a decrease in decimation factor.*

3.4 A Test Bed for Damage Detection[2]

We now study a time series algorithm that detects unplanned changes in the experiment that occur when the system is perturbed either externally or internally. This section will illustrate a direct comparison between model-based analysis and the time series embedding procedure described in the previous section. The issue of drifting conditions for a dynamical system is known as nonstationarity and has received a lot of attention of late. Proposed methods for detection include recurrence plots, recurrence quantification analysis, space-time separation plots and their associated probability distribution, metadynamical recurrence

[2]Portions of this section have appeared in [37].

plots (a statistical test using the information of the distribution of points in the reconstructed phase space), cross-correlation sum analysis, and nonlinear cross prediction analysis. (See [10, 36, 4, 21] for examples.) These methods detect slow changes in the conditions of an experiment during a measurement period and may include just parameter drift or even a state transition in the operating regime. The method we propose is related to methods of detecting nonstationarity such as the one presented in [10].

To fix ideas, we focus on detecting minute changes that occur in a system parameter. The drift in these parameters may indicate internal changes such as an evolving damage condition. An example of such damage evolution is material softening, which results in slowly changing frequencies. The algorithm treats the system under consideration as a black box and indirectly detects minute changes occurring in the system by measuring a dynamical quantity along a system trajectory. Our tested hypothesis is that minute changes in the system will affect the local statistics of a dynamical quantity measured along a system trajectory.

Our method proceeds by measuring a dynamical quantity along a trajectory, which requires a reconstruction of phase dynamics. We chose to measure the local expansion rate along trajectories, a quantity which indicates how fast trajectories diverge locally from each other. This quantity is especially suitable when the system operates in a chaotic regime. From the local expansion rates (LERs) we construct a histogram. When a parameter changes, we notice changes in the shape of this histogram. The area under the histogram changes as well as the height of the peak of this histogram. We take the area under this histogram as our detection measurement. To quantify the changes occurring when drift in parameters is present we compare the detection measurements against a control group. To establish the control group, we first take several trajectories at the same parameter value and record the area under the histogram for all of these trajectories. That will give the range of behavior that is expected in the absence of drift. Any change outside of this control group or marginal to this control group is conjectured to signify a drift in the parameter.

We apply our algorithm to a mechanical system formed by a flexible linear beam coupled with a pendulum [11, 38, 12]. The entire structure is forced externally with a periodic drive amplitude. When the beam is rigid enough the system behaves very much like a damped driven pendulum. If we allow the beam to become flexible, a rich dynamical behavior is noticed, which includes a sudden transition to chaos which can have more than one positive Lyapunov exponent (hyperchaos). We solved this system by using spectral decomposition in space and considered the ordinary differential equation model obtained when only one spectral mode is retained. The full model containing 21 spectral modes will be considered in a forthcoming paper. The one-mode model carries a lot of the features of the full model, is four-dimensional, and contains small amplitude and large amplitude chaotic oscillations as well as regions of hyperchaos, so it is complex enough to illustrate the performance of our method. By applying our method, small changes in a system parameter will be observed on chaotic time series that would be otherwise indistinguishable using linear methods.

The methods cited above that detect nonstationarity do not consider operating inside the chaotic regime. Emphasis was on detecting state transitions, bifurcation points, or easily observable qualitative changes in behavior. One reason it is important to check for nonstationarity inside a chaotic regime is that while operating chaotically, rich dynamics are excited. One may choose (by using control of chaos) to select any one of the orbits embedded in the chaotic regime and induce a desired behavior to the system. As shown in [38] and using techniques derived from [2], a large number of modes will be excited spatially. It was found that the mechanical system we consider exhibits twelve Karhunen–Loève modes while operating chaotically and only four while operating periodically. (See Figure 3.14 below.) These modes excite motion over the entire spatial domain, which consists of a linear

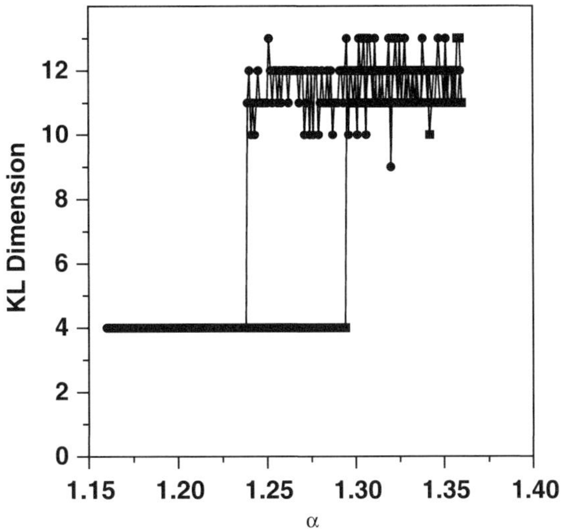

Figure 3.14. *The number of the most energetic Karhunen–Loève modes* [8, 9] *as a function of forcing amplitude. The dynamics of the nonlinear oscillator sampled at each external forcing period as a function of the forcing amplitude. Notice that when the entire system is chaotic, the number of modes increases by a factor of about* 4.

structure plus a nonlinear oscillator. Therefore, changes occurring over the entire spatial domain will be reflected in the dynamics we sample, making these changes easier to detect.

If instead of chaos the system would be operating periodically, areas of the spatial domain which possess changes may escape detection. So inducing chaos briefly may be a prerequisite to applying our method. In contrast, driving a linear structure without any attached nonlinear oscillator will not result in any coupling between modes. In the absence of any nonlinearity in the system, it is impossible to excite chaos. Periodic driving of a linear structure, therefore, will result in periodic behavior which in turn will excite motion at a specific wavelength. Therefore, different drive frequencies are required for each wavelength of the structure. Chaos, which requires a nonlinearity supplied by either an external oscillator or nonlinear material, allows many wavelengths to be sampled in parallel, making the procedure quite effective in detecting small changes at many wavelengths simultaneously.

In this section, we present the algorithm as it applies to trajectories obtained accurately from solving an analytical model of a system and show the performance of the algorithm on exact numerical solutions of the mechanical model. We also present the same method when applied to time series obtained from the same model. That is, we take the numerical solution of the mechanical model which now is treated as a black box (we no longer rely on knowing the equations) and create an approximate geometrical model using these numerical solutions. The geometrical model will represent our dynamical system. This geometrical model preserves the differential structure of the original phase space, which may be unknown in an experimental situation. In particular from this approximate geometrical model the local expansion rates for our algorithm can be measured. Working with time series will correspond to the situation when a model is difficult to obtain or one wants to work directly with data obtained experimentally.

The model system is one in which a linear structure, such as a rod, shell, or plate, is attached to a nonlinear oscillator. Generally, the equations are a set of coupled ODE-PDE equations describing the displacement of the linear structure,

$$\begin{aligned} L_\mu W(\xi, t) &= F_\mu(\xi, t), \\ \frac{d^2\theta}{d\tau^2} &= -[1 + G(W_{\tau\tau})]\sin\theta - 2\,\zeta_p \frac{d\theta}{d\tau} \end{aligned} \tag{3.9}$$

with $B(W(0,\tau), W(1,\tau)) = 0$. $W(\xi, t)$ is a function of space and time, as well as the angular displacement of the nonlinear oscillator, $\theta(t)$. F denotes a forcing function that is applied externally, and L_μ is a differential operator depending on parameter μ, which may change slowly over time. The main problem is then to detect when μ has changed sufficiently from a pristine value. The theory also will apply to other parameters, such as external forcing amplitudes and frequencies.

We use a cantilevered beam for our analysis of the coupled system. The equations and boundary conditions are

$$\begin{aligned} \frac{d^2\theta}{d\tau^2} &= -[1 + W_{A,\tau\tau}(\tau,\alpha)]\sin\theta - 2\zeta_p \frac{d\theta}{d\tau}, \\ L_\mu(W(\xi,\tau)) &= \mu^2\kappa_1^4 W_{,\tau\tau\tau\tau}(\xi,\tau) + W_{,\xi\xi\xi\xi}(\xi,\tau) + 2\zeta_b \mu W_{,\tau\xi\xi\xi}(\xi,\tau), \end{aligned} \tag{3.10}$$

$$W(0,\tau) = 0, \quad W_{,\xi}(0,\tau) = 0, \quad W_{,\xi\xi}(1,\tau) = 0, \tag{3.11}$$

$$W_{,\xi\xi\xi}(1,\tau) = \mu^2\beta\kappa_1^4\left[1 - T\left(\theta\,\frac{d\theta}{d\tau},\tau\right)\cos\theta\right], \tag{3.12}$$

$$T\left(\theta\,\frac{d\theta}{d\tau},\tau\right) = \left(\frac{d\theta}{d\tau}\right)^2 + \left[1 - \frac{d^2 W_A(\tau,\alpha)}{d\tau^2}\right]\cos\theta. \tag{3.13}$$

The physical scalings and constant definitions are listed in [11]. The important parameters are the forcing amplitude, α, and μ, which is a measure of the ratio of the natural pendulum frequency to the frequency of the first linear mode of the beam.

The dynamics of (3.10)–(3.12) generate complex spatiotemporal behavior, which is evident in Figure 3.14. The forcing and frequency parameters are chosen such that the system is in resonance. The advantage is that when the system goes chaotic, many more modes are excited since energy is distributed through the system. Therefore, when chaos is operating in resonance, all spatial scales are sampled and contain information relating to system health.

3.4.1 Is Linear Detection of Parameter Change Possible?

Before testing nonlinear schemes to detect changes in parameters, let's consider linear schemes. We assume that the parameters change slowly in time. Furthermore, we assume there exists a sampling of the system dynamics in a pristine state which we control. To make sure energy is spread throughout the structure, we assume chaos is the observed motion with a large number of modes. (It should be noted that one can have nonresonance chaos where the dynamics is low-dimensional and samples only a few modes.)

In Figure 3.15, power spectra reveal for a control case at $\alpha = 0.48$ and a changed case, $\alpha = 0.55$, that the dynamics has a broadband spectrum. Such a spectrum is indicative of chaos, and since the power spectra are quite similar, correlation tests, such as cross-correlation and autocorrelation, will not reveal any change between the two cases by virtue of the Wiener–Khinchin theorem. Therefore, we look to other means of detecting small changes in parameters by examining the dynamics explicitly.

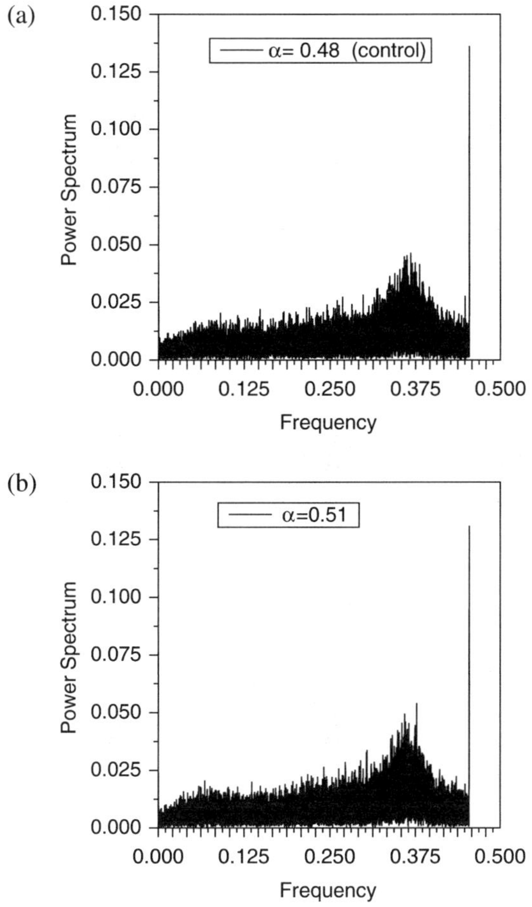

Figure 3.15. *Two power spectra for two different control parameters of forcing amplitude.*

3.4.2 Detecting Parameter Change Using Nonlinear Methods

We now assume that we have a model exhibiting slow dynamic changes in a parameter. We wish to devise a statistical method which measures local changes in phase space that may be compared to statistics of the dynamics at a controlled measurement. Our method revolves around the examination of the local expansion rates (LERs) of the measured dynamics. To see how this works, we consider the following model-based approach first.

Model-Based Methods

Given a general evolution model

$$\frac{dX}{dt} = F(X, t, \mu\,\alpha),$$

we compute a solution of the system, $\Phi(X_0, t, \mu\,\alpha)$, as well as its linear variation, $Y(t)$, along a solution, which is defined by

$$\frac{dY}{dt} = \frac{\partial F(\Phi(X_0, t, \mu\,\alpha), t, \mu\,\alpha)}{\partial X} Y, \quad Y(0) = I. \tag{3.14}$$

The maximum LER at time $t_n$is given by $\max\{|\lambda_i| \mid Y(t_n)v_i = \lambda_i v_i\}$. The maximum LERs are a measure of the finite time expansion or contraction. We typically measure the discrete times at each period of the forcing frequency. The resulting measurements of the LER yield a picture in phase space where the system is expanding and contracting, rather than the average expansion, which is sometimes called the finite time Lyapunov exponent. That is, if the points in phase space are expanding locally, we expect small perturbations due to parameter changes to be magnified when compared to the control group near the same region of phase space.

Figure 3.16 shows the results in histogram form for the control at $\alpha = 0.48$ in (a), and the adiabatic cases sampled at $\alpha = 0.49, 0.55, 0.57$. Notice the change in the peak and width near the neutral expanding value of LER in Figure 3.16(b). To formulate a metric with respect to the change in LER histograms, we compute the area above a threshold and below the histograms to highlight the changes in dynamics due to the parameters changes. Since there is almost no contribution in the tails of the distribution, thresholding above a value of 0.05 reduces the weight of the tails, while highlighting the effect near the peaks.

Statistically, we assume that the control variations obey a normal distribution. We then compare the thresholded areas against the normal distribution to make a decision based on the hypothesis that there is no change in the dynamics. If we find that Prob(Threshold Area > Area_cutoff) > 0.05, then our hypothesis fails and we have detection. That is, our metric is measure of the tail of the distribution. An example is shown in Figure 3.17 and Figure 3.18. The statistics of the control determine the mean and standard deviation to compute a cutoff based on a rejection of the null hypothesis.

Time Series–Based Methods

In applying the LER theory to experiments, the main problem in measuring the LER is that it is based on a phase space picture of the dynamics. Therefore, in the absence of a good model, the dynamics of phase space must somehow be recreated. An excellent and deep theorem due to David Ruelle and Floris Takens addresses this issue constructively. The theorem states (roughly) that for a given time series of physical system, a picture of the dynamics which is topologically equivalent to the original phase space may be constructed via the method of delay embedding. The method is described in detail in [36, 4] for a different detection application. Once the method of delays has been used to construct an analogue to the original phase space, linear least squares analysis is used to estimate the local linear maps from one point to the next in phase space. Then the LER may be computed from the eigenvalues of the local linear maps once the phase space is constructed. In our case, we measure the time series at the beam tip and reconstruct the phase space using the method of delays for both internal and external parameter changes. (Full details will be presented elsewhere.)

As a final example, we consider an internal parameter change of frequency. That is, we assume the internal frequency of the first mode of the beam changes adiabatically, and compare the dynamics of both model and time series measurement cases. The results are presented in Figure 3.19 for changes in μ. For Figure 3.19(a), we have used Prob(Threshold Area > Area cutoff) > 0.05 since the computations were done exactly with (3.14), while the time series method used in Figure 3.19(b) required Prob(Threshold Area > Area cutoff) > 0.20 to achieve decent detection.

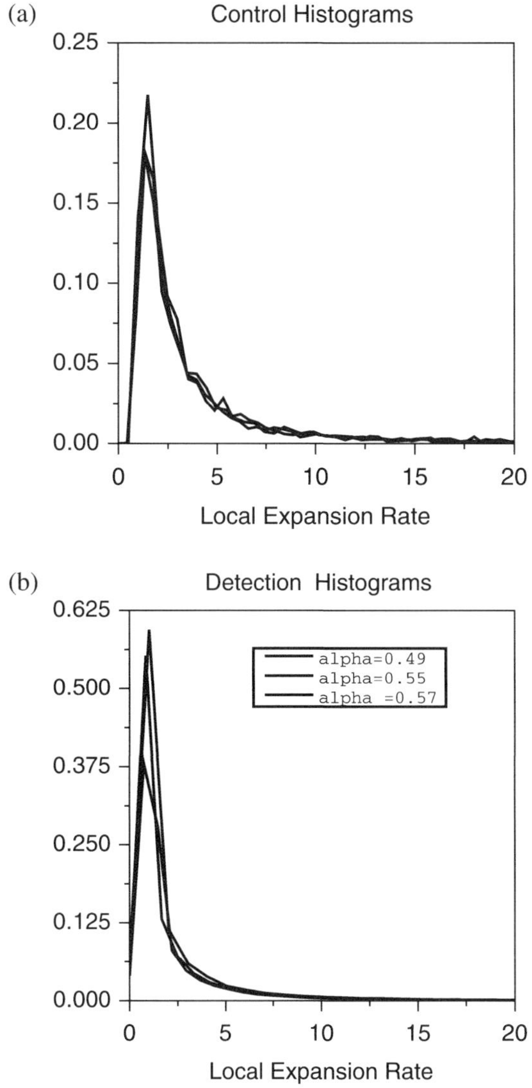

Figure 3.16. (a) *represents the histograms for three trials of LER measurements at a fixed parameter of* α. (b) *The same as in* (a), *but the differences are due to changes in the control parameter,* μ.

3.5 Discussion

We have presented analysis of two areas where applied mathematics can be applied in biochemical sensors. The sensors we have considered, the CFI sensor and cell-based biosensor, are examples chosen because of their current use in practice and the need for a performance that is superior to that of the current situation.

For the CFI sensor, we have only considered the input of the flow as either steady flow through capillary, or pulsatile flow which is periodic. The results can be analyzed further in that certain input flows may improve the situation if they are chaotic around a base

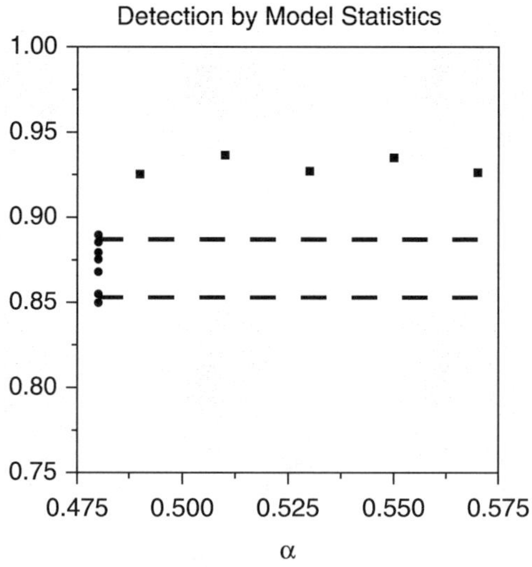

Figure 3.17. *A metric yielding the results of detecting a small parameter change in the forcing amplitude of the drive. The dotted lines denote the cutoff values based on measured control areas. Any values outside the dashed lines represent a possible detection of parameter change. The parameter used was the external drive amplitude of* α.

frequency, or stochastic. By considering such alternative flows, one may exploit a mechanism shown in Figure 3.4 which shows where the maximal signal detection occurs. That is, the signal, which is caused by reaction at the boundary and then transported downstream, is enhanced by the pulsatile flow due to stopping and reacting and flushing. That is, periodic pressures cause an increase in reaction intervals, allowing more of the reactions to occur for fluorescence to take place. Although the model in the periodic flow case predicts an increase in signal enhancement, it should be possible to optimize the problem with respect to the input controls based on reaction rates, diffusivities, and mean flow rates. Currently, experiments are ongoing to see if current predictions do indeed generate enhanced sensitivity of the CFI sensor.

Due to the nonlinear nature of spiking in excitable media, such as individual cells and tissues, detecting changes in time of biological media has required the use of new tools developed in the field of nonlinear science and statistics. From a given measured time series, we have developed new tools to detect small dynamic changes in a set of neurons perturbed by the addition of $CdCl_2$, namely, dynamically reconstructed local area expansion rates and spike area distributions.

For dynamical systems, one of the evolutionary ways of examining data in the form of time series, developed by David Ruelle and Floris Takens, is that of time series embedding. Time series embedding allows the construction of a picture of the phase space of the dynamics without writing down the equations of a model. This technique has an advantage over standard methods because no model will ever capture all of the dynamics precisely in a predictive manner. A model may miss essential properties of sensitivity to certain parametric changes. For statistical analysis, new measurement methods over the years have allowed very precise and fast data collection, allowing spike shape and area to be computed accurately, making statistical comparisons more precise.

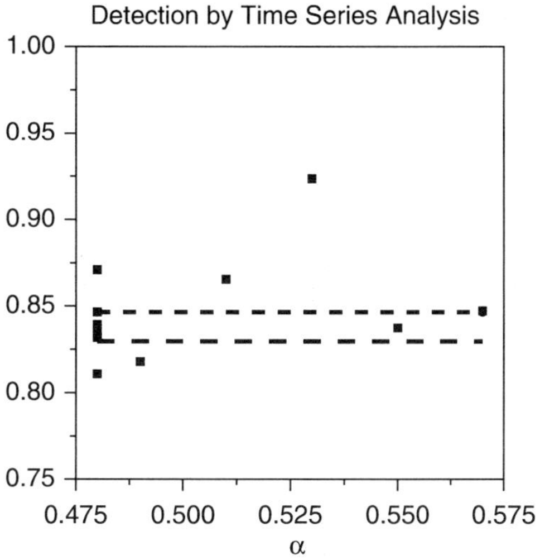

Figure 3.18. *The same metric as in Figure* 3.17, *except that phase space was reconstructed from the time series measured at the beam tip. In addition the* $Prob(Threshold\ Area > Area\ cutoff) > 0.20$ *to allow for inaccuracies in the least square fit of the local linear maps.*

Both nonlinear and statistical techniques, when compared against each other, are clearly not equivalent when predicting changes in cells due to the presence of $CdCl_2$. In point of fact, the spike area distribution method does much better in both driven and spontaneous cells, while the local expansion technique fairs well in only the spontaneous cases.

In conclusion, we have pursued the task of developing a time series embedding method for the detection of minute concentrations of a channel blocking agent present in a neuron when there are few spiking events in the time series data. When the neurons fire spontaneously, the technique based upon local expansion rates correlates with the concentration. In contrast, driven neurons show no such correlation, and the method distinctively fails to detect the concentration in any normal fashion. In contrast, the threshold areas under the spikes were statistically analyzed. Although the spike areas did not exactly correlate with concentration, they did perform much better in predicting the presence of $CdCl_2$ in both spontaneous and driven cases.

One of the major unknowns in the nonlinear dynamic portion of this work, as well as applying embedding methods to other time series problems, is that of determining the proper dimension to develop the statistics. The results of this work show that for real neurons, one requires an embedding dimension of at least 6 in order to get reasonable fidelity in constructing the local statistics. The dimension used here is higher than that for a model based on a Hodgkin–Huxley approach, which requires only 4 dimensions. Although results were excellent for the local expansion rate approach applied to the spontaneous case, it is not clear why it fails for the driven cases. It is true that for the driven neurons they are roughly slaved, modulo spike dropouts, to the period of the drive. As an example, we show the similarity of the embedded dynamics in Figure 3.20. It could be that since there is no deterministic component to a local expanding direction, the technique measures contributions to noise only. This would yield statistics that would be noise dominant instead of signal dominant. More work in understanding the differences between the two cases remains.

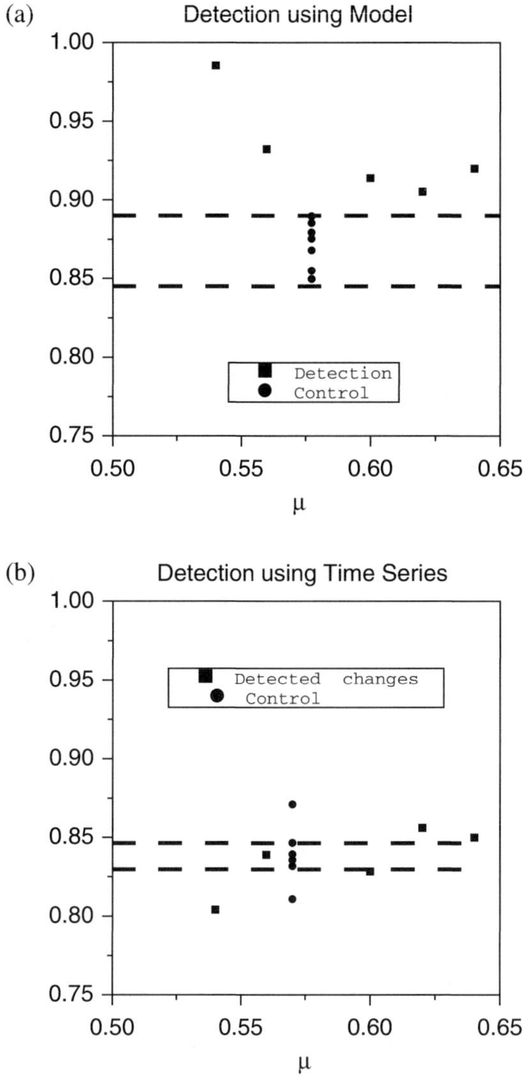

Figure 3.19. (a) *Model-based detection for parameters changes in* μ. (b) *Time series–based detection for parameter changes in* μ.

On the other hand, spike area distributions are not dependent upon dynamic determinism since the technique of detection relies solely on measuring areas under the spikes. Therefore, given the use of heavy oversampling of data (1 ms), the measurement of the statistics is straightforward, since it does not involve such machinations as delay embedding methods used for the local expansion analysis. It should be noted, however, that no optimization of threshold areas for the local expansion distribution was done, which will probably improve the results. The new method introduced for optimizing the spike area distributions can be used in the local expansion rates method, and this will be done in the future.

One big drawback of the current analysis is that it has not been tried in a high-dimensional theoretical model. As part of the discussion, we illustrated the technique when applied to discovering external and internal changes to a model of a continuum mechanical

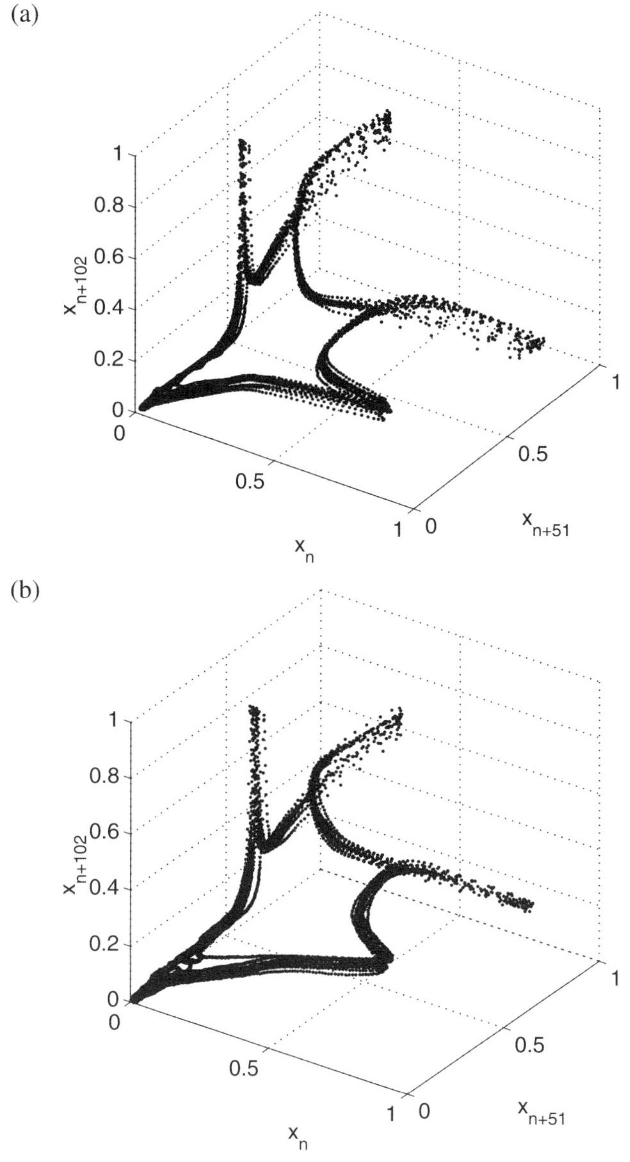

Figure 3.20. *An example of the embedded time series for* (#15 *in Table* 3.1) *in three dimensions. Notice the small difference dynamical behavior between* (a) *the control and* (b) *the* $CdCl_2$ *case for this driven cell. This may help explain why the local expansion rates method is not effective for this type of firing.*

system, where we saw how the time series embedding procedure compared with an exact computation.

Finally, this work suggests that the described techniques should have implications for the design of systems using cells for rapid detection of drugs and toxins where long streams of mammalian data are not available.

Acknowledgments

This research was supported by the Office of Naval Research and DARPA's cell- and tissue-based biosensor programs.

Bibliography

[1] H. ABARBANEL, *Analysis of Observed Chaotic Data*, Springer-Verlag, New York, 1996.

[2] N. AUBREY, R. GUYONNET, AND R. LIMA, *Spatiotemporal analysis of complex signals—theory and applications*, J. Statist. Phys., 64 (1991), pp. 683–739.

[3] J. P. BALTANÁS AND J. M. CASADO, *Bursting behaviour of the Fitzhugh-Nagumo neuron model subject to quasi-monochromatic noise*, Phys. D, 122 (1998), pp. 231–240.

[4] L. BILLINGS, I. B. SCHWARTZ, J. J. PANCRAZIO, AND J. M. SCHNUR, *Dynamic and geometric analysis of short time series: A new comparative approach to cell-based biosensors*, Phys. Lett. A, 286 (2001), pp. 217–224.

[5] W. H. CALVIN AND P. C. SCHWINDT, *Steps in production of motoneuron spikes during rhythmic firing*, J. Neurophysiol., 35 (1972), p. 297.

[6] J.-P. ECKMANN, S. O. KAMPHORST, D. RUELLE, AND S. CILIBERTO, *Liapunov exponents from time series*, Phys. Rev. A, 34 (1986), pp. 4971–4979.

[7] J.-P. ECKMANN AND D. RUELLE, *Ergodic theory of chaos and strange attractors*, Rev. Modern Phys., 57 (1985), pp. 617–656.

[8] P. FAURE AND H. KORN, *A nonrandom dynamic component in the synaptic noise of a central neuron*, Proc. Natl. Acad. Sci. USA, 94 (1997), pp. 6506–6511.

[9] R. FITZHUGH, *Impulses and physiological states in theoretical models of nerve membrane*, Biophys. J., 1 (1961), p. 445.

[10] J. B. GAO, *Detecting nonstationarity and state transitions in a time series*, Phys. Rev. E, 63 (2001), pp. 63–67.

[11] I. T. GEORGIOU AND I. B. SCHWARTZ, *Dynamics of large scale coupled structural/mechanical systems: A singular perturbation/proper orthogonal decomposition approach*, SIAM J. Appl. Math., 59 (1999), pp. 1178–1207.

[12] I. T. GEORGIOU, I. B. SCHWARTZ, E. EMACI, AND A. VAKAKIS, *Interaction between slow and fast oscillations in an infinite degree-of-freedom linear system coupled to a nonlinear subsystem: Theory and experiment*, Trans. ASME, 86 (1999), pp. 448–459.

[13] J. GUCKENHEIMER, R. HARRIS-WARRICK, J. PECK, AND A. WILLMS, *Bifurcation, bursting, and spike frequency adaptation*, J. Comp. Neuro., 4 (1997), pp. 257–277.

[14] R. GUTTMAN AND R. BARNHILL, *Oscillation and repetitive firing in squid axons, comparison of experiments with computations*, J. Gen. Physiol., 55 (1970), p. 104.

[15] O. P. HAMILL, A. MARTY, E. NEHER, B. SAKMANN, AND F. J. SIGWORTH, *Improved patch-clamp techniques for high-resolution current recording from cells and cell-free membrane patches*, Pflugers Archiv, 391 (1981), pp. 85–100.

[16] R. Harris-Warrick, L. Coniglio, S. Geuron, and J. Guckenheimer, *Dopamine modulation of 2 subthreshold currents produces phase-shifts in activity of an identified motoneuron*, J. Neurophys., 74 (1995), pp. 1404–1420.

[17] A. L. Hodgkin and A. F. Huxley, *A quantitative description of membrane current and its application to conduction and excitation in nerve*, J. Physiol., 117 (1952), pp. 500–544.

[18] D. B. Holt, *Advances in flow displacement immunoassay design*, Rev. Anal. Chem., 18 (1999), pp. 107–132.

[19] D. B. Holt, P. R. Gauger, A. W. Kusterbeck, and F. S. Ligler, *Fabrication of a capillary immunosensor in polymethyl methacrylate*, Biosensors and Bioelectronics, 17 (2002), pp. 95–103.

[20] D. B. Holt, A. W. Kusterbeck, and F. S. Ligler, *Continuous flow displacement immunosensors: A computational study*, Anal. Biochem., 287 (2000), pp. 234–242.

[21] H. Kantz and T. Screiber, *Nonlinear Time Series Analysis*, Cambridge University Press, Cambridge, UK, 1997.

[22] W. Y. Kao, Q. Y. Liu, W. Ma, G. D. Ritchie, J. Lin, A. F. Nordholm, J. Rossi, J. L. Barker, D. A. Stenger, and J. J. Pancrazio, *Inhibition of spontaneous GABAergic transmission by trimethylolpropane phosphate*, Neurotoxicology, 20 (1999), pp. 843–849.

[23] E. Kreyszig, *Advanced Engineering Mathematics*, 2nd edition, Wiley, New York, 1967.

[24] A. Longtin, *Stochastic resonance in neuron models*, J. Statist. Phys., 70 (1993), pp. 309–327.

[25] W. Ma, J. J. Pancrazio, M. Coulombe, J. Dumm, R. Sathanoori, J. L. Barker, V. C. Kowtha, D. A. Stenger, and J. J. Hickman, *Neuronal and glial epitopes and transmitter-synthesizing enzymes appear in parallel with membrane excitability during neuroblastoma x glioma hybrid differentiation*, Dev. Brain Res., 106 (1998), pp. 155–163.

[26] J. D. Murray, *Mathematical Biology*, Springer-Verlag, Berlin, 1989.

[27] J. J. Pancrazio and J. M. Schnur, The full calibration analysis of the NG108-15 cell survey will appear elsewhere.

[28] D. M. Racicot and A. Longtin, *Interspike interval attractors from chaotically driven neuron models*, Phys. D, 104-204 (1997), p. 184.

[29] F. Rieke, D. Warland, R. van Steveninck, and W. Bialek, *Spikes Exploring the Neural Code*, MIT Press, Boston, 1997.

[30] J. Rinzel, *Models in neurobiology*, in Mathematical Aspects of Physiology, Lectures in Appl. Math. 19, F. C. Hoppensteadt, ed., AMS, Providence, RI, 1981, pp. 281–297.

[31] J. Rinzel, S. M. Baer, and H. Carillo, *Analysis of an autonomous phase model for neuronal parabolic bursting*, J. Math. Biol., 33 (1995), pp. 309–333.

[32] M. Sano and Y. Sawada, *Measurement of the Lyapunov spectrum from a chaotic time series*, Phys. Rev. Lett., 55 (1985), pp. 1082–1085.

[33] T. Sauer, *Reconstruction of dynamical systems from interspike intervals*, Phys. Rev. Lett., 72 (1994), pp. 3811–3814.

[34] T. Sauer, J. A. Yorke, and M. Casdagli, *Embedology*, J. Statist. Phys., 65 (1991), pp. 579–616.

[35] E. Schneidman, B. Freedman, and I. Segev, *Ion channel stochasticity may be critical in determining the reliability and precision of spike timing*, Neural Comp., 10 (1998), pp. 1679–1703.

[36] I. B. Schwartz, L. Billings, J. J. Pancrazio, and J. M. Schnur, *Methods for short time series analysis of cell-based biosensor data*, Biosensors and Bioelectronics, 16 (2001), pp. 503–512.

[37] I. B. Schwartz and I. Triandaf, *Detecting parametric changes in structures using resonant chaos*, in High Performance Structures and Composites, C. A. Brebbia and W. P. De Wilde, eds., WIT Press, Southampton, UK, 2002, pp. 523–533.

[38] I. B. Schwartz and I. T. Georgiou, *Instant chaos and hysteresis in coupled linear-nonlinear oscillators*, Phys. Lett. A, 242 (1998), pp. 307–312.

[39] M. M. Segal, *Epileptiform activity in microcultures containing one excitatory hippocampal neuron*, J. Neurophys., 65 (1991), pp. 761–770.

[40] R. S. Skeen, W. S. Kisaalita, B. J. Vanwie, S. Fung, and C. D. Barnes, *Evaluation of neuron-based sensing with the neurotransmitter serotonin*, Biosensors and Bioelectronics, 5 (1990), pp. 491–510.

[41] R. S. Skeen, B. J. Vanwie, S. Fung, and C. D. Barnes, *Effects of temperature and analyte application technique on neuron-based chemical sensing*, Biosensors and Bioelectronics, 7 (1992), pp. 91–101.

[42] F. Takens, *Detecting strange attractors in turbulence*, in Dynamical Systems and Turbulence, D. A. Rand and L. S. Young, eds., Springer-Verlag, Berlin, 1981, pp. 366–381.

Chapter 4

The Distribution of Interpoint Distances

Marco Bonetti, Laura Forsberg,* Al Ozonoff,* and Marcello Pagano**

4.1 Introduction

Health surveillance systems are designed to collect data continuously, analyze them, and report the results in order to prevent and control diseases. Such a system that concentrates on patients presenting at an emergency department of a hospital or group of hospitals, with certain syndromes associated with biological weapons, could prove useful as an early warning system of a bioterrorist attack. If on any particular day or sequence of days, the number exceeds a predetermined amount, an alarm may be raised. Since the number of patients arriving at a hospital may be modeled as a random process, this alarm threshold may be set according to the usual hypothesis testing paradigm which is concerned with the two errors: raising false alarms and missing the raising of a warranted alarm.

Typically, the number of patients arriving at a hospital is influenced by such covariates as the season of the year, and the day of the week (see [1] and references therein, for example) and these are indeed important in order to place any set of numbers in their proper context, but in this chapter we focus on additional aspects of the data and assume a simple model for the arrival of patients. Consider the model where the number of patients is a Poisson random variable with mean λ, and suppose this represents the number of individuals we expect on any given day. To further simplify matters, suppose we wish to set up a system that raises the alarm if there are too many individuals observed on any particular day. Thus, if λ is sufficiently large to accurately use the normal approximation, one might raise the alarm if more than $\lambda + 1.645\sqrt{\lambda}$ patients arrive on any particular day. A one-sided alarm system like this runs a 5% false-positive rate. The associated power curve is easy to calculate.

Such a system would seem optimal in possible early detection of a disturbance that impacts uniformly across the whole area under surveillance. Alternatively, if the disturbance is due to a single emitting source (such as happened in the anthrax catastrophe in Sverdlovsk [2]) or a number of such sources, so that the disturbance to the system is geographically localized

*Department of Biostatistics, Harvard School of Public Health, 655 Huntington Avenue, Boston, MA 02115.

(for example, a triangular plume downwind from the offending source, as in Sverdlovsk), then this geographical information, if available (such as in the addresses of the patients), should prove useful. Similarly, if the disturbance to the system is some contagious agent, then one would again expect some geographic clustering amongst the patients. Both these instances argue for a surveillance system that not only looks at the number of patients but simultaneously looks at the location of where the patients were afflicted. Such systems are the motivation for this study, and we present some initial thoughts on the subject in this chapter.

We look at the distribution of distances between individuals as a summary of information on patient locations. In the cases we have investigated, we have found that this distribution does not seem to be affected by time or even season, so that it can form the basis for normalcy. Unfortunately this distribution is not easy to characterize because it differs for every different geographic distribution, and intuition is often foiled when attempting to estimate this function. We first look at some simple examples that make this point. Subsequently, we look at the empirical distribution of a statistic to measure the deviations from normalcy of such a distribution.

Consider the locations $X_1, X_2, \ldots,$ of patients arriving at random and indexed by the order in which they arrive. Denote by $D(X_i, X_j)$ the geographical distance between individuals at X_i and X_j. Consider too the distribution function $F(d) = Pr(D(X_i, X_j) \leq d)$ for nonnegative d, and assume that F is independent of i and j and is constant over time. Suppose that we have a long history of steady-state behavior of a hospital admission system so that we may estimate F by its empirical counterpart with confidence and act as if F is known, and equal to this estimate.

Now consider a disturbance to the system that may be reflected in the locations from which the patients come. For example, a point of toxic emissions might infect a neighborhood and result in a large increase of patients coming to the hospital from that particular neighborhood. We may phrase the problem by asking whether the distance distribution associated with the latest group of patients is given by F. We show in [3] how to test hypotheses about the distribution of distances with power to detect unusual clustering amongst patients.

In our current context, suppose we check on a daily basis not only whether there are too many patients, but also whether there is unusual clustering amongst the patients. Consider one particular day and denote by n the number of patients arriving on that day. Denote by F_n the empirical cumulative distribution function (ecdf) based on the n patients; i.e., if the patients are located at $X_1, \ldots, X_n$ and their interpoint distances are $D(X_i, X_j), i, j = 1, \ldots, N$, then

$$F_n(d) = \frac{1}{n^2} \sum_{i=1}^{n} \sum_{j=1}^{n} I(D(X_i, X_j) \leq d) \quad \forall d \geq 0. \tag{4.1}$$

Define a statistic T_n to measure the distance between F and F_n—below we consider a number of these statistics. In order to test the hypothesis that $X_1, \ldots, X_n$ is a random sample from the steady-state distribution of patients arriving at the hospital, we need the null joint distribution of the couplet, (n, T_n). The rest of this chapter considers the interpoint distance distribution and how it can be used to detect deviations from the steady-state distribution.

4.2 The Interpoint Distance Distribution

4.2.1 The Continuous Case

Consider a point process such that the observations can appear anywhere inside some bounded region, S. Let the distribution P of points in S be absolutely continuous, and define a nonnegative function d of pairs of observations in this region. Henceforth we generically refer to such a function as a distance, even though in subsequent developments we do not make use of the triangle inequality a distance function must satisfy. The cumulative distribution function (cdf) $F(\cdot)$ of the interpoint distance D between two independent points selected according to P is then $F(d) = \mathcal{E}1(d(X_1, X_2) \leq d)$, where $1(\cdot)$ is the indicator function and $\mathcal{E}$ denotes expectation with respect to the $P \times P$ distribution; thus, on average, $F(d)$ is the proportion of distances less than or equal to d.

As an example in which the spatial distribution is analytically known, consider the case of a mixture of K bivariate normal distributions f_i on the plane; i.e., let

$$f_i(x) = N_2\left(\begin{pmatrix} \mu_{1_i} \\ \mu_{2_i} \end{pmatrix}, \sigma^2 I_2\right) = N_2(\mu_i, \sigma^2 I_2).$$

It is easy to show that in this case the interpoint distance $Y = d(X_1, X_2)$ between two points randomly generated from $f(x) = \sum_{i=1}^{K} \pi_i f_i(x)$ is distributed as the square root of a mixture of chi-square densities. If each individual has probability π_i of belonging to each of the distributions f_i, then the density $f_{Y^2}(\cdot)$ can be written as $f_{Y^2}(\cdot) = \sum_{i=1}^{K} \sum_{j=1}^{K} g(i, j)\pi_i \pi_j$, where $g(i, j)$ is the density function of the $\chi^2(2, \frac{1}{4\sigma^2}((\mu_{1_i} - \mu_{1_j})^2 + (\mu_{2_i} - \mu_{2_j})^2))$ distribution.

With an extension of the usual definition of empirical distribution for random samples we define the ecdf $F_n(.)$ of the interpoint distances associated with a random sample $X_1, \ldots, X_n$ as defined in (4.1).

As an illustration, Figure 4.1 below shows the smoothed interpoint distance density function estimated on all the $\binom{n}{2}$ dependent distances obtained from $n = 100$ points generated from such a mixture of three bivariate normal distributions (top histogram), and the density function estimated on the 10,000 distances computed from 10,000 independent *pairs* of points from that same distribution (lower left graph). The histogram and smooth density estimates illustrate the closeness of the interpoint distance distribution of the dependent distances to the empirical distribution of the interpoint distance between two randomly chosen points, and the closeness of the latter density function to the theoretical density $f_D(\cdot)$. The ecdf (see (4.1)) of the dependent interpoint distances among n points in the plane thus is a well-defined and well-behaved summary of a configuration of observations.

The definition, and use, of the interpoint distribution function given above does not require that the point process be stationary, but if it is, a number of theoretical results follow. On the plane, Bartlett [10] reports the distribution of the interpoint distances for randomly distributed points on the unit square and on the unit circle (results originally due to Borel [11]). The latter distribution can be shown to be equal to

$$f_D(d) = \frac{4d}{\pi}\left\{\cos^{-1}\frac{d}{2} - \frac{d}{2}\sqrt{1 - \frac{d^2}{4}}\right\}, \qquad d \in [0, 2].$$

Bartlett in [10] suggests computing a chi-square test to measure the deviation between the observed and the expected frequencies over a grid. He also recognizes that distributional problems arise because the observed distances do not constitute a sample of independent observations.

For fixed d, $F_n(d)$ is a V-statistic (see, for example, [16, p. 172]). The scaled distribution of $F_n(d)$ computed at a finite set of values, $d_1, \ldots, d_m$, converges to a multivariate normal distribution as $n \to \infty$. Also, the quantity $\sqrt{n}(F_n(d) - F(d))$, considered as a stochastic process indexed by d, converges weakly to a Gaussian process [15, 3]. The use of this result, however, requires knowledge of the underlying spatial distribution, since the

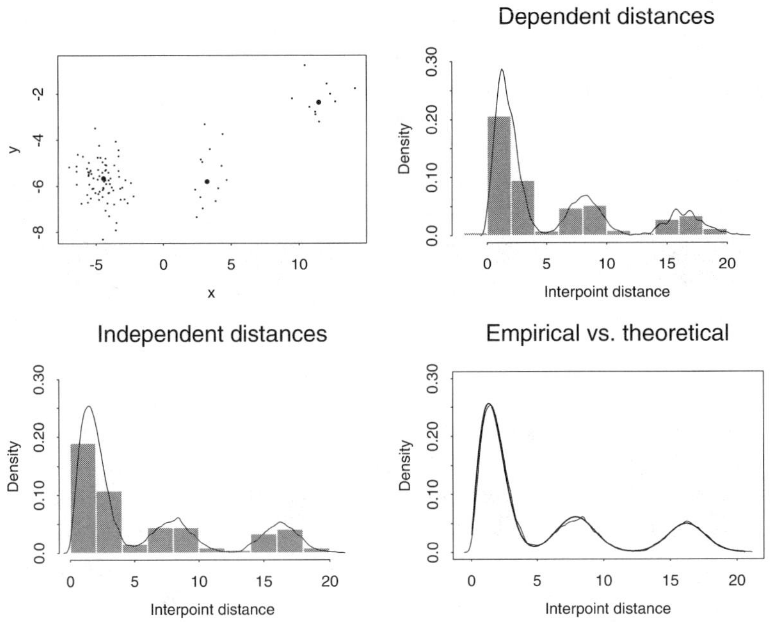

Figure 4.1. *An example of a realization of* 100 *points, top left, of a point process. The histogram from* 10,000 *dependent distances is given in the top right, and independent distances in the bottom left. The bottom right shows the comparison of these two smoothed estimates.*

covariance function of the associated Gaussian process $GP(d)$ is equal to

$$\mathrm{cov}(GP(d_1), GP(d_2)) = E[1(d(X_1, X_2) \le d_1)1(d(X_1, X_3) \le d_2)] - F(d_1)F(d_2).$$

If that is available, then the sampling distribution of the empirical interpoint distance distribution function can be obtained. Otherwise, it can be estimated via resampling methods or via simulation methods, as we do below. In lower-dimensional settings one could estimate the intensity function via kernel methods, in which case, as an alternative to the methods discussed here, it is also possible to compare intensity functions [13].

4.2.2 The Discrete Case

Often continuous data is not available, so consider the case of a fixed population distribution with population centers $l_1, \ldots, l_k$ wherein live $N_1, \ldots, N_k$, individuals, respectively. For example, these may be the centers of census tracts or, on a smaller scale, houses. Let N be the total population size ($N = \sum_{i=1}^{k} N_i$). Let the random variable D represent the distance between two individuals chosen at random (with replacement) from this population. Formally, let $p_i = N_i/N$, $i = 1, \ldots k$ and $p = (p_1, \ldots, p_k)$, and let $N \to \infty$. Let d_{ij} be the distance between locations l_i and l_j. The random variable D then takes on the value d_{ij} with probability $p_i p_j$. The distribution function of this nonnegative random variable is thus

$$F(d) = F(d; p) = \sum_{i=1}^{k} \sum_{j=1}^{k} p_i p_j 1(d_{ij} \le d). \tag{4.2}$$

Consider a random sample $n_1, \ldots, n_k$ of individuals distributed over the same geographic region, and let $n = \sum_{i=1}^{k} n_i$. Consider all the $\binom{n}{2}$ distances between the individuals in the sample, and compute the function $F_n(d) = F(d; \widehat{p})$, where $\widehat{p} = (\widehat{p}_1, \ldots, \widehat{p}_k)$ and for $i = 1, \ldots, k$, $\widehat{p}_i = n_i/n$. These definitions of $F(d; p)$ and $F(d; \widehat{p})$ are the discrete analogues and are equivalent to those of $F(d)$ and $F_n(d)$ given above for the continuous case.

If one is interested in the distribution of the distances between individuals and does not wish to make assumptions or inference about the value of the sample size, n, one may condition on it and then use the distribution of the distances obtained by choosing samples of size n at locations l_i with probabilities p_i, $i = 1, \ldots, k$, as the null distribution (see [4]). Then the null hypothesis of random sampling from the population distribution corresponds to the null hypothesis that the n_i are a multinomial sample with probabilities $p = (p_1, \ldots, p_k)$. Since the $\widehat{p}_i$ are strongly consistent estimators of the p_i (as $n \to \infty$), then for any fixed real d, $F(d; \widehat{p})$ is a strongly consistent estimator of $F(d; p)$. Some measure of the difference between $F(d; \widehat{p})$ and $F(d; p)$ can thus be used as a gauge of the null hypothesis of spatial randomness.

Note that in this discrete setting (as opposed to the continuous case) one can expect the underlying population distribution to be known. For a fixed value d the empirical cdf $F(d; \widehat{p})$ has $\sqrt{n}$-convergence to $\mathcal{E}(d(X_1, X_2) \le d)$. Moreover, there is convergence to a multivariate normal distribution when one computes the cdf at the finite set of values $d_1, d_2, \ldots, d_m$, and the covariance structure of the limiting distribution can be expressed analytically [3].

4.2.3 Two Discrete Examples of Interpoint Distance Distribution

Example 1: Two Points

Let n individuals be assigned either to location $A = (0, 0)$ or to location $B = (1, 0)$, with probabilities p_A and $p_B = 1 - p_A$, respectively. Denote by n_A and n_B the number of individuals assigned to the two locations. The matrix of the n^2 interpoint distances between individuals thus contains only the two values zero and one. In particular, a total of $n_A^2 + n_B^2$ zero distances are observed, and a total of $2n_A n_B$ distances equal to one. The relative frequency distribution of the interpoint distances is thus defined by the two proportions $P_0 = (n_A^2 + n_B^2)/n^2$ and $P_1 = 2n_A n_B/n^2$. The expected values of P_0 and P_1 are equal to

$$
\begin{aligned}
EP_0 &= \frac{En_A^2 + En_B^2}{n^2} = \frac{(np_A(1 - p_A) + n^2 p_A^2) + (np_B(1 - p_B) + n^2 p_B^2)}{n^2} \\
&= \frac{2p_A p_B}{n} + p_A^2 + p_B^2, \\
EP_1 &= \frac{2E(n_A n_B)}{n^2} = \frac{2E(n_A(n - n_A))}{n^2} = \frac{2(nE(n_A) - E(n_A^2))}{n^2} \\
&= 2p_A p_B - \frac{2p_A p_B}{n},
\end{aligned}
$$

since n_A and n_B are binomial random variables. Note how the expected value of P_1 follows immediately from that of P_0, as $P_1 = 1 - P_0$. As n tend to infinity the expected values converge to the distribution of the interpoint distance $D = D(X_1, X_2)$ between two independent points X_1 and X_2, i.e., to the two probabilities $P(D = 0) = p_A^2 + p_B^2$ and $P(D = 1) = 2p_A p_B$.

The terms of the variance-covariance matrix of (P_0, P_1) for fixed n are equal to

$$\begin{aligned}
\operatorname{var}(P_0) &= \frac{1}{n^4}\operatorname{var}(n_A^2 + n_B^2) = \frac{1}{n^4}(En_A^4 + En_B^4 + En_A^2 n_B^2 - (En_A^2 + En_B^2)^2),\\
\operatorname{var}(P_1) &= \operatorname{var}(1 - P_0) = \operatorname{var}(P_0),\\
\operatorname{cov}(P_0, P_1) &= E(P_0 P_1) - E(P_0)E(P_1) = E(P_0(1 - P_0)) - E(P_0)E(1 - P_0) = -\operatorname{var}(P_0).
\end{aligned}$$

From differentiation of the moment generating function of the binomial $\psi_{\mathbf{X}}(\mathbf{t}) = (p_A \exp(t_A) + p_B \exp(t_B))^n$ one obtains all necessary moments, and after some algebra the result

$$\begin{aligned}
\operatorname{var}(P_0) &= -\frac{4p_A}{n^3}(6p_A^3 - 10np_A^3 + 4n^2 p_A^3 - 12p_A^2 - 12np_A + 7p_A - 1\\
&\qquad + 20np_A^2 - 8n^2 p_A^2 + 2n - n^2 + 5n^2 p_A),\\
\operatorname{var}(P_1) &= \operatorname{var}(P_0),\\
\operatorname{cov}(P_0, P_1) &= -\operatorname{var}(P_0).
\end{aligned}$$

Note how the last two expressions also follow immediately from the fact that $P_1 = 1 - P_0$. Thus we can focus on P_0 alone. For $p_A = p_B = 1/2$ the expected value of P_0 is equal to $(1 + 1/n)/2$, which tends to 1/2 as $n \to \infty$. Also, the expression for the variance of P_0 becomes

$$\operatorname{var}(P_0) = \frac{1}{2}\frac{n-1}{n^3}.$$

Thus $\operatorname{var}(\sqrt{n}P_0) \to 0$ as $n \to \infty$. Rescaling by n instead of $\sqrt{n}$ yields that

$$\operatorname{var}(nP_0) \to \frac{1}{2} \quad \text{as } n \to \infty.$$

In fact, this happens if and only if $p_A = 1/2$, since in the expression of $\operatorname{var}(P_0)$ one needs the condition $4p_A^3 - 8p_A^2 + 5p_A - 1 = 0$ to be satisfied, and $1/2$ is the only solution in $(0, 1)$. As a final remark, note how $1/2$ is *not* the variance of the binomial random variable with probability of success (i.e., of falling into the value $D = 0$) equal to $1/2$.

From the familiar results about U-statistics one might expect that $\sqrt{n}P_0$ has a nondegenerate asymptotic distribution, while from the expressions above it is clear that this does not happen. In fact, one would need to normalize P_0 by multiplication by n and not by $\sqrt{n}$ to converge to nonzero variances. The reason why this happens is worth discussing in detail. Below we refer for simplicity to one-dimensional U-statistics, but with minor changes the same considerations apply to V-statistics such as those considered here.

This phenomenon is due to the fact that one of the requirements for the usual asymptotic normality results of U-statistics is not satisfied whenever $p_A = p_B$. In fact, consider the m-order U-statistic

$$U_n = \frac{1}{n_{(m)}}\sum h(X_{i_1}, \ldots, X_{im})$$

with (symmetric) kernel $h(X_1, \ldots, X_m)$ (such that $E[h^2] < \infty$). The summation is taken over all $n_{(m)} = n(n-1)\cdots(n-m+1)$ m-tuples $(i_1, \ldots, i_m)$ of distinct elements from $\{1, \ldots, n\}$. It is well known (see, for example, [16, p. 162]) that if one defines the auxiliary functions $h_d(x_1, \ldots, x_d) = E[h(x_1, \ldots, x_d, X_{d+1}, \ldots, X_m)]$ and the parameters $\zeta_0 = 0$ and, for $1 \le d \le m$, $\zeta_d = \text{var}(h_d(X_1, \ldots, X_d))$, then the variance of U_n can be expressed as

$$\text{var}(U_n) = \binom{n}{m}^{-1} \sum_{d=1}^{m} \binom{m}{d}\binom{n-m}{m-d}\zeta_d$$
$$= \frac{m^2\zeta_1}{n} + O(n^{-2}).$$

In our case above we have $m = 2$, $h(X_1, X_2) = 1(d(X_1, X_2) = d)$ for $d = 0$. (A symmetric argument with $d = 1$ would be used for X_1.) Then one has

$$\zeta_1 = \text{var}(h_1(X_1)) = \text{var}(E[h(x_1, X_2)])$$
$$= \text{var}(P(d(x_1, X_2) = d))$$

and

$$P(d(x_1, X_2) = 0) = \begin{cases} p_A, & x_1 = (0,0), \\ p_B, & x_1 = (1,0), \end{cases}$$

so that $p_A = p_B = 1/2 \Rightarrow \zeta_1 = 0$. In fact, here this is indeed a necessary and sufficient condition for the convergence in probability of $\sqrt{n}P_0$ to $P(D = 0) = 1/2$.

For a connection with related problems in genetics we refer the reader to [17].

This example is a degenerate one in the sense that one deals with only one (binomial) random variable. We now present another example that is constructed on four points and allows for the interpoint distance to take on four different values.

Example 2: Four Points

Consider the situation of each of n individuals being assigned to one of the four points $A = (0, 2)$, $B = (1, 2)$, $C = (2, 0)$, and $D = (0, 0)$. The interpoint distance between two individuals can take on one of the four values $\{0, 1, 2, \sqrt{5}\}$. If one calls D_{ij} the interpoint distances D_{ij} between individuals i and j, one can then define the quantities $X_d = \sum_{i=1}^{n}\sum_{j=1}^{n} 1(D_{ij} = d)/n^2$, i.e., the proportion of interpoint distances equal to d. In particular,

$$P_0 = \frac{1}{n^2}(n_A^2 + n_B^2 + n_C^2 + n_D^2),$$
$$P_1 = \frac{2}{n^2}(n_A n_B + n_C n_D),$$
$$P_2 = \frac{2}{n^2}(n_A n_D + n_B n_C),$$
$$P_{\sqrt{5}} = \frac{2}{n^2}(n_A n_C + n_B n_D),$$

where n_A, n_B, n_C, and n_D are the numbers of individuals at the four location, which we assume are assigned according to the multinomial distribution with parameters n and $\mathbf{p} = (p_A, p_B, p_C, p_D)$ for the four locations. Similarly to what was done in Example 1 one can derive the expressions for the expected values and for the variance-covariance matrix of the vector $\mathbf{P} = [P_0, P_1, P_2, P_{\sqrt{5}}]$. The expressions for the expected values are as follows:

$$\begin{aligned}
EP_0 &= (1 + (n-1)(p_A^2 + p_B^2 + p_C^2 + p_D^2))/n,\\
EP_1 &= 2(p_A p_B + p_D p_C)\frac{n-1}{n},\\
EP_2 &= 2(p_A p_D + p_B p_C)\frac{n-1}{n},\\
EP_{\sqrt{5}} &= 2(p_A p_C + p_B p_D)\frac{n-1}{n}.
\end{aligned}$$

The expression of the variance-covariance matrix for general $\mathbf{p}$ is quite messy.

As $n \to \infty$ the expected value of $\mathbf{P}$ converges to the vector of probabilities $P(D = d)$, where D is the distance between two individuals placed at random, i.e.,

$$\begin{aligned}
P(D = 0) &= p_A^2 + p_B^2 + p_C^2 + p_D^2,\\
P(D = 1) &= 2(p_A p_B + p_C p_D),\\
P(D = 2) &= 2(p_A p_D + p_B p_C),\\
P(D = \sqrt{5}) &= 2(p_A p_C + p_B p_D).
\end{aligned}$$

For the special case of $\mathbf{p} = [.25, .25, .25, .25]$ the expected value $E(\mathbf{P})$ becomes

$$E\begin{bmatrix} P_0 \\ P_1 \\ P_2 \\ P_{\sqrt{5}} \end{bmatrix} = \frac{1}{4}\begin{bmatrix} 1 + 3/n \\ 1 - 1/n \\ 1 - 1/n \\ 1 - 1/n \end{bmatrix},$$

and the variance-covariance matrix takes the simple form

$$\frac{1}{8}\left(\frac{1}{n^2} - \frac{1}{n^3}\right)\begin{bmatrix} 3 & -1 & -1 & -1 \\ -1 & 3 & -1 & -1 \\ -1 & -1 & 3 & -1 \\ -1 & -1 & -1 & 3 \end{bmatrix}.$$

Here, too, as $n \to \infty$ a phenomenon similar to that observed in Example 1 above occurs, namely the fact that $\mathrm{var}(\sqrt{n}\mathbf{P}) \to \mathbf{c}$ with $\mathbf{c} = [.25, .25, .25, .25]'$ and that $\mathrm{var}(n\mathbf{P}) \to [0, 0, 0, 0]'$.

4.2.4 Test Statistics

It is easier and more informative to visualize the differences between the two cdfs $F_n(d)$ and $F(d)$ if we define the scaled first difference function $f(\mathbf{d})$. This is defined as a vector $f_n(\mathbf{d}) = (f_n(d_1), \ldots, f_n(d_m))$ of values

$$f_n(d) = \frac{1}{\epsilon}[F_n(d + \epsilon/2) - F_n(d - \epsilon/2)]$$

computed at the values $d_1, \ldots, d_m$ such that $d_j - d_{j-1} = \epsilon$ for $j = 1, \ldots, m$ and m some positive integer. We set $d_1 = \epsilon/2$, and for definiteness we define $f_n(d_1) = F_n(\epsilon)/\epsilon$ so that it includes the origin. The population equivalent of $f_n(\mathbf{d})$ is the vector $f(\mathbf{d}) = (f(d_1), \ldots, f(d_m))$ computed as $f_n(\mathbf{d})$ and at the same values $d_1, \ldots, d_m$, but replacing $F_n(\cdot)$ with $F(\cdot)$. (The constant ϵ may be any size and can be made as small as the accuracy with which the distances are measured.)

Because of its linear relationship with $F_n(\cdot)$, the first difference function $f_n(d)$ has $\sqrt{n}$-convergence to the expected value, $\mathcal{E}(1(d - \epsilon/2 < d(X_1, X_2) \leq d + \epsilon/2)$, and that for a fixed d, $n^{1/2} f(d; \widehat{p})$ has an asymptotically normal distribution. The joint asymptotic distribution for multiple values of d also follows immediately.

Several test statistics can be defined to measure the distance between $\widehat{F}_n(\cdot)$ and $F(\cdot)$, and thus allow the testing for deviations from the null spatial distribution. The asymptotic normality noted above suggests the use of the following statistic to measure the distance between the two vectors $f(\mathbf{d})$ and $f_n(\mathbf{d})$:

$$M(f_n(\mathbf{d}), f(\mathbf{d})) = (f_n(\mathbf{d}) - f(\mathbf{d}))^t S^- (f_n(\mathbf{d}) - f(\mathbf{d})), \tag{4.3}$$

where S^- is the (Moore–Penrose) generalized inverse of the sample covariance estimator computed on the samples. Note that the parameter ϵ needs to be set for M to be defined. The asymptotic distribution of NM can be shown to be chi-squared [3].

Note that we have defined the statistic in terms of the "densities," $f(\mathbf{d})$ and $f_n(\mathbf{d})$, but that we could equally well define the M statistic directly in terms of the cdfs computed at the same values $d_1, \ldots, d_m$. The two forms with, of course, appropriate definitional changes in the covariance matrix yield identical results.

The cutoffs at which the interpoint distance distribution is evaluated can also be chosen so that between each two subsequent cutoffs one has, say, 10% of the probability mass. Such a choice can be expected to be more robust at the extremes.

Figure 4.2 shows the QQ plots for the null distribution of the M statistic for varying numbers of points versus a chi-squared random variable. The QQ plots are based upon 1000 realizations of the M statistic with points randomly distributed in the unit circle in the plane. The degrees of freedom used were the number of bins minus one. When N is 10, one of the bins is always empty, due to the smaller number of distances in this case, so we used eight degrees of freedom for the chi-squared distribution. These show that, except for the extreme right tails, the asymptotic results given above provide a good approximation to the distribution of M, even for small values of N.

Note that if one uses equal probability histogram, then the ecdf is such that the area of the empirical histogram of the frequency distribution of the interpoint distances converges to 0.1 within each bin. However, it is *not* true in general that the centered and scaled histogram converges in distribution to a multinomial distribution. The two discrete examples above illustrate this and show that in situations of symmetry the rate of convergence of $F_n(d)$ to $F_D(d)$ may not be the one that one would expect.

4.3 Cluster Detection

To study the power properties of various detection statistics we designed a simple study with points generated at random on a unit circle in the plane. The number of points generated were determined by a Poisson distribution with mean 25. This represents the steady state. Outbreaks were then superimposed on this null distribution in the form of clusters of various sizes and locations. The power calculations are based on 1000 repetitions.

We first looked at detection based solely on observing clusters based on the interpoint distances. One can compare the distributions using the classical Kolmogorov–Smirnov test statistic or the Wilcoxon test statistic (see, for example, [5]). The Kolmogorov–Smirnov test considers the largest difference between the two sample cdfs. The Wilcoxon test is rank-based and considers ranking within a combined sample of the two populations. The statistic is given by

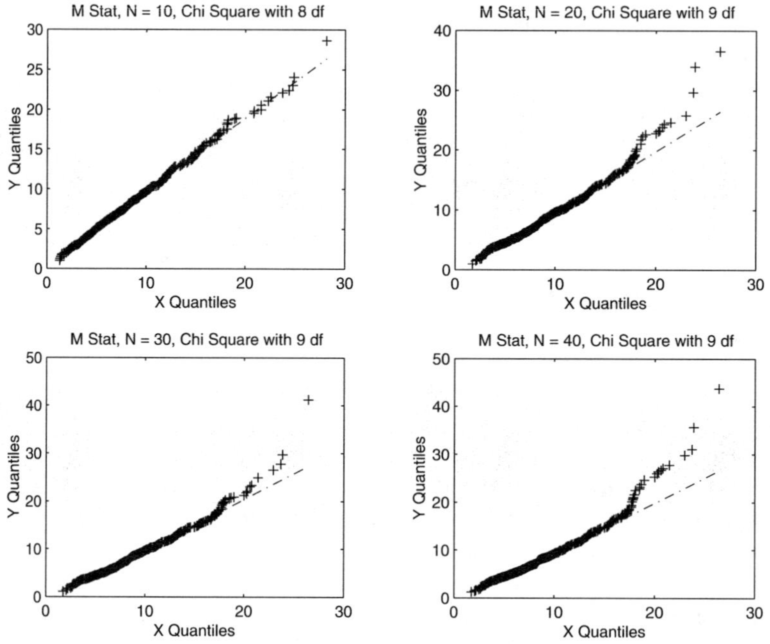

Figure 4.2. *QQ plots of the null distribution of the M statistic versus the chi-squared distribution when points are randomly distributed in the unit circle, based on* 1000 *realizations of the statistic.*

$$W = \sum_{j=1}^{N} S_j,$$

where the S_j are the ranks for one of the samples and N is the total number of distances in the combined sample. Because of the dependencies in the distances between patients, we cannot rely on published tables to determine the p-values of the tests, so we turned to simulations to derive empirical null distributions of the statistics (see [4]). In our example we used 1000 samples to generate this distribution and determine a cutoff value corresponding to a type I error rate of 0.05.

The tests that utilized the Wilcoxon and Kolmogorov–Smirnov statistics give very similar results for the problem at hand. Both tests are quite sensitive to the location of the cluster. In our example, we see that the power for detecting a cluster declines as the cluster moves farther from the origin. Figure 4.3 illustrates the results for the Wilcoxon test. These tests are sensitive to cluster location because of the impact the location has on the sample cdf. When a cluster is placed at the center of the circle the number of very small distances increases, but as the cluster moves to a more extreme location on the circle, the number of larger distances also increases. As a result, the alternative cdf becomes bimodal, and as neither of these tests has large power against such a bimodal distribution, they fail to detect such clustering.

To address this lack of power, we choose to discretize the cdfs by placing the sorted interpoint distances into ten equiprobable bins. These bins are determined by the empirical distribution of distances of the null distribution. Subsequently, we can compare the numbers falling in each bin. To this end consider the M statistic as defined in (4.3). When using the

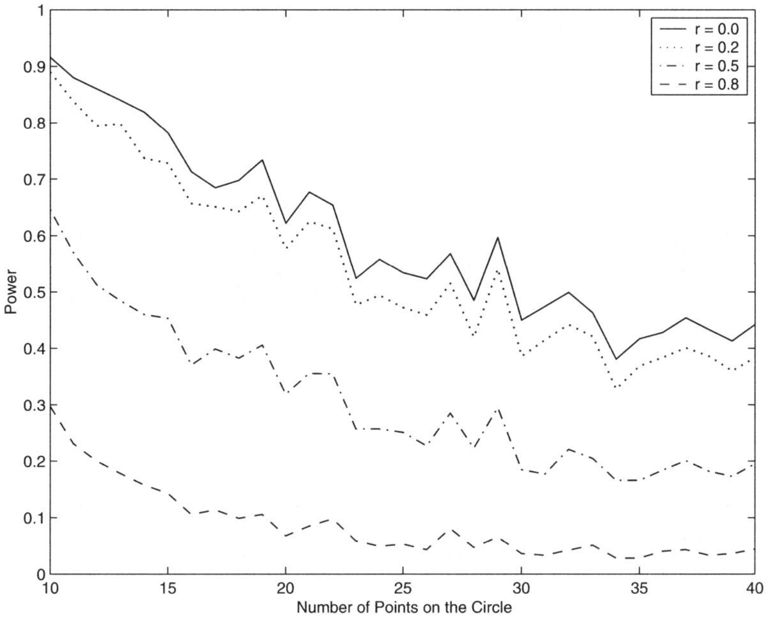

Figure 4.3. *Powers for the Wilcoxon test with varying locations of the cluster. Cluster size* = 5, *radius* = 0.001, *location* = 0.8.

M statistic in this example, we see an increase in the power to detect a cluster, regardless of its location on the circle (Figure 4.4), in contrast to the other two test statistics previously considered.

The Wilcoxon statistic seems to fail in the case where the cluster is at an extremity on the circle because the sum of the ranks for the two groups appear similar. This is because the statistic for the group with the cluster will tend to be composed of the ranks of the lower and higher points in the combined sample while the other group will be mainly composed of the intermediate ranks (the bimodality effect). Therefore the overall sums of the ranks will be very similar.

The Kolmogorov–Smirnov statistic only considers the maximal difference between the cdfs. This summary ignores substantial information about the overall behavior of the groups in relation to one another. When a cluster is added to an extreme location on the circle, the difference between the two cdfs is divided between differences that occur in the smaller distances and the larger distances. Therefore the maximal difference does not appear to be so extreme and only captures one of the aberrations created by this case. The M statistic does not suffer from either of these shortcomings as illustrated above.

In what follows we consider some of the properties of the M statistic in order to better understand its distributional properties so as to enhance our capabilities for inference.

4.3.1 Bivariate Power Calculations

We have seen the effectiveness of the M statistic at detecting clusters generated on a unit circle. As described previously, our aim is to be able to combine this test with a test of a Poisson process in order to more accurately and rapidly detect aberrations in a system that considers the number of events occurring as well as the spatial clustering of those events.

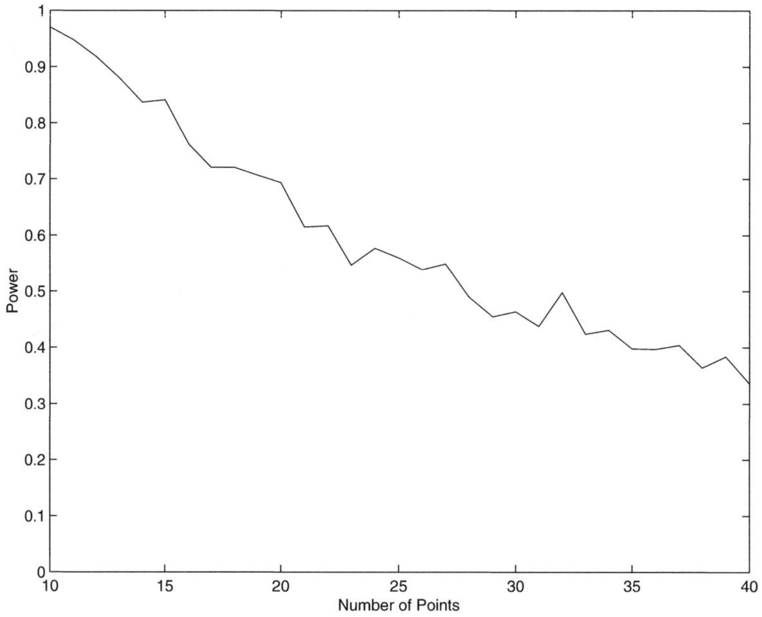

Figure 4.4. *Power of the M statistic to detect a cluster. Cluster size* $= 5$, *radius* $= 0.001$, *and various locations indicated by r in the legend; r varies from* 0.0 *to* 0.8.

Clearly, the number of cases of a disease are the determinant of whether an outbreak has occurred or not. When there is value in early detection of an outbreak, other useful information can be incorporated to facilitate more rapid detection. So the task at hand is to *simultaneously* consider the number of cases *and* the location of these cases. Combining these two streams of information should improve our ability to detect outbreaks.

A very simple first step analysis is to consider that N, the number of cases, follows a Poisson distribution with parameter λ. Then one can superimpose an outbreak on this.

We know that asymptotically, conditional on n, as $n \to \infty$, nM is distributed as a chi-squared variable [3] whose parameters are independent of n. This motivates us to think of n and nM as independent. This, in turn, suggests a bivariate rejection region such as shown in Figure 4.5 where the two pieces of information are combined so that the overall type I error rate is 0.05. This compromise region protects against either too many cases or too much clustering individually, or in concert.

We continue with the example previously described, but now additionally assume that the number of sample points follow a Poisson distribution with a mean of 25. Table 4.1 gives a summary of the bivariate powers when the Poisson and M statistic are combined and contrasts these to the situation when we use only the M statistic. One can see that this method of combining the two statistics is not optimal. Indeed, an optimal boundary may depend on where the cluster is placed. But as a compromise boundary we see that the power is not too influenced by the location of the cluster.

4.3.2 Use of Nearest Neighbors Distances

Traditional methods for using distances to detect clustering have involved the nearest neighbor (see [6], for example). The M statistic above uses all distances, so the question naturally

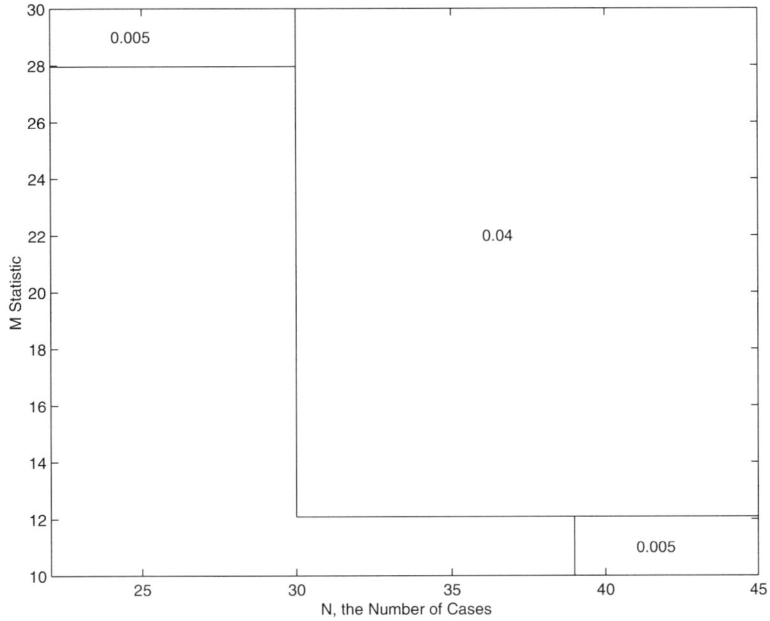

Figure 4.5. *Critical region for the bivariate test, a combination of a region where n is too large, where M is too large, and where both are too large simultaneously. The probabilities of lying in each region under the null hypothesis are shown.*

Table 4.1. *Power to detect clusters at various locations in the unit circle when the clusters are superimposed on a Poisson (mean 25) number of points. The columns headed M statistic are the powers when considering the M statistic alone. The columns headed Joint are the powers when using the bivariate statistic.*

Cluster Location	Cluster size 5		Cluster size 8	
	Poisson power = 0.24		Poisson power = 0.45	
	M statistic	Joint	M statistic	Joint
0.0	0.7798	0.6156	0.9254	0.9107
0.2	0.5877	0.5680	0.8861	0.9497
0.5	0.4974	0.5602	0.8554	0.9071
0.8	0.4731	0.5021	0.8504	0.9054
alpha	0.05	0.0474	0.05	0.0474

arises as to whether one loses any information by only looking at the closest neighbor. We report on a small study that looked at the power of detecting a cluster based on looking at the k-nearest neighbors. This tends to simplify the problem and minimize the amount

Table 4.2. *Bivariate power calculations when using nearest neighbors.*

Number of near neighbors	Power
1	0.2054
2	0.2078
3	0.5031
4	0.4774
5	0.6645
6	0.5834
7	0.6279
8	0.6226
9	0.5387
All	0.5602

of dependencies in the data, but it suffers from other weaknesses. Table 4.2 illustrates the results from doing this with our usual problem. Here the cluster is of size five, located at radius 0.5 and of radius 0.001. We see that there is no advantage to using near neighbor data; in fact it might compromise power, depending on the size of the cluster. Since we would not anticipate knowing the size of the cluster at the time of testing, this method does seem to be optimal.

Therefore, for this simple case, we have a method of detecting spatial and quantitative aberrations in a system. By utilizing the M statistic to detect clustering and combining it with information gained from the Poisson distribution, one can powerfully detect an outbreak. It is possible to extend this technique for use in more complicated situations, such as in populations where the cases are not expected to occur uniformly over the area of consideration. One also might consider the case where multiple addresses are used, as will be described later, or the case when exact addresses are not available.

4.3.3 Discretization of Addresses

Often the only data that a medical institution will release are zip codes of the patient. The following simulates such a situation by discretizing the points generated on a uniform circle into sectors. Two methods for doing this are described below.

Single Circle

The addresses were discretized into eight sectors in this simulation study. The sectors created are pie-shaped and all cases falling in a given sector are reported as coming from a point that is central to the sector. All of these points lie a distance of 0.5 from the center and are equally spaced (the upper graph in Figure 4.6). The M statistic and the Poisson, as described in the previous section, were used to calculate the powers that are shown in Table 4.3. We considered three different locations of clusters: First we placed a cluster at the origin, then we centered a cluster on a boundary between two sectors, and finally we placed the cluster entirely in one sector. The cluster at the center had the potential of being split amongst any of the sectors. The cluster on a boundary had its impact split between two sectors. Thus the third placement was expected to have the most noticeable impact, as indeed it did.

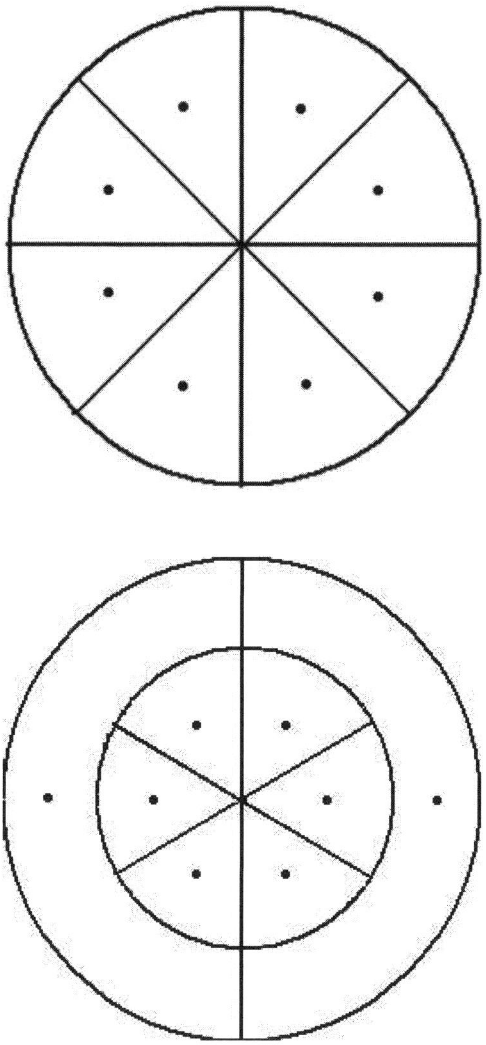

Figure 4.6. *The two graphs show the regions broken down into separate sectors. Anyone within a particular sector is reported as coming from the point displayed within that sector.*

There is a clear decline in power due to the discretization of the addresses. This may be partly due to the regular design we used that also greatly diminished the distinct number of distances. It is unusual that the powers for the cluster placed at the origin, where the points in the cluster could potentially fall into any one of the eight sectors, would be equivalent to the case when the cluster is entirely contained within one sector.

Two-Step Discrete

One might also consider a case similar to an urban setting, where the population density is largest at the center and diminishes as we move away from the center. One way to portray this is as follows. The circle is divided into two concentric circles, with the radius of the

Table 4.3. *Bivariate powers for the simple discrete case.*

Cluster Center	Cluster size 5	Cluster size 8
Poisson power =	0.24	0.45
0.0	0.2721	0.4880
On boundary	0.2449	0.3076
Contained in One Sector	0.2796	0.6842
alpha	0.0435	0.0435

Table 4.4. *Bivariate powers for the two-step discrete case.*

Cluster Center	Cluster Size 5	Cluster Size 8
0.0	0.3101	0.5188
On boundary		
Inner Circle	0.3202	0.6582
Outer Circle	0.4140	0.6656
In one Sector		
Inner Circle	0.3101	0.6761
Outer Circle	0.2493	0.5482
alpha	0.0483	0.0483

inner circle being 0.7. Six sectors are created in the inner circle while two are used on the outer circle (see the lower figure in Figure 4.6). We considered five different cluster locations as described in Table 4.4.

We were drawn to considering what happens when an individual's address is coarsely recorded, such as knowing only the zip code, possibly for privacy reasons. One cannot draw too many general conclusions because of the very special design we used, with possibly too many regularities. These regularities restrict the number of distinct values the distances can take and thus the distribution of distances is now quite discrete with some large steps. But it is instructive to see what a large impact discretizing the data can have. More work needs to be done in this area.

4.4 The Use of Multiple Distances

Conceptually, the consideration of multiple addresses is appealing because it brings more information to bear on the problem. The simulations reported above are useful if the formation of the clusters has some connection to the point at which individuals were infected.

Practically, we can use the individuals' home addresses to serve as proxies for the points of infection, admitting, of course, that this may not be the best proxy. Thus, if more than one address is available per individual, such as an individual's home address and/or work address, then it may prove beneficial to use these multiple addresses. Building on the statistical framework that we have already developed, we now consider this problem of incorporating multiple addresses and other data into the distribution of distances.

More formally, suppose for each individual we record a set of data, taken from a sample space $\mathcal{D}$. We have seen examples where $\mathcal{D}$ was discrete (section 4.2) and where $\mathcal{D} = \mathbb{R}^2$ (the address of the individual, as in sections 4.2 and 4.3). We now allow $\mathcal{D}$ to be a higher-dimensional space, perhaps the product of several copies of $\mathbb{R}^2$ corresponding to multiple addresses, or a combination of discrete and continuous data. We will call any map $d : \mathcal{D} \times \mathcal{D} \to \mathbb{R}$ a *distance function*, understanding that this is an abuse of terminology (indeed, we do not require that our "distance function" take only positive values, satisfy the triangle inequality, or place any other restrictions on d).

In actual practice, the space $\mathcal{D}$ will be determined by the available data. The distance function d should make some attempt to model proximity in the context of the data and the problem at hand. We give some examples below.

1. Let $\mathcal{D} = \mathbb{R}^2 \times \mathbb{R}^2$. Each component indicates a separate recorded address. Possible distance functions between two data points include taking the minimum of the (Euclidean) distance between first address and second address; taking the minimum of all possible address comparisons; or taking a weighted average of distance between first address and second address.

2. Let $\mathcal{D} = \mathbb{R}^2 \times \mathbb{R}$. The first component indicates a recorded address, and the second component a time of event. An appropriately chosen distance function can measure various spatial-temporal clustering effects (see [8, 9, 7] and references therein, for example). As one example, given two data points (x_1, y_1, t_1) and (x_2, y_2, t_2), set $\Delta x = |x_1 - x_2|$, $\Delta y = |y_1 - y_2|$, and $d = |t_1 - t_2| \cdot \sqrt{(\Delta x)^2 + (\Delta y)^2}$. This measures proximity as a product of spatial and temporal distance.

3. Let $\mathcal{D} = \mathbb{R}^2 \times S$, where $S = \{1, 2, \ldots, n\}$ is a finite or discrete countable set. Given two data points (x_1, y_1, s_1) and (x_2, y_2, s_2), set $d = \sqrt{(\Delta x)^2 + (\Delta y)^2} + \alpha_1 \cdot \delta(s_1, s_2)$. Here, the delta function $\delta(s_1, s_2)$ takes the value 1 if $s_1 = s_2$, and 0 otherwise. The parameter α_1 is a real constant.

To illustrate the importance of using all of the available information, we ran a series of power calculations based on the last example above. Our simulated data consist of individuals with address coordinates generated uniformly from the unit circle. Each quadrant of the circle corresponded to a hypothetical school district, and so each individual was assigned to a school according to the quadrant containing the individual's address. Now $\mathcal{D} = \mathbb{R}^2 \times \{1, 2, 3, 4\}$.

We then simulated disease outbreak in the following way. We generated a random sample of size between 10 and 40. We then reassigned five of these individuals to a broad region in the first quadrant (uniformly distributed throughout $R = \{(\rho, \theta) | 0.1 \leq \rho \leq 0.7, 0 \leq \theta \leq \frac{\pi}{2}\}$). The intention was to simulate an undetected attack at a school; because the residential addresses of affected children would be widely dispersed, usual methods of cluster detection may lack sensitivity to this pattern of disease.

The interpoint distances were calculated, first using the Euclidean distance only and ignoring the school district. We then used a chi-squared test (three degrees of freedom) on the expected and actual counts of individuals in each school district, ignoring the address.

Table 4.5. *Power to detect a cluster of size five in the first quadrant when the others in the sample are uniformly distributed in the unit circle. The contrast is between using solely the home address, only the school address, or both addresses.*

Sample size	10	15	20	25	30	35	40
Address only	0.40	0.21	0.20	0.14	0.13	0.12	0.10
District only	0.51	0.31	0.23	0.19	0.19	0.17	0.14
Address and district	0.71	0.41	0.34	0.23	0.19	0.19	0.18

Finally, we used both available components of the data, computing the interpoint distances with a distance function d as described in Example 3 above. The univariate test statistic used on the interpoint distance distributions was the M statistic.

The power to detect these events of five individuals in one school district was greatest when both components of data were utilized. Gain in power was on the order of 25–75%. Table 4.5 summarizes the power results.

When working with the distribution of interpoint distances, the outcome of any test statistic will depend on the choice of the distance function d. For statistics such as the M statistic (section 4.3) which rely on a binning procedure based on quantiles, there are equivalence classes of distance functions that leave the statistic invariant. Call two distance functions $d_1, d_2 : \mathcal{D} \to \mathbb{R}$ *monotonically equivalent* if $d_1 = \phi \circ d_2$ for some monotonic (increasing or decreasing) function $\phi : \mathbb{R} \to \mathbb{R}$. Then we have the following.

Fact. The M statistic is invariant across each equivalence class of monotonically equivalent distance functions.

Indeed, suppose $X_1 \dots X_n$ are observed interpoint distances distributed according to a cdf F_X. Write $q_i = F^{-1}(i/100)$ for the quantiles of F_X. Suppose ϕ is a monotonically increasing (decreasing) function from $\mathbb{R}$ to $\mathbb{R}$. Then for any particular observation X_i that lies between q_j and q_{j+1}, monotonicity guarantees that $\phi(q_j) \leq \phi(X_i) \leq \phi(q_{j+1})$ (or reverse the inequalities for ϕ decreasing). Then bin the distances X_i into deciles of F_X as described above. This yields the same bin counts as when we bin the distances $\phi(X_i)$ and use the deciles of $F_{\phi(X)}$. Hence the value of M remains unchanged after transformation $X \mapsto \phi(X)$.

Example. Let $\mathcal{D} = \mathbb{R}^2$, and let $d : \mathcal{D} \to \mathbb{R}$ be the ordinary Euclidean distance. Let $\phi_1 = \frac{1}{x} : \mathbb{R} \to \mathbb{R}$ and $\phi_2 = \frac{\exp(1/x)}{1+\exp(1/x)} : \mathbb{R} \to [0, 1]$. Then for a fixed set of data, the M statistic will take the same value whether computed using distance function d, $d_1 = \phi_1 \circ d$, or $d_2 = \phi_2 \circ d$.

We find that using the transformation suggested by d_2 above is convenient for maintaining a normalized measure of proximity, taking values in $[0, 1]$. Note that we have inverted the scale of values, so that proximity close to 1 indicates two individuals that are very similar, while a value close to 0 indicates to individuals that are dissimilar. Depending on the model, we may gain some interpretability using this similarity measure. One possible interpretation might be $d_2 =$ the probability that two individuals became infected from the same source of disease. In practice we would first determine the available data and formulate an appropriate method of measuring proximity that captures the essence of the problem at hand. Call this distance function d, and applying ϕ_2 to d gives a proximity or similarity measure that takes values between 0 and 1.

4.5 Conclusions

We have attempted to show how to use the distribution of the interpoint distance between two randomly selected observations as a summary of a spatial distribution. It can be used in biosurveillance if it provides a sufficiently stable constancy in order to define normal behavior against which deviations can be spotted. That has been our experience. We have observed the distributions of distances between patients arriving at the Emergency Department with flulike symptoms at a children's hospital in Boston, and they display a remarkable constancy over time [18].

The distribution can be estimated from the dependent distances between observations in a random sample generated from an underlying spatial distribution. It can then serve as the null distribution against which the deviations of future samples can be compared. The M statistic is an example of a derived statistic that is specifically designed to detect these deviations.

We show that the combination of the M statistic (or of another statistic based on the interpoint distance distribution) with a statistical test for the presence of an excessively high number of cases of some disease in a given time frame (in a day or week, for example) allows for an increase in the power to detect outbreaks caused by naturally occurring or deliberately released agents.

Given the many possible routes of infection in the event of an outbreak, the possibility of extending the concept of interpoint distance to a dissimilarity measure—in particular, through the use of the multiple addresses usually associated with each individual—allows for a straightforward generalization of the methods described here. Further extensions also include the fitting of models for the interpoint distance, also from dependent quantities obtained from all pairs of observations.

The development of surveillance systems that collect real-time information on health-related events (such as flulike symptoms in pediatric emergency room admissions) and that use detection methods such as the ones introduced here should become a priority for both its public health and its national security implications.

Acknowledgments

This research was funded in part by National Institutes of Health grants RO1AI28076 and T32AI07358.

Bibliography

[1] B. Y. Reis, M. Pagano, and K. D. Mandl, *Using temporal context to improve biosurveillance*, Proc. Natl. Acad. Sci., 100 (2003), pp. 1961–1965.

[2] M. Meselson, J. Guillemin, M. Hugh-Jones, A. Langmuir, I. Popova, A. Shelokov, O. Yampolskaya, *The Sverdlovsk anthrax outbreak of* 1979, Science, 266 (1994), pp. 1202–1207.

[3] M. Bonetti and M. Pagano, *The interpoint distance distribution as a descriptor of point patterns: An application to cluster detection*, to be submitted, 2003.

[4] M. Dwass, *Modified randomization tests for nonparametric hypotheses*, Ann. Math. Statist., 28 (1957), pp. 181–187.

[5] M. Hollander and D. A. Wolfe, *Nonparametric Statistical Methods*, Wiley, New York, 1999.

[6] N. A. C. Cressie, *Statistics for Spatial Data*, Wiley–Interscience, New York, 1991.

[7] G. M. Jacquez, *A k nearest neighbour test for space-time interaction*, Statist. Med., 15 (1996), pp. 1935–1949.

[8] G. Knox, *The detection of space-time interactions*, Appl. Statist., 13 (1964), pp. 25–29.

[9] N. Mantel, *The detection of disease clustering and a generalized regression approach*, Cancer Res., 27 (1967), pp. 209–220.

[10] M. S. Bartlett, *The spectral analysis of two-dimensional point processes*, Biometrika, 5 (1964), pp. 299–311.

[11] E. Borel, *Traité du Calcul des Probabilités et de ses Applications*, I, Gauthier-Villars, Paris, 1925.

[12] V. de la Peña and E. Giné, *Decoupling*, Springer-Verlag, New York, 1999.

[13] J. E. Kelsall and P. J. Diggle, *Non-parametric estimation of spatial variation in relative risk*, Statist. Med., 14 (1995), pp. 2335–2342.

[14] W. K. Newey and D. McFadden, *Large sample estimation and hypothesis testing*, in Handbook of Econometrics, Vol. IV, R. F. Engle and D. McFadden, eds., Elsevier–North-Holland, Amsterdam, 1994, pp. 2113–2241.

[15] B. W. Silverman, *Limit theorems for dissociated random variables*, Adv. Appl. Probab., 8 (1976), pp. 806–819.

[16] A. W. van der Vaart, *Asymptotic Statistics*, Cambridge University Press, Cambridge, UK, 1998.

[17] B. S. Weir, *Genetic Data Analysis* II, Sinauer Associates, Sunderland, MA, 1996.

[18] K. L. Olson, M. Bonetti, M. Pagano, and M. D. Mandl, *Syndromic surveillance: A population-adjusted, stable geospatial baseline for outbreak detection*, in Proceedings of the AMIA Fall Symposium 2003, to appear.

Chapter 5

Epidemiologic Information for Modeling Foot-and-Mouth Disease

*Thomas W. Bates,** *Mark C. Thurmond,*† *and Tim E. Carpenter*‡

5.1 Introduction

Foot-and-mouth disease (FMD) is widely regarded as one of the most economically important livestock diseases in the world. Although FMD has not affected US livestock since 1929 [1], the disease is endemic in many areas around the world, which has created a constant threat to US herds and flocks. Changes in modern livestock management, coupled with continued international trade of livestock and livestock products and the potential for FMD to be intentionally introduced as a means of economic bioterrorism, have elevated concerns that FMD may reenter the US. A recent study estimated that if an FMD epidemic occurred in California economic losses totaling $4.3 to $13.5 billion could be expected [2]. Improved mathematical modeling techniques, data collection and management tools, and analysis techniques may permit evaluation of improved means of controlling an FMD epidemic.

5.2 History

FMD was first described by Fracastorius during an Italian outbreak in 1514 and has since been diagnosed sporadically, or become endemic, in many countries around the world [3]. Epidemics were reported in Germany, France, and Italy during the seventeenth and eighteenth centuries, although the first reports of FMD outside Europe did not occur until the early nineteenth century, when countries in Asia, Africa, Australia, North America, and South America reported outbreaks [4]. The US has remained free of FMD for over 70 years.

*Medical Physics and Biophysics Division, Lawrence Livermore National Laboratory, P. O. Box 808, L-174, Livermore, CA 94550.

†Department of Medicine and Epidemiology, School of Veterinary Medicine, University of California at Davis, Davis, CA 95616.

‡Department of Medicine and Epidemiology, School of Veterinary Medicine, University of California at Davis, Davis, CA 95616.

Prior to legislation introduced in 1930 creating livestock import restrictions, however, FMD epidemics occurred in the US during 1870, 1881, 1884, 1902, 1908, 1914, 1924–1925, and 1929. Also, in 1978 an accidental release of the virus occurred at the Plum Island Animal Disease Research Center in New York [5].

FMD has not been recently reported in North America. The last Canadian FMD outbreak occurred in 1952 and the last Mexican outbreak occurred in 1954 [4, 6, 7]. Other countries, however, have been less fortunate in remaining FMD-free (Table 5.1). FMD remains endemic in many countries in South America, Africa, and Asia and is a constant threat to countries free of the disease. In 1997 an FMD outbreak in Taiwan affected 6,147 swine herds, devastating its swine producing economy. By some estimates, costs of FMD in Taiwan may have ranged from $1.6 to $3.6 billion US dollars [8, 9, 10]. FMD outbreaks also occurred in 2000 in South Korea and Japan, which had been free of the disease since 1934 and 1908, respectively [11, 12, 13].

Table 5.1. *Selected foot and mouth disease epidemics.*

Country	Year	Number of herds infected	Number of herds depopulated
Argentina	2001	N/A	1,105
Canada	1951–1952*	29	42
France	2001*	2	2
Ireland	2001	4	400
Japan	2000*	4	4
Netherlands	2001*	26	2,600
Taiwan	1997	6,147	6,147
United Kingdom	1967–1968	2,364	2,364
	2001*	2,030	9,318
United States	1924–1925	N/A	948
	1929*	5	33

*Year of most recently reported FMD outbreak.

Few FMD outbreaks were reported in Europe during the 1990s after the disease was eradicated by use of vaccination [14, 15]. Outbreaks in Italy and Greece, however, are notable exceptions [16]. Also, in 2001 after being free of the disease for 20 years, the UK was affected by a large FMD epidemic that led to the destruction of 9,318 herds, including 600,000 cattle, 3,190,000 sheep, 147,000 pigs, 2,000 goats, 1,000 deer, and 200 other animals [17]. England, Wales, Scotland, and Northern Ireland were affected by the disease, which later spread to the Republic of Ireland, France, and The Netherlands. The 2001 FMD epidemic in the UK greatly exceeded the number of animals affected during the "Great UK Cattle Epidemic" in 1967, which affected 2,339 livestock facilities and resulted in the slaughter of 443,000 animals before the epidemic ended in 1968 [18].

5.3 Description of the Agent

FMD is caused by the *aphthovirus* of the family Picornaviridae. There are seven immunologically distinct FMD virus (FMDV) serotypes, namely A, C, O, Asia 1, and SAT (Southern African Territories) 1, 2, 3 [19]. Molecular epidemiological investigations have found the

FMDV genome, which contains approximately 8,500 nucleotides [20], to be stable and does not rapidly change antigenic characteristics to evade host immunity [20]; however, there have been reports of vaccine program failures, which were attributed to genetic drift [21, 22].

5.4 Clinical Signs

FMD can affect all cloven-hoofed animals, including cattle, sheep, goats, and swine and cloven-hoofed wildlife species, such as antelope, bison, deer, elk, and moose. FMD can be subclinical or manifest as clinical disease. Small ruminants, such as goats and sheep, often display only mild signs and have been described as having iceberg infections because their milder signs may lead to misdiagnosis of FMD, potentially leading to widespread transmission among susceptible populations [23]. Typical clinical signs include excessive salivation and development of vesicles in the oral cavity and on the tongue, muzzle, teats, coronary bands, and interdigital space of the hoof [24]. Clinically affected animals will often smack their lips in response to formation of vesicles, causing drooling or frothing around the mouth. Painful erosions on the coronary band of the hoof, and possibly sloughing of the hoof, are common and may result in dehydration or loss of condition because it is too painful for the animal to walk to water or feed.

FMD is not generally considered a zoonotic disease, although there have been more than 40 human confirmed cases reported [25] and over 400 reports of possible cases [26] since the disease was first discovered. The virus, typically type O or C [25], has been recovered from the nose, throat, saliva, and air expelled during coughing, sneezing, talking, and breathing of people exposed to animals infected with FMDV [27]. When humans have become infected, only mild signs and symptoms have been reported. For example, a man infected in the UK in 1966 displayed vesicles on the mouth and reported having a headache and sore throat, but recovered uneventfully [28]. Nasal isolates of FMDV from man have been experimentally shown to cause subsequent infection in livestock, indicating that humans can act as a carrier host, with the potential to transmit the virus [29].

5.4.1 Pathogenesis

Within 20 to 24 hours, an infected animal may develop a viremia that disseminates the virus to secondary sites, which are predominately epithelium tissues on the tongue, oral cavity, and feet. Internal tissues, however, also have been known to be affected, including pituitary and mammary glands, lymphoid and myocardial tissues, small and large intestines, and rumen [30, 31]. Additional vesicles may develop at the secondary sites and can rupture within a few hours, exposing raw tissue. Vesicles will typically heal with little or no residual effect; however, secondary bacterial infections may occur, resulting in mastitis or pneumonia or long-term damage, such as permanent deformities of the hoof [30]. It also has been estimated that production losses of 25% or more can be expected of infected and recovered animals, costing an estimated $50 billion annually worldwide [32].

Transmission

Transmission of FMDV occurs primarily via the respiratory route. A large amount of virus can be aerosolized during normal breathing of infected animals and transmitted directly to susceptible animals via very small, aerosolized saliva particles, also called droplet nuclei, containing one or more infective doses of virus. An infective dose (ID_{50}) is the virus concentration required to infect 50% of the exposed individuals or cell cultures [33]. Virus also can be released though excessive salivation, a common condition of FMDV-infected animals. Feces may contain large amounts of FMDV and can play a role in contaminating the environment, potentially leading to indirect transmission. The maximum ID_{50} concentrations of FMDV found in feces, urine, semen, blood, epithelium, bone marrow, and air for cattle, pigs, and sheep have been summarized [44].

When one animal in a herd becomes infected, the remaining susceptible animals will generally become infected within 2–3 weeks; however, the speed and likelihood of transmission vary by herd size, pen density, species, host condition, and viral subtype. Analysis of results from one study indicated that virus excretion from experimentally exposed cattle, pigs, and sheep was biphasic [34]. The first phase occurred within 30 minutes to 22 hours post exposure and was thought to have been caused by recirculation of FMDV inhaled during exposure. The second phase, which occurred 2 to 7 days post exposure, was thought to have been caused by virus that multiplied within the respiratory tract. The incubation period (time from infection to clinical signs) for swine, cattle, and sheep, however, was observed to be 5–8, 3–5, and 5–13 days, respectively [35]. Similarly, the mean durations of the subclinical period (duration of viral excretion before appearance of clinical signs) for swine, cattle, and sheep were 4.3, 2.2, and 2.6 days, respectively [35]. The infectious period (duration of viral excretion) generally lasts 5 to 19 days [36], although the mean infectious period of herds infected during the 1951–1952 epidemic in Canada [7] were estimated to be 16.7 days (ranging from 6 to 63) for cattle and 10.3 days (ranging from 7 to 14) for swine [37]. It has been reported that infected cattle, however, may continue to carry the virus in their esophageal-pharyngeal region for at least 12 months, indicating that transmission could occur long after clinical signs have subsided [38].

Direct Contact FMDV is commonly transmitted from animal to animal via the respiratory and fecal-oral routes through close contact of animals. FMDV also has been isolated from bull semen and raw milk [39, 40, 41], indicating that horizontal transmission by coitus and nursing are possible. A calf, however, may be protected from postnatal infection for up to 12 weeks if FMDV-specific antibodies were passed in colostrum from the dam [30].

Indirect Contact by Personnel Indirect contact has been the most often implicated means of spreading disease [42] and is an important consideration for potential FMDV transmission. FMDV has been isolated from the human respiratory track up to 48 hours after contact with FMDV-infected livestock [27, 29]. The survivability of FMDV in clothing material, such as cotton and wool, also has been documented, and FMDV particles in materials potentially can be aerosolized and inhaled by susceptible animals [43]. Because affected animals may excrete massive quantities of FMDV into the environment, including approximately 79,000 ID_{50} per milliliter of urine and approximately 316,000 ID_{50} per gram of feces [44], an important means of indirect transmission may be by personnel who did not adequately disinfect themselves before visiting a susceptible herd. As an example, the Canadian FMD outbreak in 1951–1952 was thought to have been introduced by a farmhand coming from an FMDV-infected region in Germany. Nearly all secondary outbreaks in that epidemic also were thought to have been caused by neighbors or farmhands that had indirect contact with the index farm and other nearby livestock [45].

Indirect Contact by Vehicles Mechanical transmission by contact of animals with material deposited by vehicles is thought to be possible because FMDV survives well in the common herd environments. Contaminated feces, slurry, and feed can remain infectious in the environment for several weeks [46] and can be transferred from herd to herd via a vehicle acting as a fomite. In one report, FMDV was recovered from air sampled from a stall vacated by livestock with FMD [47], suggesting that reintroduction of a susceptible animal into the stall up to 24 hours after being vacated could result in infection. Results of that study have potential implications for FMDV transmission via the air in livestock trailers that are commonly used in the US to move animals from herd to herd. Investigation into potential transmission by other types of vehicles, however, can be more difficult to confirm. When investigations of FMD outbreaks found no clear epidemiologic link between two infected herds, transmission was often categorized as "local spread." Because vehicles traveling from herd to herd may visit several more herds in a single day, their routes are difficult to trace during investigations, although they are thought to contribute to "local spread."

Consumption of Contaminated Food Products Transmission of FMDV also can occur by consumption of contaminated food products, such as raw or processed pork, lamb, or beef. Dairy products, such as milk, cream, butter, and whey, also have been shown to contain viable virus, even after processing [40, 41, 48, 49]. Swine are much more susceptible to FMDV transmission via the oral route than cattle and other cloven-hoofed species; however, all cloven-hoofed species are susceptible, given sufficient viral doses. The ID_{50} for oral infection of swine is $10^{3.9}$ but is much higher in cattle ($10^{5.8}$) [50]. Because livestock are susceptible to FMDV infection via the oral route, human food waste (e.g., food from ships and airplanes) must be cooked for 30 minutes, or 120 minutes if garbage is from a foreign country, at 212 degrees Fahrenheit to attempt to eliminate any exotic disease agents before consumption by swine is allowed [51, 52]. Inquires were inconclusive as to whether feeding uncooked swill to pigs was the source of FMDV introduction into the UK in 2001 [53].

Airborne Transmission Livestock are highly susceptible to infection via the respiratory route; intradermal inoculation is the only other route with sensitivity nearing that of the respiratory route [50]. Cattle and sheep can become infected by inhaling 10–25 ID_{50}, while swine require $10^{2.6}$ ID_{50} [50]. Airborne movement of FMDV is thought to be possible when certain environmental conditions are present. The minimum relative humidity required for FMDV to survive without dehydration is 55–60% [54, 55]. Although FMDV is sensitive to ultraviolet light, at least one study indicates that the amount of ultraviolet light found in sunlight has little effect on the virus [56]. Reports of the amount of FMDV released during normal breathing have been inconsistent. One report found that pigs shed $10^{6.08}$ ID_{50} of FMDV per 24-hour period [57] and cattle and sheep shed $10^{4.6}$ ID_{50}, while other reports indicated that $10^{5.4}$ ID_{50}were shed from cattle and sheep, and $10^{8.6}$ ID_{50} from swine [58, 59].

5.5 Prevention and Control

In non–FMD-endemic countries there are three primary ways to prevent or control FMDV transmission. First, limit disease transmission by restricting livestock movement and not permitting importation of animals and livestock products from areas known to have FMD. Second, vaccinate susceptible livestock after the disease has been introduced. Third, interrupt the life cycle of FMDV by destroying the host [24, 60]. In predominantly hot, dry countries, where risk of airborne transmission is minimal, intraherd FMDV transmission also was shown to have been limited by segregation of pens with infected and susceptible animals [61].

5.5.1 Livestock and Livestock Product Import Restrictions

In the US, livestock and livestock products from FMD-affected countries have been strictly regulated since 1930, when legislation such as the Smoot–Hawley Tariff Act, section 306(a) [62], was introduced. To ensure compliance, US Department of Agriculture (USDA) Animal and Plant Health Inspection Service (APHIS) veterinarians and more than 1,800 nationwide inspectors conduct point-of-entry inspections, quarantine live animal and bird imports, and restrict potential introduction of foreign animal diseases by inspecting cargo, controlling aircraft and ship garbage, and verifying germplasm certificates [63].

5.5.2 Emergency Response Preparedness

To prepare for highly infectious foreign animal disease (FAD) outbreaks, the US has developed the following animal health emergency management objectives: prevent the introduction of foreign animal diseases, be prepared to manage an outbreak, develop appropriate response strategies for control and eradication of the disease, and take active measures to recover from the animal health emergency [63]. As part of its FAD preparedness, USDA supports a team of emergency response personnel, termed the Regional Emergency Animal Disease Eradication Organization (READEO). During an actual FMD outbreak, READEO members would be expected to (1) create a command center to coordinate eradication efforts; (2) quarantine affected premises and create a 10-km high-risk zone around the known affected herds and a 20–40-km buffer zone around the affected herds; (3) identify and destroy sick and exposed animals; (4) dispose of carcasses and contaminated items; and (5) clean and disinfect the affected premises [43, 64, 65]. Every few years the READEO has been activated to conduct mock outbreak eradication exercises. In a recent exercise in 1998, simultaneous outbreaks were considered at locations in eastern and western regions of the US [66], and a follow-up exercise in 2000 included decision-makers from Canada and Mexico [67].

A comprehensive electronic emergency response system capable of predicting potential spread of FMD, with the ability to analyze potential control strategies, does not yet exist in the US. Other countries, however, have developed such systems. For example, the EpiMan system was developed in New Zealand to provide FMD outbreak decision support and simulation modeling tools [68, 69]. EpiMan considers herd location and size information from a continuously maintained herd database that is linked to a geographical information system and includes simulation models that consider data from a study of contact among livestock facilities in New Zealand [69, 70], in addition to data derived from the 1967–1968 epidemic in the UK. A project to adapt EpiMan for use within the EU (EpiMan-EU) was initiated in 1995, using modeling parameters derived from an analysis of results from a herd contact study conducted among livestock facilities in The Netherlands [71, 72, 73].

5.5.3 Prophylactic Vaccination

Prophylactic vaccination of susceptible animals can be used to protect livestock from clinical disease and reduce transmission, and was largely responsible for the eradication of FMD from Europe by the early 1990s [14, 15]. The first successful FMD vaccines were developed in 1925 [4], and by 1988 it was estimated that the annual worldwide demand for FMD vaccines exceeded $1 billion, surpassing demand for any other vaccine, even human vaccines [32]. Most FMD vaccines are chemically inactivated and use an adjuvant, such as Al(OH)3/saponin or an incomplete oil-based formulation [74]. Successful use of a genetically altered, live-attenuated FMD vaccine, however, also has been demonstrated [75, 76].

Analysis of results from tests of highly potent emergency vaccines with up to five times more viral antigen than traditional vaccines have shown that they can provide an early (3–4-day) immune response against challenge, even after only one dose [77, 78, 79, 80]. There are, however, disadvantages to vaccine use: (1) minimal cross-protection among FMDV serotypes necessitates that the vaccine contain a strain antigenically similar, if not identical to, the outbreak strain [20]; (2) the vaccine should lack nonstructural viral proteins (e.g., 3ABC) so that diagnostic tests are able to differentiate vaccinated from naturally infected animals [81]; (3) although transmission of FMDV is reduced, vaccination does not always prevent replication of FMDV in the esophageal-pharyngeal tissues, indicating that airborne transmission could potentially still occur [82, 83]; (4) depending on dosage, FMD vaccines can have a short "shelf-life" and they must be kept cold (2–6°C) or they can lose immunogenicity [84]; (5) the vaccine may cause allergic reactions, which may temporarily decrease milk production [85]; (6) there are potential severe economic trade ramification for countries that vaccinate for the disease because they lose their "FMD-free" trading status [86]; and (7) in rare instances, vaccines have been shown to have caused FMD outbreaks [87].

Reports of vaccine effectiveness have not been widely published. In one report, FMD immunity among dairy herds prophylactically vaccinated in Saudi Arabia varied between 81 and 98% [88]. The vaccine also was considered to confer protection for approximately 2.5 months; however, results from a susceptible-latent-infectious-removed (SLIR) simulation model indicated that the protective antibody half-life was 43 days, leading the authors to conclude that prophylactic vaccination would be inappropriate because the required intravaccination period of 43 days or less was unreasonably short [89]. Evaluations of South American FMD control programs have reported herd protection levels of at least 80% [90], and a similar vaccine effectiveness level (89.3%) was observed during the Mexican outbreak in 1952 [91]; however, all reports were observational and the true level of herd FMDV exposure, if any, was not known.

5.5.4 Emergency Vaccination

Herd vaccination alone is generally considered inadequate to control an epidemic of FMD. Instead it is used as an adjunct to herd slaughter because exposed, vaccinated animals can potentially act as FMDV carriers [39]. Due to the expense, international trade ramifications, and potential ineffectiveness of FMD vaccines, FMD eradication guidelines for the US recommend that vaccination be considered only if an FMD epidemic persists for six months, 1 million susceptible animals have been slaughtered, wildlife in multiple states have endemic infection, or the cost-benefit ratio favors vaccination [43]. The policy, however, would likely be reviewed on a case-by-case basis because analysis of results from recent high-potency vaccine trials indicated that immunologic response may be sufficient to prevent infection from FMD challenge in less than 4 days [78, 79, 80]. Also, a recent outbreak of FMD among herds in The Netherlands was promptly eradicated by means of an eradication strategy that included vaccination [92].

If the decision were made to vaccinate susceptible livestock during an FMD epidemic, vaccines could be made available through international vaccine banks. The US is a member of the North American Vaccine Bank (NAVB) located at the Plum Island Animal Disease Research Laboratory in New York state, which maintains stored antigen sufficient to produce 2.5–8 million doses and has established contracts with vaccine producers to formulate, fill, and package vaccine upon request [93, 94]. The vaccine bank is supported by the US, Canada, and Mexico, who contribute money toward operating expenses based on their percent of the total North American cattle population, which, in 1991, was 72%, 8%, and 20%, respectively

[93]. The International Vaccine Bank at Pirbright, UK, which supports the UK, Australia, Finland, Ireland, New Zealand, Norway, and Sweden, also stores concentrated antigen, but, in contrast to the NAVB, it maintains facilities to rapidly manufacture the final vaccine product. In the UK, manufacturer vaccine production estimates are 100,000–175,000 doses per 24-hour period and could be initiated within 3–4 days, whereas production by vaccine production contractors of the NAVB may require up to 7 days [95]. The NAVB has chosen not to have large quantities of vaccine on hand because of expenses related to storage, quality assurance, and the potentially short "validity period," which can be as little as 12 months, depending on potency [96].

5.5.5 Herd Slaughter

Herd slaughter, often referred to as "stamping-out," [97] and burial, burning, or rendering of infected animal carcasses are commonly practiced as a means of stopping FMDV transmission [24, 60]. The benefit of rapid slaughter of infected herds has recently been quantified in a report that used a simulation model to compare the epidemic size when considering the actual day of slaughter to the predicted epidemic size if herds were slaughtered on the day they were diagnosed [98]. It was speculated that if herds affected during the 1967–1968 UK FMD epidemic had been slaughtered the day they were diagnosed, the 1967–1968 FMD epidemic in the UK may have been reduced by 588 herds (from 2,339 to 1,751) [98]. Important considerations before slaughtering herds, including methods of humane euthanasia and disposal, have been reported elsewhere [43].

5.6 Modeling FMDV Transmission

Models are often used to analyze incomplete, complex, or voluminous data and are especially useful as tools for simulating and analyzing a natural process, such as disease transmission. Numerous models have been developed to analyze various aspects of FMD; however, most have been developed to investigate the possibility of airborne transmission [99, 100, 101, 102, 103, 104, 105, 106] or to aid cost-effectiveness analyses considering use of vaccine [2, 86, 107, 108, 109, 110, 111, 112, 113].

5.6.1 Nonspatial Models

State-transition models [114, 115] have been extensively used to estimate the potential impact of various infectious diseases, including FMD [2, 72, 108, 116, 117, 118, 119]. By modeling the number of herds that become latently infected, infectious, and immune, state-transition models have been found to produce distributions of cases that mimic known epidemic distributions [116]. An important parameter in state-transition models is the probability of effective contact, which is the probability that a susceptible herd had sufficient direct or indirect contact with one or more infectious herds to cause at least one animal to become infected. The probability of effective contact is difficult to estimate directly because it depends on numerous factors, including the virus strain; herd size, location, and density; animal species; and the types and numbers of direct and indirect contacts that occur among the study herds. Instead of directly estimating contact rates, investigators typically will consider the numbers of herds and dates that herds became infected during previous FMD epidemics to calculate dissemination rates (DRs) (also called effective contact rates [37]), which represent the number of herds expected to be infected in the next time period by each current time period infectious herd. DRs then can be included in equations to estimate the probability of effective contact, which is a required parameter in most epidemic state-transition models. One report estimated weekly DRs of 0.53 to 4.52 by using data from the large UK FMD epidemic in 1967–1968 and then estimated the probability that a susceptible herd had effective contact with an infectious herd as $1 - e^{-\mathrm{DR}\times\mathrm{Pa}}$, where Pa is the proportion of herds affected during the previous time period and thus infectious in the current time period [116]. The DR at time $= 0$ is similar to the reproductive number (R_o), which is commonly used to compare the infectiousness of different disease agents. R_o represents the number of secondary cases that one case would produce in an otherwise completely susceptible population [120] and was recently estimated to be as high as 73 among initial FMDV-infected herds in Saudi Arabia [89] and to be 38.4 and 36 among herds initially infected in the 1967–1968 and 2001 epidemics in the UK, respectively [119, 121]. After FMD has been diagnosed in a region and control polices are in place, however, much smaller R_o values (< 2) are frequently observed [116, 122].

5.6.2 Spatial Models (Nonairborne)

Epidemic models that incorporate spatial information (i.e., herd locations) are not new. Their applications in investigation of spatial disease transmission, however, is still not well developed because of the considerable computational power required to run the complex models and because spatial herd data sets have been rare [70]. Spatial epidemic models have been used to consider potential nonairborne FMDV transmission. For example, after investigating potential spatial transmission among herds affected during a large 1967–1968 FMD epidemic in the UK, the author of one report indicated that approximately 25% of the transmission could have been attributed to milk trucks [123]. The report, however, was later challenged after analysis of results from a simulation model indicated that 21% of the risk could have been due to chance, leaving 4% (or less, given possible modeling variation) attributable to the milk trucks [124]. An FMD spatial diffusion model estimated the carrying capacity required to sustain FMD in the Australian feral pig population at 2.3–14 pigs/km and the potential velocity of propagation at 2.8 km/day [102, 125]. Results from a prototype spatial, stochastic FMD epidemic simulation model developed in The Netherlands indicated that, on average, 39 (SD $=$ 31.9) herds would become infected if high-risk contact and infected herds were slaughtered as an eradication strategy [72]. The average epidemic size from that model was higher, but not significantly, than a recently observed FMD epidemic size of 26 herds, which occurred in The Netherlands in 2001 [92]. A strategy that included herd slaughter and emergency vaccination, however, was used to control the epidemic in 2001. Survival analysis also has been used to calculate risk estimates of FMDV transmission by considering data from the 1967–1968 FMD outbreak in the UK [126]. Analysis of these results indicated that during the first three days of an outbreak, herds located within 3 km of an infectious herd were at 3.7- to 10-fold higher risk of infection than those located 3 to 5 km from the same herd.

More recently, a comprehensive spatial-temporal, stochastic simulation model was developed to assess potential FMDV-eradication strategies in a three-county area of California [127, 128]. Because of the long duration since FMD had been diagnosed in California and unique aspects of California livestock husbandry practices, herd sizes, and herd types, many modeling inputs were derived from an animal contract study and expert opinion surveys, as opposed to, for example, assuming that transmission among contemporary herds in California would be similar to that of herds in other countries or historic herds in California during a previous outbreak. Analysis of results suggested that ring vaccination was the most successful strategy to pursue if herds could be rapidly vaccinated. The second most effective strategy was one to slaughter "high-risk" herds, which required the slaughter of infected herds and the selective slaughtering of noninfected herds that were at highest risk of infection. The calculation of "high risk" considered the proximity of susceptible herds to infectious herds, and the expected numbers of direct and indirect contact on both the infected and susceptible herds.

During the recent FMD epidemic in the UK [129], several independent teams of investigators reported independent results from epidemic models developed to predict the epidemic size, while also considering various alternatives for eradication [121, 130, 131]. One team used a mathematical model to represent disease transmission and considered the potential impact of delays in slaughtering herds [121, 122]. The team's approach was to consider information from the first two months of the epidemic, including numbers, types, and locations of herds affected, to develop transmission risk estimates, and to parameterize a model to predict the size and duration of the epidemic. However, because the epidemic model was developed to simulate epidemics in large regions of the UK with over 45,000 susceptible herds, detailed representation of how individual herd-to-herd transmission occurred (e.g., direct contact, airborne, etc.) was limited. When compared to the observed epidemic size, however, analysis of results indicated that the model was reasonably accurate and capable of predicting epidemic sizes, as well as effectiveness of potential eradication strategies. The recommended strategy was to continue slaughtering infected and nearby herds. Use of vaccination also was considered, although it was predicted to be less effective than slaughter. The authors, however, reported that the actual number of susceptible herds in the UK was 131,000, but they did not discuss how their modeling results may have differed had the entire susceptible livestock population in the UK been considered instead.

Another group used a previously developed computer simulation model that emphasized the biological mechanisms of transmission instead of complex mathematical representations [130]. The model contained approximately 54 input parameters; however, the parameter values (or distributions) considered by the model were not presented and so critical review of the modeling results is not possible. The researchers considered strategies to increase the number of preemptively slaughtered herds, to increase the time to slaughter, to use an FMD vaccine, and to use a combination of strategies. When the model considered the actual control strategies used during the epidemic, the group was able to successfully predict the epidemic size and duration. They concluded that 1,800 to 1,900 farms would be infected and the epidemic would end between July and October 2001, which was close to the actual results. The epidemic ended September 30. The epidemic appeared to have ended after 2,030 herds were infected [132]. They also concluded that use of vaccine would have decreased the epidemic size only by about 100 farms. The model, however, was programmed to consider vaccination in a highly unorthodox manner. For example, sheep were not vaccinated and only herds in geographic "bands" stretching east-west across the UK (to prevent transmission between regions) were considered, instead of the more traditional ring vaccination strategy or a strategy to vaccinate all herds.

A third group used a spatial-temporal stochastic simulation model to evaluate retrospectively potential eradication strategies, including various combinations of herd slaughter and vaccination [131, 133]. The modeling kernel was derived from analyses of the historic data available from the 2001 UK epidemic and parameters of susceptibility and transmissibility of individual animals were derived from previous analyses of the 1967–1968 epidemic [134]. The authors concluded that had slaughter of only infected herds been used to control the epidemic, the size may have been approximately ninefold larger; however, pursuing the actual policies followed by UK decision-makers but adding prompt slaughter of confirmed infected herds would have resulted in an epidemic approximately 57% smaller than the observed size. A policy to slaughter only infected herds and to vaccinate herds within 3 km of an infected herd was found to be ineffective (epidemic size increased by 784%), but vaccination strategies that would augment the preemptive slaughter eradication policy used in the UK were somewhat beneficial. Various strategies prevented 3–30% of the expected cases. Stochastic modeling results, however, were based on simulations of only 50 epidemics, and it was unclear how additional simulations would have altered the modeling results.

5.6.3 Airborne Transmission Models

The rapidity with which FMDV has spread among livestock facilities with no identifiable contact, other than the facilities were downwind from an index herd, suggests the possibility of long-distance FMDV transmission via aerosolized virus particles. The potential for airborne transmission has been noted during numerous FMD epidemics, particularly the 1967–1968 FMD epidemic in the UK [135]. One hypothesis proposed to explain the unique distribution of affected herds was that animals became infected by aerosolized viral particles that traveled distances of 20 km or more [136]. Results from spatial analyses of that study indicated herds were grouped in three clusters, with each subsequent cluster downwind from the previous cluster. It was postulated that the virus traveled downwind in an atmospheric phenomenon referred to as Lee waves [136], which help explain how material can be carried aloft, circulated in the air, and deposited downwind at estimable intervals. Under light wind conditions, the calculated length of the Lee waves, and thus intervals of potential viral deposition, were approximately 20 km, which corresponded with the distance between the three herd clusters.

More advanced airborne diffusion models were developed in the late 1970s and have spawned resurgence in the examination of potential airborne FMDV transmission. One report described an airborne prediction model that was tested retrospectively against the Northumberland and Hampshire, UK outbreaks of 1966 and 1967 [104]. Analysis of modeling results supported the hypothesis that airborne transmission was possible in those outbreaks and that the model could be used to assist decision-makers in future outbreaks in the UK. Additional research considered the possibility of long distance transport of FMDV over the sea during epidemics that occurred in Denmark and Sweden in 1966 and the Channel Islands in 1974 [105]. During the epidemics, risk of FMDV transmission via human mechanical vectors was discounted because no contact between the two locations was uncovered after extensive investigations. Wind direction, however, was consistent with possible airborne transmission. Analysis of results from a different airborne prediction model indicated that spread of FMDV 250 km across the British Channel from Brittany, France, to Jersey and the Isle of Wight, UK was possible because (1) abundant meteorological evidence indicated that the transmission was feasible; (2) the FMDV strain-specific serology among animals at affected locations was identical; and (3) no other contacts could be identified between the two locations [99]. The model later considered the possibility of airborne transmission of FMDV from herds in Jordan across the border to herds in Israel [106], and analysis of modeling results indicated that if a herd of pigs were infected near the border in Jordan, there may have been enough aerosolized virus and sufficient meteorological evidence to support the hypothesis that transborder airborne transmission occurred.

An airborne prediction model was used retrospectively to analyze potential airborne spread of FMD to herds whose likely source of infection during the 1951–1952 FMD epidemic in Saskatchewan, Canada was otherwise unknown [137]. It was concluded that herd-to-herd airborne transmission might have occurred, and in some instances, sufficient virus could have been transported up to 20 km downwind. A predictive model also identified areas at high risk of potential airborne FMDV transmission during the 1993 Italian FMD outbreak [101]. Analysis of the modeling results indicated that there was a low probability for long-distance airborne FMDV transmission, suggesting that emergency response resources may be better focused on localized FMDV transmission.

Analysis of results from an alternative Gaussian plume dispersion model that considered airborne transmission for two FMD epidemics and two Aujeszky's disease epidemics that occurred in the UK indicated that simulated airborne transmission matched actual outbreak occurrences [100, 138]. An updated version of the same model (Rimpuff) [139] was used for further analysis of the France/UK FMD outbreaks in 1981 and Denmark/Germany outbreaks in 1982 [103, 140]. Modeling results were consistent with the original hypothesized means of airborne FMDV transmission, if the actual number of infected animals were slightly above the official numbers, which was considered probable.

5.7 Conclusion

In summary, FMD is a complex disease that affects multiple animal species, consists of multiple virus serotypes, and has a wide variety of routes whereby transmission can occur. To perform a realistic analysis of disease FMDV transmission and its control, mathematical or simulation-based epidemic models should reflect this diversity and should be able to (1) incorporate parameters for multiple animal species and herd sizes, various virus serotypes, and multiple routes of virus transmission and (2) evaluate spatial disease transmission and control. The potential benefit of the development and application of appropriate models is that optimal control and eradication programs may be designed and evaluated prospectively, thereby potentially resulting in a substantial reduction of economic losses as well as increased animal welfare.

Bibliography

[1] J. R. Mohler and R. Snyder, *The* 1929 *Outbreak of Foot-and-Mouth Disease in Southern California*, US Department of Agriculture, Washington, DC, 1930.

[2] J. Ekboir, *Potential Impact of Foot-And-Mouth Disease in California: The Role and Contribution of Animal Health Surveillance and Monitoring Services*, Agricultural Issues Center, Division of Agriculture and Natural Resource, University of California at Davis, Davis, CA, 1999.

[3] W. Bullock, *Foot-and-mouth disease in the sixteenth century*, J. Comparative Pathology and Therapeutics, 40 (1927), p. 75.

[4] W. Henderson, *An historical review of the control of foot and mouth disease*, British Veterinary J., 134 (1978), pp. 3–9.

[5] Anonymous, *Cattle virus escapes from a P4 lab*, Science, 202 (1978), p. 290.

[6] M. Loring, *Foot-and-mouth disease in Mexico*, Agribusiness Worldwide, 3 (1982), pp. 22–26.

[7] R. F. Sellers and S. M. Daggupaty, *The epidemic of foot-and-mouth disease in Saskatchewan, Canada*, 1951–1952, Canadian J. Veterinary Res., 54 (1990), pp. 457–464.

[8] A. I. DONALDSON, *Foot-and-mouth disease in Taiwan*, Veterinary Record, 140 (1997), p. 407.

[9] P. C. YANG, R. M. CHU, W. B. CHUNG, AND H. T. SUNG, *Epidemiological characteristics and financial costs of the* 1997 *foot-and-mouth disease epidemic in Taiwan*, Veterinary Record, 145 (1999), pp. 731–734.

[10] B. J. CHEN, W. H. T. SUNG, AND H. K. SHIEH, *Managing an animal health emergency in Taipei China: Foot and mouth disease*, Rev. Sci. Tech. Off. Internat. Epizooties, 18 (1999), pp. 186–192.

[11] ANONYMOUS, *Suspected Foot and Mouth Disease, South Korea, March* 2000, Impact Worksheet, Center for Emerging Diseases, Animal and Plant Health Inspection Service, US Department of Agriculture, Hyattsville, MD; available online from http://www.aphis.usda.gov/vs/ceah/cei/fmdkoreaext.htm; accessed September 14, 2001.

[12] ANONYMOUS, *Suspected Foot and Mouth Disease, Japan, March* 2000, Impact Worksheet, Center for Emerging Diseases, Animal and Plant Health Inspection Service, US Department of Agriculture, Hyattsville, MD; available online from http://www.aphis.usda.gov/vs/ceah/cei/fmdjapanext.htm; accessed September 14, 2001.

[13] G. MURRAY, *Foot and mouth disease hits both Japan and South Korea*, Australian Veterinary J., 78 (2000), p. 303.

[14] Y. LEFORBAN, *Prevention measures against foot-and-mouth disease in Europe in recent years*, Vaccine, 17 (1999), pp. 1755–1759.

[15] A. I. DONALDSON AND T. R. DOEL, *Foot-and-mouth disease—the risk for Great Britain after* 1992, Veterinary Record, 131 (1992), pp. 114–120.

[16] R. KITCHING, *Foot-and-mouth disease in the European Union* Boletín del Centro Panamericano de Fiebre Aftosa, 61 (1995), pp. 51–56.

[17] ANONYMOUS, *Statistics on Foot and Mouth Disease*, UK Department for Environment, Food and Rural Affairs, London; available online from http://www.defra.gov.uk/footandmouth/cases/statistics/generalstats.asp; accessed November 12, 2001.

[18] W. DAVIES, G. LEWIS, AND H. RANDALL, *Some distributional features of the foot and mouth epidemic*, Nature, 219 (1968), p. 1968.

[19] J. J. CALLIS, P. D. MCKERCHER, AND J. H. GRAVES, *Foot-and-mouth disease—a review*, J. Veterinary Med. Assoc., 153 (1968), pp. 1798–1802.

[20] R. P. KITCHING *The application of biotechnology to the control of foot-and-mouth disease virus*, British Veterinary J., 148 (1992), pp. 375–388.

[21] A. RAI AND A. C. GOEL, *Emergence of an antigenic variation in FMD virus type O strains in India*, Rev. Sci. Tech. Off. Internat. Epizooties, 2 (1983), pp. 161–170.

[22] A. C. GOEL AND A. RAI, *Antigenic variation in FMD virus type O strains in India during* 1967–1982, Indian J. Comparative Microbio!. Immunology and Infectious Diseases, 5 (1984), pp. 108–115.

[23] S. K. Sharma, P. P. Singh, and D. K. Murty, *Foot and mouth disease in sheep and goats: An iceberg infection*, Indian Veterinary J., 58 (1981), pp. 925–928.

[24] J. Blackwell, *Internationalism and survival of foot-and-mouth disease virus in cattle and food products*, J. Dairy Sci., 63 (1980), pp. 1019–1030.

[25] K. Bauer, *Foot-and-mouth disease as zoonosis*, Arch. Virology Suppl., 13 (1997), pp. 95–97.

[26] N. S. Hyslop, *Transmission of the virus of foot and mouth disease between animals and man*, Bull. World Health Org., 49 (1973), pp. 577–585.

[27] R. F. Sellers, A. I. Donaldson, and K. A. J. Herniman, *Inhalation, persistence and dispersal of foot-and-mouth disease virus by man*, J. Hygiene, 68 (1970), pp. 565–573.

[28] Anonymous, *Report on the Animal Health Services in Great Britain*, Ministry of Agriculture, Fisheries and Food Department of Agriculture and Fisheries for Scotland, London, 1968.

[29] R. F. Sellers, K. A. J. Herniman, J. A. Mann, *Transfer of foot-and-mouth disease virus in the nose of man from infected to non-infected animals*, Veterinary Record, 89 (1971), pp. 447–449.

[30] A. Andersen, E. Kurstak, and C. Kurstak, *Picornaviruses of Animals: Clinical Observations and Diagnosis*, Academic Press, New York, 1981.

[31] H. Smith, T. C. Jones, and R. D. Hunt, *Veterinary Pathology*, 4th ed., Lea and Febiger, Philadelphia, 1972.

[32] J. Glosser, *Regulation and application of biotechnology products for use in veterinary medicine*, Rev. Sci. Tech. Off. Internat. Epizooties, 7 (1988), pp. 223–237.

[33] P. Sutmoller and D. J. Vose, *Contamination of animal products: The minimum pathogen dose required to initiate infection*, Rev. Sci. Tech. Off. Internat. Epizooties, 16 (1997), pp. 30–32.

[34] R. F. Sellers, K. A. J. Herniman, and I. D. Gumm, *The airborne dispersal of foot-and-mouth disease virus from vaccinated and recovered pigs, cattle and sheep after exposure to infection*, Res. Veterinary Sci., 23 (1977), pp. 70–75.

[35] R. Burrows, *Excretion of foot-and-mouth disease virus prior to the development of lesions*, Veterinary Record, 82 (1968), pp. 387–388.

[36] R. L. Sanson, *The epidemiology of foot-and-mouth disease: Implications for New Zealand*, New Zealand Veterinary J., 42 (1994), pp. 41–53.

[37] M. Garner, *Modeling foot-and-mouth disease in Australia*, in Proceedings of the National Symposium on FMD, Canberra, 1992, pp. 157–175.

[38] R. Hedger, *The isolation and characterization of foot-and-mouth disease virus from clinically normal herds of cattle in Botswana*, J. Hygiene, 66 (1968), pp. 27–36.

[39] R. F. Kahrs, E. P. J. Gibbs, and R. E. Larsen, *The search for viruses in bovine semen, a review*, Theriogenology, 14 (1980), pp. 151–165.

[40] J. H. BLACKWELL, *Survival of foot-and-mouth disease virus in cheese*, J. Dairy Sci., 59 (1976), pp. 1574–1579.

[41] J. H. BLACKWELL, *Persistence of foot-and-mouth disease virus in butter and butter oil*, J. Dairy Res., 45 (1978), pp. 283–285.

[42] W. B. FORD, *Disinfection procedures for personnel and vehicles entering and leaving contaminated premises*, Rev. Sci. Tech. Off. Internat. Epizooties, 14 (1995), pp. 393–401.

[43] ANONYMOUS, *Foot-and-Mouth Disease Emergency Disease Guidelines*, Animal and Plant Health Inspection Service, US Department of Agriculture, Hyattsville, MD, 1991.

[44] R. SELLERS, *Quantitative aspects of the spread of foot-and-mouth disease*, Veterinary Bull., 41 (1971), pp. 431–439.

[45] N. HYSLOP, *The epizootiology and epidemiology of foot and mouth disease*, Adv. Veterinary Sci. Comparative Med., 14 (1970), pp. 261–307.

[46] G. E. COTTRAL, *Persistence of foot-and-mouth disease virus in animals, their products and the environment*, Rev. Sci. Tech. Off. Internat. Epizooties, 71 (1969), pp. 549–568.

[47] R. F. SELLERS, K. A. J. HERNIMAN, AND A. I. DONALDSON, *The effects of killing or removal of animals affected with foot-and-mouth disease on the amounts of airborne virus present in looseboxes*, British Veterinary J., 127 (1971), pp. 358–365.

[48] J. BLACKWELL, *Foreign animal disease agent survival in animal products: Recent developments*, J. Amer. Veterinary Med. Assoc., 184 (1984), pp. 675–679.

[49] J. H. BLACKWELL, *Potential transmission of foot-and-mouth disease in whey constituents*, J. Food Protection, 41 (1978), pp. 631–633.

[50] A. I. DONALDSON, *Foot-and-mouth disease: The principal features*, Irish Veterinary J., 41 (1987), pp. 325–327.

[51] US CODE OF FEDERAL REGULATIONS, *Swine Health Protection*, Title 9, Subchapter L, Section 166.7, 1982.

[52] CALIFORNIA CODE OF REGULATIONS, *Food and Agricultural Code*, Title III, Section 10952; available online from http://www.leginfo.ca.gov/calaw.html; accessed August 14, 2001.

[53] ANONYMOUS, *Source of the Outbreak*, UK Department for Environment, Food, and Rural Affairs, London, 2001; available online from http://www.defra.gov.uk/footandmouth/about/current/source.asp; accessed January 4, 2002.

[54] D. F. BARLOW, *The aerosol stability of a strain of foot-and-mouth disease virus and the effects on stability of precipitation with ammonium sulphate, methanol or polyethylene glycol*, J. Gen. Virology, 15 (1972), pp. 17–24.

[55] A. I. DONALDSON, *The influence of relative humidity on the aerosol stability of different strains of foot-and-mouth disease virus suspended in saliva*, J. Gen. Virology, 15 (1972), pp. 25–33.

[56] A. I. Donaldson and N. P. Ferris, *The survival of foot-and-mouth disease virus in open air conditions*, J. Hygiene (London), 74 (1975), pp. 409–416.

[57] R. F. Sellers and J. Parker, *Airborne excretion of foot-and-mouth disease virus*, J. Hygiene (Cambridge), 67 (1969), pp. 671–677.

[58] A. I. Donaldson, *Quantitative data on airborne foot-and-mouth disease virus: Its production, carriage and deposition*, Philos. Trans. Roy. Soc. London Ser. B, 302 (1983), pp. 529–534.

[59] A. I. Donaldson, N. P. Ferris, and J. Gloster, *Air sampling of pigs infected with foot-and-mouth disease virus: Comparison of Litton and cyclone samplers*, Res. Veterinary Sci., 33 (1982), pp. 384–385.

[60] R. F. Sellers, *The nature and control of foot-and-mouth disease*, Soc. Dairy Tech. J., 22 (1969), pp. 90–93.

[61] A. M. Hutber and R. P. Kitching, *The role of management segregations in controlling intra-herd foot-and-mouth disease*, Tropical Animal Health and Production, 32 (2000), pp. 285–294.

[62] United States Code, *Tariff Act of* 1930, Title 19, Chapter 4, Section 1306(a); available online from http://uscode.house.gov/download/19C4.doc; accessed January 4, 2001.

[63] Q. P. Bowman and J. M. Arnoldi, *Management of animal health emergencies in North America: Prevention, preparedness, response and recovery*, Rev. Sci. Tech. Off. Internat. Epizooties, 18 (1999), pp. 76–103.

[64] J. C. Gordon and S. Bech-Nielsen, *Biological terrorism: A direct threat to our livestock industry*, Military Med., 151 (1986), pp. 357–363.

[65] Anonymous, *National Emergency Response to a Highly Contagious Animal Disease*, Executive Summary, Animal and Plant Health Inspection Service, United States Department of Agriculture, Hyattsville, MD, 2001.

[66] Anonymous, *Results of* 1998 *NIMBY Exercise*, Veterinary Services, Animal and Plant Health Inspection Service, United States Department of Agriculture, Hyattsville, MD, 1998.

[67] Anonymous, *Tripartite Exercise* 2000, Animal and Plant Health Inspection Service, United States Department of Agriculture, Hyattsville, MD; available online from http://www.aphis.usda.gov/oa/tripart/index.html; accessed October 1, 2001.

[68] R. L. Sanson, H. Liberona, and R. S. Morris, *The use of a geographical information system in the management of a foot-and-mouth disease epidemic*, Preventive Veterinary Med., 11 (1991), pp. 309–313.

[69] R. L. Sanson, G. Struthers, P. King, J. F. Weston, and R. S. Morris, *The potential extent of transmission of foot-and-mouth disease: A study of the movement of animals and materials in Southland, New Zealand*, New Zealand Veterinary J., 41 (1993), pp. 21–28.

[70] R. L. Sanson and A. Pearson, *Agribase: A national spatial farm database*, Epidemiologie et Sante Animale, 31–32 (1997), pp. 12.16.11–12.16.13.

[71] A. W. Jalvingh, M. Nielen, A. A. Dijkhuizen, and R. S. Morris, *A computerized decision support system for contagious animal disease control*, Pig News and Inform., 16 (1995), pp. 9N–12N.

[72] A. W. Jalvingh, M. Nielen, M. P. M. Meuwissen, A. A. Dijkhuizen, and R. S. Morris, *Economic evaluation of foot-and-mouth disease control strategies using spatial and stochastic simulation*, Epidemiologie et Sante Animale, 31–32 (1997), pp. 10.22.11–10.22.13.

[73] M. Nielen, A. W. Jalvingh, H. S. Horst, A. A. Dijkhuizen, H. Maurice, B. H. Schut, L. A. Van Wuijckhuise, and M. F. De Jong, *Quantification of contacts between Dutch farms to assess the potential risk of foot-and-mouth disease spread*, Preventive Veterinary Med., 28 (1996), pp. 143–158.

[74] F. Sobrino, M. Saiz, M. A. Jimenez-Clavero, J. I. Nunez, and M. F. Rosas, *Foot-and-mouth disease virus: A long known virus, but a current threat*, Veterinary Res., 32 (2001), pp. 1–30.

[75] C. C. Brown, M. E. Piccone, P. W. Mason, T. S. C. McKenna, and M. J. Grubman, *Pathogenesis of wild-type and leaderless foot-and-mouth disease virus in cattle*, J. Virology, 70 (1996), pp. 5638–5641.

[76] P. W. Mason, M. E. Piccone, S. C. McKenna, J. Chinsangaram, and M. J. Grubman, *Evaluation of a live-attenuated foot-and-mouth disease virus as a vaccine candidate*, Virology, 227 (1997), pp. 96–102.

[77] A. I. Donaldson and U. Kihm, *Research and technological developments required for more rapid control and eradication of foot and mouth disease*, Rev. Sci. Tech. Off. Internat. Epizooties, 15 (1996), pp. 863–873.

[78] S. J. Cox, P. V. Barnett, P. Dani, and J. S. Salt, *Emergency vaccination of sheep against foot-and-mouth disease: Protection against disease and reduction in contact transmission*, Vaccine, 17 (1999), pp. 1858–1868.

[79] T. R. Doel, L. Williams, and P. V. Barnett, *Emergency vaccination against foot-and-mouth disease—rate of development of immunity and its implications for the carrier state*, Vaccine, 12 (1994), pp. 592–600.

[80] J. S. Salt, P. V. Barnett, P. Dani, and L. Williams, *Emergency vaccination of pigs against foot-and-mouth disease: Protection against disease and reduction in contact transmission*, Vaccine, 16 (1998), pp. 746–754.

[81] I. E. Bergmann, V. Malirat, E. Neitzert, E. Beck, N. Panizzutti, C. Sanchez, and A. Falczuk, *Improvement of a serodiagnostic strategy for foot-and-mouth disease virus surveillance in cattle under systematic vaccination: A combined system of an indirect ELISA-3ABC with an enzyme-linked immunoelectrotransfer blot assay*, Arch. Virology, 145 (2000), pp. 473–489.

[82] C. F. Gibson, A. I. Donaldson, and N. P. Ferris, *Response of sheep vaccinated with large doses of vaccine to challenge by airborne foot and mouth disease virus*, Vaccine, 2 (1984), pp. 157–161.

[83] R. P. KITCHING, *A recent history of foot-and-mouth disease*, J. Comparative Pathology, 118 (1998), pp. 89–108.

[84] A. J. M. GARLAND, *Vital elements for the successful control of foot-and-mouth disease by vaccination*, Vaccine, 17 (1999), pp. 1760–1766.

[85] I. YERUHAM, H. YADIN, M. HAYMOVICH, AND S. PERL, *Adverse reactions to FMD vaccine*, Veterinary Dermatology, 12 (2001), pp. 197–201.

[86] E. H. MCCAULEY, N. A. AULAQI, J. C. NEW, W. B. SUNDQUIST, AND W. M. MILLER, EDS., *A Study of the Potential Economic Impact of Foot-and-Mouth Disease in the United States*, University of Minnesota, St. Paul, MN, 1979.

[87] E. BECK AND K. STROHMAIER, *Subtyping of European foot-and-mouth disease virus strains by nucleotide sequence determination*, J. Virology, 61 (1987), pp. 1621–1629.

[88] A. M. HUTBER, R. P. KITCHING, AND D. A. CONWAY, *Control of foot-and-mouth disease through vaccination and the isolation of infected animals*, Tropical Animal Health and Production, 30 (1998), pp. 217–227.

[89] M. E. J. WOOLHOUSE, D. T. HAYDON, A. PEARSON, AND R. P. KITCHING, *Failure of vaccination to prevent outbreaks of foot-and-mouth disease*, Epidemiology and Infection, 116 (1996), pp. 363–371.

[90] ANONYMOUS, *Coordenacao do Combate a Febre Aftosa*, Ministerio Agricultura Avaliacao do Plano Nacional de Combate a Febre Aftosa-PNCFA, Embaixador, SCS, Brasil, 1971–1976; cited in [109].

[91] J. J. CALLIS, *National and international foot-and-mouth disease control programmes in Panama, Central and North America*, British Veterinary J., 134 (1978), pp. 10–15.

[92] ANONYMOUS, *Disease Information Bulletin*, Office International des Epizooties, Paris; available online from http://www.oie.int/eng/info/hebdo/AIS_01.htm#Sec1; accessed November 29, 2001.

[93] C. MEBUS, *Vaccine banks and emergency stocks of vaccines against exotic diseases: The Plum Island (USA) model*, in Quality Control of Veterinary Vaccines in Developing Countries: Proceedings of the Expert Consultation Held at FAO, Rome, 2–6 December 1991, N. Mowat, ed., Animal Production and Health Paper 116, Food and Agriculture Organization of the United Nations, Rome, 1993, pp. 301–304.

[94] P. THORNBER, *Zoning for FMD Infrastructure and Surveillance—Canada and USA Attitudes to Zoning and Regionalisation*, unpublished report from the 13th Conference of the OIE Regional Commission for the Americas, Havana, 1996, Office International des Epizooties, Paris, 1996; cited in [95].

[95] M. G. GARNER, R. T. ALLEN, AND S. SHORT, *Foot-and-Mouth Disease Vaccination: A discussion paper on its use to control outbreaks in Australia*, 1996, Bureau of Resource Sciences, Kingston, ACT, Australia, 1997.

[96] T. DOEL, *International vaccine banks*, in Quality Control of Veterinary Vaccines in Developing Countries: Proceedings of the Expert Consultation Held at FAO, Rome, 2–6 December 1991, N. Mowat, ed., Animal Production and Health Paper 116, Food and Agriculture Organization of the United Nations, Rome, 1993, pp. 305–309.

[97] H. S. Horst, C. J. de Vos, F. H. M. Tomassen, and J. Stelwagen, *The economic evaluation of control and eradication of epidemic livestock diseases*, Rev. Sci. Tech. Off. Internat. Epizooties, 18 (1999), pp. 367–379.

[98] S. C. Howard and C. A. Donnelly *The importance of immediate destruction in epidemics of foot and mouth disease*, Res. Veterinary Sci, 69 (2000), pp. 189–196.

[99] A. I. Donaldson, J. Gloster, L. D. Harvey, and D. H. Deans, *Use of prediction models to forecast and analyse airborne spread during the foot-and-mouth disease outbreaks in Brittany, Jersey and the Isle of Wight in* 1981, Veterinary Record, 110 (1982), pp. 53–57.

[100] J. Casal, J. M. Moreso, E. Planas Cuchi, and J. Casal, *Simulated airborne spread of Aujesky's disease and foot-and-mouth disease*, Veterinary Record, 140 (1997), pp. 672–676.

[101] F. Moutou and B. Durand, *Modeling the spread of foot-and-mouth disease virus*, Veterinary Research, 25 (1994), pp. 279–285.

[102] R. P. Pech and J. C. McIlroy, *A model of the velocity of advance of foot and mouth disease in feral pigs*, J. Appl. Ecol., 27 (1990), pp. 635–650.

[103] J. H. Sorensen, D. K. J. Mackay, C. O. Jensen, and A. I. Donaldson, *An integrated model to predict the atmospheric spread of foot-and-mouth disease virus*, Epidemiology and Infection, 124 (2000), pp. 577–590.

[104] J. Gloster, R. M. Blackall, R. F. Sellers, and A. I. Donaldson, *Forecasting the airborne spread of foot-and-mouth disease*, Veterinary Record, 108 (1981), pp. 370–374.

[105] J. Gloster, R. F. Sellers, and A. I. Donaldson, *Long distance transport of foot-and-mouth disease virus over the sea*, Veterinary Record, 110 (1982), pp. 47–52.

[106] A. L. Donaldson, M. Lee, and A. Shimshony, *A possible airborne transmission of foot and mouth disease virus from Jordan to Israel—a simulated computer analysis*, Israel J. Veterinary Med., 44 (1988), pp. 92–96.

[107] M. G. Garner and M. B. Lack, *An evaluation of alternate control strategies for foot-and-mouth disease in Australia—a regional approach*, Preventive Veterinary Med., 23 (1995), pp. 9–32.

[108] P. B. M. Berentsen, A. A. Dijkhuizen, and A. J. Oskam, *A dynamic model for cost-benefit analyses of foot-and-mouth disease control strategy*, Preventive Veterinary Med., 12 (1992), pp. 229–243.

[109] V. M. Astudillo and P. Auge de Mello, *Cost and effectiveness analysis of two foot-and-mouth disease vaccination procedures*, Bol. Centro Panamericano de Fiebre Aftosa, 37–38 (1980), pp. 49–63.

[110] G. Davies, *Risk assessment in practice: A foot and mouth disease control strategy for the European Community*, Rev. Sci. Tech. Off. Internat. Epizooties, 12 (1993), pp. 1109–1119.

[111] A. D. JAMES AND P. R. ELLIS, *Benefit-cost analysis in foot and mouth disease control programmes*, British Veterinary J., 134 (1978), pp. 47–52.

[112] M. VAN HAM AND Y. ZUR, *Estimated damage to the Israeli dairy herd caused by foot and mouth disease outbreaks and a cost/benefit analysis of the present vaccination policy*, Israel J. Veterinary Med., 49 (1994), pp. 13–16.

[113] R. J. LORENZ, *A cost-effectiveness study on the vaccination against foot-and-mouth disease (FMD) in the Federal Republic of Germany*, Acta Veterinary Scandinavian Suppl., 84 (1988), pp. 427–429.

[114] T. E. CARPENTER, *Microcomputer programs for Markov and modified Markov chain disease models*, Preventive Veterinary Med., 5 (1988), pp. 169–179.

[115] T. E. CARPENTER, *Stochastic epidemiologic modeling using a microcomputer spreadsheet package*, Preventive Veterinary Med., 5 (1988), pp. 159–168.

[116] W. MILLER, *A state-transition model of epidemic foot-and-mouth disease*, in A Study of the Potential Economic Impact of Foot-and-Mouth Disease in the United States, E. H. McCauley, N. A. Aulaqi, J. C. New, W. B. Sundquist, and W. M. Miller, eds., University of Minnesota, St. Paul, MN, 1979, pp. 113–131.

[117] M. G. GARNER AND M. B. LACK, *Modelling the potential impact of exotic diseases on regional Australia*, Australian Veterinary J., 72 (1995), pp. 81–87.

[118] B. DURAND AND O. MAHUL, *An extended state-transition model for foot-and-mouth disease epidemics in France*, Preventive Veterinary Med., 47 (2000), pp. 121–139.

[119] D. T. HAYDON, M. E. J. WOOLHOUSE, AND R. P. KITCHING, *An analysis of foot-and-mouth-disease epidemics in the UK*, IMA J. Math. Appl. Med. Biol., 14 (1997), pp. 1–9.

[120] K. DIETZ, *The estimation of the basic reproduction number for infectious diseases*, Statist. Methods Med. Res., 2 (1993), pp. 23–41.

[121] M. N. FERGUSON, C. A. DONNELLY, AND R. M. ANDERSON, *The foot-and-mouth epidemic in Great Britain: Pattern of spread and impact of interventions*, Science, 292 (2001), pp. 1155–1160.

[122] M. N. FERGUSON, C. A. DONNELLY, AND R. M. ANDERSON, *Transmission intensity and impact of control policies on the foot and mouth epidemic in Great Britain*, Nature, 413 (2001), pp.542–548.

[123] P. S. DAWSON, *The involvement of milk in the spread of foot-and-mouth disease: An epidemiological study*, Veterinary Record, 87 (1970), pp. 543–548.

[124] M. E. HUGH JONES, *A simulation spatial model of the spread of foot-and-mouth disease through the primary movement of milk*, J. Hygiene (London), 77 (1976), pp. 1–9.

[125] R. P. PECH AND J. HONE, *A model of the dynamics and control of an outbreak of foot and mouth disease in feral pigs in Australia*, J. Appl. Ecol., 25 (1988), pp. 63–77.

[126] R. L. SANSON AND R. S. MORRIS, *The use of survival analysis to investigate the probability of local spread of foot-and-mouth disease: An example study based on the United Kingdom epidemic of* 1967–1968, Kenya Veterinarian, 18 (1994), pp. 186–188.

[127] T. Bates, M. Thurmond, and T. Carpenter, *Evaluation of strategies to control foot and mouth disease virus by use of an epidemic simulation model, Part* II: *Simulation modeling results*, Amer. J. Veterinary Res., in press, 2002.

[128] T. Bates, M. Thurmond, and T. Carpenter, *Evaluation of strategies to control foot and mouth disease virus by use of an epidemic simulation model, Part* I: *Model description and parameters*, Amer. J. Veterinary Res., in press, 2002.

[129] Anonymous, *Foot-and-mouth disease confirmed in the UK*, Veterinary Record, 148 (2001), p. 222.

[130] R. S. Morris, J. W. Wilesmith, M. W. Stern, R. L. Sanson, and M. A. Stevenson, *Predictive spatial modelling of alternative control strategies for the foot-and-mouth disease epidemic in Great Britain*, 2001, Veterinary Record, 149 (2001), pp. 137–144.

[131] M. J. Keeling, M. E. J. Woolhouse, D. J. Shaw, L. Matthews, M. Chase-Topping, D. T. Haydon, S. J. Cornell, J. Kappey, J. Wilesmith, and B. T. Grenfell, *Dynamics of the* 2001 *UK foot and mouth epidemic: Stochastic dispersal in a heterogeneous landscape*, Science, 294 (2001), pp. 813–817.

[132] Anonymous, *Statistics on Foot and Mouth Disease*, UK Department for Environment, Food, and Rural Affairs, London; available online from http://www.defra.gov.uk/footandmouth/cases/statistics/generalstats.asp; accessed December 30, 2001.

[133] M. J. Keeling, M. E. J. Woolhouse, D. J. Shaw, L. Matthews, M. Chase-Topping, D. T. Haydon, S. J. Cornell, J. Kappey, J. Wilesmith, and B. T. Grenfell, *Supplementary material for "Dynamics of the* 2001 *UK foot and mouth epidemic: Stochastic dispersal in a heterogeneous landscape*," Science Online; available online from http://www.sciencemag.org/cgi/content/full/1065973/DC1/1; accessed November 1, 2001.

[134] M. E. Hugh Jones, *Epidemiological studies on the* 1967–68 *foot and mouth epidemic: Attack rates and cattle density*, Res. Veterinary Sci., 13 (1972), pp. 411–417.

[135] M. E. Hugh Jones and P. B. Wright, *Studies on the* 1967–8 *foot-and-mouth disease epidemic: The relation of weather to the spread of disease*, J. Hygiene (London), 68 (1970), pp. 253–271.

[136] R. Tinline, *Lee wave hypothesis for the initial pattern of spread during the* 1967–68 *foot and mouth epizootic*, Nature, 227 (1970), pp. 860–862.

[137] S. M. Daggupaty and R. F. Sellers, *Airborne spread of foot-and-mouth disease in Saskatchewan, Canada*, 1951–1952, Canadian J. Veterinary Res., 54 (1990), pp. 465–468.

[138] J. Casal, E. Planascuchi, J. M. Moreso, and J. Casal, *Forecasting virus atmospherical dispersion—studies with foot-and-mouth disease*, J. Hazardous Materials, 43 (1995), pp. 229–244.

[139] Anonymous, *Risø Mesoscale PUFF Model*, Risø National Laboratory, Roskilde, Denmark, 1999.

[140] S. French and J. Smith, *Using monitoring data to update atmospheric dispersion models with an application to the Rimpuff model*, Radiation Protection Dosimetry, 50 (1993), pp. 317–320.

Chapter 6

Modeling and Imaging Techniques with Potential for Application in Bioterrorism

H. T. Banks, David Bortz,† Gabriella Pinter,‡ and Laura Potter§*

6.1 Introduction

In this paper we present a survey of several recent and emerging ideas and efforts on modeling and system interrogation in the presence of uncertainty that we feel have significant potential for applications related to bioterrorism. The first focuses on physiologically based pharmacokinetic (PBPK)–type models and the effects of drugs, toxins, and viruses on tissue, organs, individuals and populations wherein both intra- and interindividual variability are present when one attempts to determine kinetic rates, susceptibility, efficacy of toxins, antitoxins, etc., in aggregate populations. Methods combining deterministic and stochastic concepts are necessary to formulate and computationally solve the associated estimation problems. Similar issues arise in the human immunodeficiency virus (HIV) infectious models we also present below.

A second effort concerns the use of remote electromagnetic interrogation pulses linked to dielectric properties of materials to carry out macroscopic structural imaging of bulk packages (drugs, explosives, etc.) as well as test for presence and levels of toxic chemical compounds in tissue. These techniques also may be useful in functional imaging (e.g., of brain and central nervous system activity levels) to determine levels of threat in potential adversaries via changes in dielectric properties and conductivity.

The PBPK and cellular level virus infectious models we discuss are special examples of a much wider class of population models that one might utilize to investigate potential agents for use in attacks, such as viruses, bacteria, fungi, and other chemical, biochemical or radiological agents. These include general epidemiological models such as susceptible-

*Center for Research in Scientific Computation, North Carolina State University, Raleigh, NC 27695-8205.

†Department of Mathematics, University of Michigan, Ann Arbor, MI 48109.

‡Department of Mathematical Sciences, University of Wisconsin at Milwaukee, Milwaukee, WI 53201.

§Center for Research in Scientific Computation, North Carolina State University, Raleigh, NC 27695-8205. Current address: Scientific Computing and Mathematical Modeling, GlaxoSmithKline, Research Triangle Park, NC 27709.

infectious-recovered (SIR) infectious spread models containing contact and susceptibility rates with structures (e.g., public vs. private transport; residence times in exposure; subnetworks of populations) as well as more general population models with heterogeneities and/or behavioral structures (e.g., social interaction, age/size dependency, spatial-temporal dependency, adaptive transient behavior, etc.). These may involve general dynamical systems, both discrete and continuous, including ordinary and/or partial differential equations and delay differential equations. Included are well known structured population models, such as those of Sinko–Streifer and McKendrick–Von Foerster. These deterministic models often must be augmented with probabilistic and/or statistical structures such as mixing distributions, random effects, etc. (see [20, 22] for discussions and references). Such models combine ideas from continuum population models with aspects of agent based models incorporating individual level effects. The results are population models encompassing intraindividual and/or interindividual variability that in some cases describe (predict) continuous population evolution that is driven by distributions of individual level mechanisms and behaviors. The models described in section 6.2 below, where the parameters are viewed as random variables, or realizations thereof, are examples of these.

The use of models such as those outlined above ultimately lead to estimation or inverse problems containing both a mathematical model and a statistical model. These are treated in a fit-to-data formulation using either experimental data or synthetic "data" simulating a desired response. The latter arises, for example, in design of a drug or therapy that will result in a sought-after response of an individual or a population to a threat. However, the rationale to support elaborate models with structures does not lie simply in the desire to better fit a data set but rather to aid in understanding basic mechanisms, pathways, behavior, etc. and to better frame population as well as individual responses to a challenge or to a prophylactic. But it is not just inverse problems that arise in the context of these models (although that is the focus in this chapter); indeed, ideas from control theory and system optimization are also important. In almost every instance, including those discussed in the examples below, fundamental mathematics, especially modeling, theoretical and computational analysis, and probability and statistics, plays a significant role.

The electromagnetic interrogation and imaging ideas discussed in section 6.3 could conceivably be a part of a surveillance technology in a first line of defense against bioterrorism. More precisely, physical detection and identification of hidden substances and agents (whether in food and water supplies, luggage, mail and packages, etc.) as a part of biodefense depends not only on the electromagnetic techniques discussed below but also on characterization of dielectric properties of specific molecules and compounds. Although we present only deterministic aspects of the problems here, it can be expected that a successful methodology will also involve probabilistic and statistical formulations as well as tools from computational molecular biology.

6.2 Inverse Problems

In these discussions we shall consider inverse or estimation problems involving aggregate data for populations which may be described by two different types of "parameter-dependent" dynamics; for the lack of better terms we shall refer to these as "individual dynamics" and "aggregate dynamics." In both cases the data and populations inherently contain variability of parameters; this variability may be intraindividual, interindividual, or both.

The problems for individual dynamics can be used to treat many examples of practical interest, including physiologically based pharmacokinetic (PBPK) models, biologically based dose response (BBDR) models, and susceptible-infectious-recovered (SIR) models of disease spread. The aggregate dynamics problems include cellular level virus models such as those for HIV growth.

In the first type of problem we consider below, one has a *mathematical model* at what we shall term (in perhaps something of a misnomer) the "individual" level. That is, the population count or density is described by a parameter dependent system. To facilitate our discussions here we use, without loss of generality, ordinary differential equation (ODE) models of the form

$$\dot{x}(t) = f(t, x(t), q), \qquad q \in Q, \tag{6.1}$$

where the parameters q (e.g., growth, mortality, fecundity, etc.) in the model vary from individual to individual across the population according to some probability distribution P on a set of admissible parameters Q. More precisely, we suppose that the population is made up of subpopulations distinguished by common values of the parameters q and whose time course is described by the solution $x(t; q)$ of (6.1) for the shared value of q. The total population count or density is then given by a weighted sum of these solutions over all possible $q \in Q$ so that the counts or densities one expects to observe at any time t are given by

$$\begin{aligned} \bar{x}(t; P) &= \mathcal{E}[x(t; q)|P] \\ &\equiv \int_Q x(t; q) dP(q). \end{aligned} \tag{6.2}$$

Experimental observations or data $\{\hat{d}_i\}$ corresponding to times $\{t_i\}$ are then given by the expected values $\bar{x}(t_i; P)$ of (6.2) plus some error ε_i so that

$$\hat{d}_i = \bar{x}(t_i; P) + \varepsilon_i.$$

Assumptions about the error $\{\varepsilon_i\}$ in the observation process constitute the basis of an associated *statistical model* for the inverse problems. For discussions in this chapter, we will simply (and perhaps naively) assume that the errors are independent identically distributed (i.i.d.) Gaussian and will use an ordinary least squares (OLS) formulation for our inverse problems. This will then be completely equivalent to the traditional maximum likelihood estimator (MLE) problems. Thus, we formulate our inverse problem in terms of seeking to minimize

$$J(P) = \sum_{i=1}^{n} |\mathcal{E}[x(t_i; q)|P] - \hat{d}_i|^2 \tag{6.3}$$

over P in the set $\mathcal{P}(Q)$ of probability measures on Q subject to $t \to x(t; q)$ satisfying (6.1) for a given $q \in Q$.

The second type of problem involves aggregate dynamics wherein one has ODEs that describe the *expected values* of the population counts or densities. Essentially one has dynamics which already have been summed over the variability in parameters resulting in *measure-dependent dynamics* (as opposed to *parameter-dependent dynamics*) given by

$$\dot{\bar{x}}(t) = g(t, \bar{x}(t), P), \qquad P \in \mathcal{P}(Q), \tag{6.4}$$

where now $\bar{x}(t; P)$ is the average or expected value of the population count or density at time t. In this case the OLS formulation takes the form of minimizing

$$J(P) = \sum_{i=1}^{n} |\bar{x}(t_i; P) - \hat{d}_i|^2 \tag{6.5}$$

over $P \in \mathcal{P}(Q)$ subject to the aggregate dynamics (6.4). As we shall note in the examples below, models such as (6.4) occur naturally and may not be readily formulated in terms of dynamics of the form (6.1) and vice versa.

In section 6.2.1 we outline a theoretical and computational framework for problems involving (6.1), (6.3) and illustrate the approach with a PBPK model for trichloroethylene (TCE). We follow this by discussing a framework for problems based on (6.4), (6.5) in the context of an inverse problem for virus dynamics (HIV in this case).

6.2.1 Inverse Problems for Individual Dynamics

Our goal is to estimate $q \in Q \subset R^m$ from solutions of $\dot{x}(t) = f(t, x(t), q)$. To do this we visualize parameters as realizations of a random variable and attempt to estimate the probability distribution function (PDF) $P \in \mathcal{P}(Q)$ where $\mathcal{P}(Q)$ is the set of all PDFs on the Borel subsets of Q. We then attempt to estimate P from given data $\hat{d}_i$, $i = 1, \ldots, n$, where

$$\begin{aligned}\hat{d}_i &\approx \mathcal{E}[x(t_i; q)|P] \\ &= \int_Q x(t_i; q) dP(q),\end{aligned}$$

which in the case of a discrete probability measure can be written as

$$\hat{d}_i \approx \sum_{j=1}^{M} x(t_i, q_j) p_j$$

for P a discrete PDF with atoms at $\{q_j\}_{j=1}^M \subset Q$ and associated probabilities $\{p_j\}_{j=1}^M$.

We can then, as noted above, define the OLS estimation problem of minimizing

$$J(P) = \sum_{i=1}^{n} |\mathcal{E}[x(t_i; q)|P] - \hat{d}_i|^2 \tag{6.6}$$

over $P \in \mathcal{P}(Q)$. To consider a theoretical and computational foundation for such problems, one needs the following items:

(i) a topology on $\mathcal{P}(Q)$;

(ii) continuity of $P \to J(P)$;

(iii) compatible compactness results (well-posedness);

(iv) computational tools (approximations, etc.).

Fortunately, probability theory offers a great start toward a possible complete, tractable computational methodology [16]. The most important tool is the Prohorov metric, which we proceed to define. Suppose (Q, d) is a complete metric space. For any closed subset $F \subset Q$ and $\varepsilon > 0$, define

$$F^\varepsilon = \{q \in Q : d(\tilde{q}, q) < \varepsilon, \tilde{q} \in F\}.$$

We then define the Prohorov metric $\rho : \mathcal{P}(Q) \times \mathcal{P}(Q) \to R^+$ by

$$\rho(P_1, P_2) \equiv \inf\{\varepsilon > 0 : P_1[F] \leq P_2[F^\varepsilon] + \varepsilon, \ F \text{ closed}, \ F \subset Q\}.$$

This can be shown to be a metric on $\mathcal{P}(Q)$ and has a number of well-known properties including the following:

(a) $(\mathcal{P}(Q), \rho)$ is a complete metric space;

(b) If Q is compact, then $(\mathcal{P}(Q), \rho)$ is a compact metric space.

We note that the definition of ρ is not intuitive. For example, what does $P_k \to P$ in ρ mean? We have the following important characterizations [16].

Theorem 1. *Given $P_k, P \in \mathcal{P}(Q)$, the following convergence statements are equivalent:*

(i) $\rho(P_k, P) \to 0$;

(ii) $\int_Q f dP_k(q) \to \int_Q f dP(q)$ *for all bounded, uniformly continuous* $f : Q \to R^1$;

(iii) $P_k[A] \to P[A]$ *for all Borel sets* $A \subset Q$ *with* $P[\partial A] = 0$.

Thus, one obtains immediately the following useful results:

- Convergence in the ρ metric is equivalent to convergence in distribution;
- Let $C_B^*(Q)$ denote the topological dual of $C_B(Q)$, where $C_B(Q)$ is the usual space of bounded continuous functions on Q with the supremum norm. If we view $\mathcal{P}(Q) \subset C_B^*(Q)$, convergence in the ρ topology is equivalent to weak* convergence in $\mathcal{P}(Q)$.

More importantly,

$$\rho(P_k, P) \to 0 \quad \text{is equivalent to} \int_Q x(t_i; q) dP_k(q) \to \int_Q x(t_i; q) dP(q),$$

and $P_k \to P$ in ρ metric is hence equivalent to

$$\mathcal{E}[x(t_i; q)|P_k] \to \mathcal{E}[x(t_i; q)|P]$$

or "convergence in expectation." This yields that

$$P \to J(P) = \sum_{i=1}^{n} |\mathcal{E}[x(t_i; q)|P] - \hat{d}_i|^2$$

is continuous in the ρ topology. Continuity of $P \to J(P)$ and compactness of $\mathcal{P}(Q)$ (each with respect to the ρ metric) allows one to assert the existence of a solution to min $J(P)$ over $P \in \mathcal{P}(Q)$.

Computational Issues and Approximation Ideas

We first note that $(\mathcal{P}(Q), \rho)$ is infinite-dimensional and hence one must use finite-dimensional approximations to obtain tractable computational algorithms. To this end, one may prove the following theorem (see [4]).

Theorem 2. *Let Q be a complete, separable metric space with metric d, $\mathcal{S}$ the class of all Borel subsets of Q and $\mathcal{P}(Q)$ the space of probability measures on $(Q, \mathcal{S})$. Let $Q_0 = \{q_j\}_{j=1}^{\infty}$ be a countable, dense subset of Q. Then the set of $P \in \mathcal{P}(Q)$ such that P has finite support in Q_0 and rational masses is dense in $\mathcal{P}(Q)$ in the ρ metric. That is,*

$$\mathcal{P}_0(Q) \equiv \left\{ P \in \mathcal{P}(Q) : P = \sum_{j=1}^{k} p_j \delta_{q_j}, k \in \mathcal{N}^+, q_j \in Q_0, p_j \text{ rational}, \sum_{j=1}^{k} p_j = 1 \right\}$$

is dense in $\mathcal{P}(Q)$ relative to ρ, where δ_{q_j} is the Dirac measure with atom at q_j.

Given $Q_d = \bigcup_{M=1}^{\infty} Q_M$ with $Q_M = \{q_j^M\}_{j=1}^M$ chosen so that Q_d is dense in Q, define

$$\mathcal{P}^M(Q) = \left\{ P \in \mathcal{P}(Q) : P = \sum_{j=1}^{M} p_j \delta_{q_j^M}, q_j^M \in Q_M, p_j \text{ rational}, \sum_{j=1}^{k} p_j = 1 \right\}.$$

Then we find

- $\mathcal{P}^M(Q)$ is a compact subset of $(\mathcal{P}(Q), \rho)$;
- $\mathcal{P}^M(Q) \not\subset \mathcal{P}^{M+1}(Q)$;
- "$\mathcal{P}^M(Q) \to \mathcal{P}(Q)$" in the ρ topology—that is, elements in $\mathcal{P}(Q)$ may be approximated arbitrarily closely in the ρ metric by elements in $\mathcal{P}^M(Q)$ for M sufficiently large.

These ideas and results can then be used to establish a type of "stability" of the inverse problem (see [4, 12]). We first define a series of approximate problems consisting of minimizing

$$J(P_M) = \sum_{x=1}^{n} |\mathcal{E}[x(t_i; q)|P_M] - \hat{d}_i|^2$$

over $P_M \in \mathcal{P}^M(Q)$. Then we have the following theorem.

Theorem 3. *Let Q be a compact metric space and assume solutions $x(t; q)$ of $\dot{x}(t) = f(t, x(t), q)$ are continuous in q on Q. Let $\mathcal{P}(Q)$ be the set of all probability measures on Q and let Q_d be a countable dense subset of Q as defined previously with $Q_M = \{q_j^M\}_{j=1}^M$. Define $\mathcal{P}^M(Q)$ as above. Suppose $P_M^*(\hat{d}^k)$ is the set of minimizers for $J(P)$ over $P \in \mathcal{P}^M(Q)$ corresponding to the data $\{\hat{d}^k\}$ and $P^*(\hat{d})$ is the set of minimizers over $P \in \mathcal{P}(Q)$ corresponding to $\hat{d}$, where $\hat{d}^k, \hat{d} \in R^n$ are the observed data such that $\hat{d}^k \to \hat{d}$. Then $dist(P_M^*(\hat{d}^k), P^*(\hat{d})) \to 0$ as $M \to \infty$ and $\hat{d}^k \to \hat{d}$. Thus the solutions depend continuously on the data and the approximate problems are method stable.*

To illustrate the above methodology with a relevant example, we present here a brief description of a PBPK-hybrid model for trichloroethylene (TCE) and indicate how one formulates and implements the corresponding estimation problems. TCE is a metal-degreasing agent that is a widespread environmental contaminant and has been linked to several types of cancer in laboratory animals and humans. This compound is highly soluble in lipids and is known to accumulate within the fat tissue. Therefore, in order to accurately predict toxicity-related measures such as the net clearance rate of TCE and the effective dose of TCE delivered to target tissues, it is important to accurately capture the transport of TCE within the fat tissue.

Physiologically based pharmacokinetic (PBPK) models are used to describe the disposition of compounds such as TCE within the tissues and organs. These models include compartments for tissues that are involved in the uptake, metabolism, elimination, and/or transport of the compound, as well as compartments for tissues that are targets of the chemical's toxic effects. See [24] for a detailed description of standard PBPK modeling techniques.

As discussed in [1, 27], the standard perfusion-limited and diffusion-limited compartmental models used in PBPK modeling are not able to describe the dynamics of TCE in fat tissue as seen in experimental data, and the assumptions for these ODE-based models do not match well with the heterogeneous physiology of fat tissue. This motivated the development of a specialized compartmental model for the fat tissue, which is then coupled with standard compartments for the remaining nonfat compartments to produce a PBPK-hybrid model.

The resulting compartmental model for the fat tissue is based on an axial dispersion model originally developed by Roberts and Rowland [28] for the transport of solutes in the liver. The underlying assumptions for the dispersion model match well with the physiology of fat tissue (see [1, 27] for details), and the geometry for the PDE-based fat model is based specifically on the known geometry of fat cells and their accompanying capillaries. A key feature of the dispersion model is its aggregate nature, using a representative "cell" to capture the transport behavior of the compound in a collection of many similar "cells" that have varying properties.

In this particular case, the representative "cell" is a unit containing three subcompartments: a single adipocyte (fat cell) together with an adjoining capillary, and the surrounding interstitial fluid. In the model, the adipocyte is represented by a sphere and the capillary is a cylindrical tube with circular cross-section; the interstitial fluid fills in the space surrounding the other two regions. The model geometry and equations are given in spherical coordinates. See [1, 27] for a complete description of the model.

It is assumed that TCE enters the capillary region of the fat compartment along with the arterial blood. The capillary equation (6.7) includes a one-dimensional convection-dispersion term together with a term based on Fick's first law of diffusion for the exchange between the capillary and the other two subcompartments. The accompanying boundary conditions (6.8) and (6.9) connect the capillary with the arterial and venous blood systems, and are based on flux balance. The adipocyte and interstitial equations (6.10) and (6.15) each contain two-dimensional diffusion terms together with terms for the exchange of TCE between the subcompartments. The boundary conditions (6.11)–(6.14) and (6.16)–(6.19) are based on standard periodic and finiteness conditions that are appropriate for diffusion on a spherical domain.

In addition to the fat compartment, there are perfusion-limited tissue compartments used in the PBPK-hybrid model to represent the brain, kidney, liver, muscle and remaining tissues. Uptake of TCE is via inhalation in the lungs, which is modeled using a standard steady-state assumption. Metabolism of TCE is described with a Michaelis–Menten term in the liver with parameters $v_{\max}$ (mg/hour) and k_M (mg/liter). The resulting equations for the PBPK-hybrid model are given by

$$\begin{aligned} V_B \frac{\partial C_B}{\partial t} &= \frac{V_B}{r_2 \sin\phi}\frac{\partial}{\partial \phi}\left[\sin\phi\left(\frac{\mathcal{D}_B}{r_2}\frac{\partial C_B}{\partial \phi} - vC_B\right)\right] \\ &\quad + \lambda_I \mu_{BI}(f_I C_I(\theta_0) - f_B C_B) + \lambda_A \mu_{BA}(f_A C_A(\theta_0) - f_B C_B), \end{aligned} \tag{6.7}$$

$$-\frac{\mathcal{D}_B}{r_2}\frac{\partial C_B}{\partial \phi}(t,\phi) + vC_B(t,\phi)\bigg|_{\phi=\varepsilon_1} = \frac{Q_c}{1000A_B}C_a(t), \tag{6.8}$$

$$-\frac{\mathcal{D}_B}{r_2}\frac{\partial C_B}{\partial \phi}(t,\phi) + vC_B(t,\phi)\bigg|_{\phi=\pi-\varepsilon_2} = \frac{Q_c}{1000A_B}C_v(t), \tag{6.9}$$

$$V_I \frac{\partial C_I}{\partial t} = \frac{V_I D_I}{r_1^2}\left[\frac{1}{\sin^2\phi}\frac{\partial^2 C_I}{\partial \theta^2} + \frac{1}{\sin\phi}\frac{\partial}{\partial \phi}\left(\sin\phi\frac{\partial C_I}{\partial \phi}\right)\right]$$

$$+ \delta_{\theta_0}(\theta)\chi_B(\phi)\lambda_I\mu_{BI}(f_B C_B - f_I C_I) + \mu_{IA}(f_A C_A - f_I C_I), \tag{6.10}$$

$$C_I(t,\theta,\phi) = C_I(t,\theta+2\pi,\phi), \tag{6.11}$$

$$\frac{\partial C_I}{\partial \theta}(t,\theta,\phi) = \frac{\partial C_I}{\partial \theta}(t,\theta+2\pi,\phi), \tag{6.12}$$

$$C_I(t,\theta,0) < \infty, \tag{6.13}$$

$$C_I(t,\theta,\pi) < \infty, \tag{6.14}$$

$$V_A\frac{\partial C_A}{\partial t} = \frac{V_A D_A}{r_0^2}\left[\frac{1}{\sin^2\phi}\frac{\partial^2 C_A}{\partial\theta^2} + \frac{1}{\sin\phi}\frac{\partial}{\partial\phi}\left(\sin\phi\frac{\partial C_A}{\partial\phi}\right)\right]$$

$$+ \delta_{\theta_0}(\theta)\chi_B(\phi)\lambda_A\mu_{BA}(f_B C_B - f_A C_A) + \mu_{IA}(f_I C_I - f_A C_A), \tag{6.15}$$

$$C_A(t,\theta,\phi) = C_A(t,\theta+2\pi,\phi), \tag{6.16}$$

$$\frac{\partial C_A}{\partial \theta}(t,\theta,\phi) = \frac{\partial C_A}{\partial \theta}(t,\theta+2\pi,\phi), \tag{6.17}$$

$$C_A(t,\theta,0) < \infty, \tag{6.18}$$

$$C_A(t,\theta,\pi) < \infty, \tag{6.19}$$

$$V_v\frac{dC_v}{dt} = \frac{Q_m C_m}{P_m} + \frac{Q_t C_t}{P_t} + \frac{Q_f C_f}{P_f} + \frac{Q_{br} C_{br}}{P_{br}} + \frac{Q_l C_l}{P_l} + \frac{Q_k C_k}{P_k} - Q_c C_v, \tag{6.20}$$

$$C_a = \frac{Q_c C_v + Q_p C_c}{Q_c + \frac{Q_p}{P_b}}, \tag{6.21}$$

$$V_m\frac{dC_m}{dt} = Q_m(C_a - C_m/P_m), \tag{6.22}$$

$$V_t\frac{dC_t}{dt} = Q_t(C_a - C_t/P_t), \tag{6.23}$$

$$V_{br}\frac{dC_{br}}{dt} = Q_{br}(C_a - C_{br}/P_{br}), \tag{6.24}$$

$$V_l\frac{dC_l}{dt} = Q_l(C_a - C_l/P_l) - \frac{v_{\max} C_l/P_l}{k_M + C_l/P_l}, \tag{6.25}$$

$$V_k\frac{dC_k}{dt} = Q_k(C_a - C_k/P_k). \tag{6.26}$$

The variables in the model are the concentrations of TCE (in mg/liter) in each of the compartments/subcompartments, and are denoted by C with subscripts corresponding to the respective tissue/region. Model parameters include tissue volumes V in liters, blood flow rates Q in liters/hour, and partition coefficients P, each with the appropriate tissue subscripts. Parameters specific to the dispersion model include the dispersion coefficient $\mathcal{D}_B$ and the diffusion coefficients D_I and D_A in m^2/hour; unbound fractions f_B, f_I, f_A; permeability coefficients μ_{BA}, μ_{IA}, μ_{BI} in liters/hour; blood flow parameters v (m/hour) and $\mathcal{F}$; and interregion transport parameters λ_I and λ_A. A complete discussion of the model equations and parameters is presented in [1, 27].

Here we utilize the TCE model (6.7)–(6.26) to illustrate parameter estimation techniques for models with individual-level dynamics that have realization-dependent derivatives. We present results for both parametric and nonparametric parameter estimation approaches, where the parameter of interest is the probability distribution of the fat dispersion coefficient $\mathcal{D}_B$ in the capillary. This parameter is an important measure of the degree of heterogeneity within an individual's fat tissue.

The parametric and nonparametric approaches each fit into the general framework presented earlier in this chapter for models with individual dynamics. We assume that the parameter $q \equiv \mathcal{D}_B \in Q$ is distributed across the population with distribution $P \in \Pi \subset \mathcal{P}(Q)$, where Π is a set of admissible probability distribution functions (possibly all of $\mathcal{P}(Q)$). Then the general objective function for the standard least squares parameter estimation problem is given by

$$J(P) = \sum_{i=1}^{n} \left| \mathcal{E}\left[x(t_i; q)|P\right] - \hat{d}_i \right|^2, \tag{6.27}$$

where, in this case, $\hat{d}_i$ represents a measurement of the spatial mean concentration of TCE in the fat cells at time t_i, and $x(t_i; q)$ is the spatial mean concentration of TCE in the adipocyte region of the fat compartment that is obtained by solving (6.7)–(6.26) with parameter q.

For the parametric approach, we assume that the probability distribution P for q is of a particular form with parameterization $\tilde{q} \in R^{N_q}$ (e.g., a normal distribution $\mathcal{N}(\mu, \sigma)$ with parameterization $\tilde{q} = (\mu, \sigma)$), so that the set Π of admissible probability distributions is defined as the set of all distributions $P_{\tilde{q}}$ of that given form. The estimation problem is then reduced to the N_q-dimensional problem of minimizing

$$J(\tilde{q}) = \sum_{i=1}^{n} \left| \mathcal{E}\left[x(t_i; q)|P_{\tilde{q}}\right] - \hat{d}_i \right|^2 \tag{6.28}$$

over $P_{\tilde{q}} \in \Pi$ for admissible parameterizations $\tilde{q} \in \tilde{Q} \subset R^{N_q}$.

A major advantage of this approach is the reduction of the original infinite-dimensional objective function (6.27) to a more tractable N_q-dimensional problem. When there is a high degree of confidence about the specific form of the probability distribution P, this method can be expected to perform reasonably well. In many cases, however, the exact form of P is unknown, making it difficult to choose the proper restriction for the set Π and the corresponding parameterization $\tilde{q}$. If an incorrect form and parameterization are chosen for the distribution function, the parametric approach is likely to provide a poor fit to the data since the "true" underlying distribution may not correspond to a distribution in the admissible set Π. Even more alarming are situations where a reasonable fit is found even though an incorrect parameteric form has been assumed (see [14] for examples). In this situation, a nonparametric approach is often more appropriate.

Instead of using a specific form for the distribution P with a finite-dimensional parameterization $\tilde{q}$, the nonparametric parameter estimation approach utilizes a discretization of the admissible parameter set Q to achieve a finite-dimensional approximation for the original objective function (6.27). The resulting family of finite-dimensional estimation problems can be solved in a straightforward manner using quadratic programming, and theoretical results established in [4, 14] guarantee that the minimizers converge to a minimizer for the infinite-dimensional problem (e.g., see Theorem 3 above).

As described earlier in this chapter, we utilize the set $Q_d = \bigcup Q_M$, a dense, countable subset of Q, together with convex combinations of Dirac delta distributions defined over Q_M, to define the following family of objective functions over $\mathcal{P}^M(Q)$:

$$J(P_M) = \sum_{i=1}^{N} \left| \mathcal{E}\left[x(t_i; q)|P_M\right] - \hat{d}_i \right|^2, \tag{6.29}$$

where $\hat{d}_i$ are observations corresponding to the expected value and P_M is a probability distribution in $\mathcal{P}^M(Q)$ as defined in Theorem 3 above.

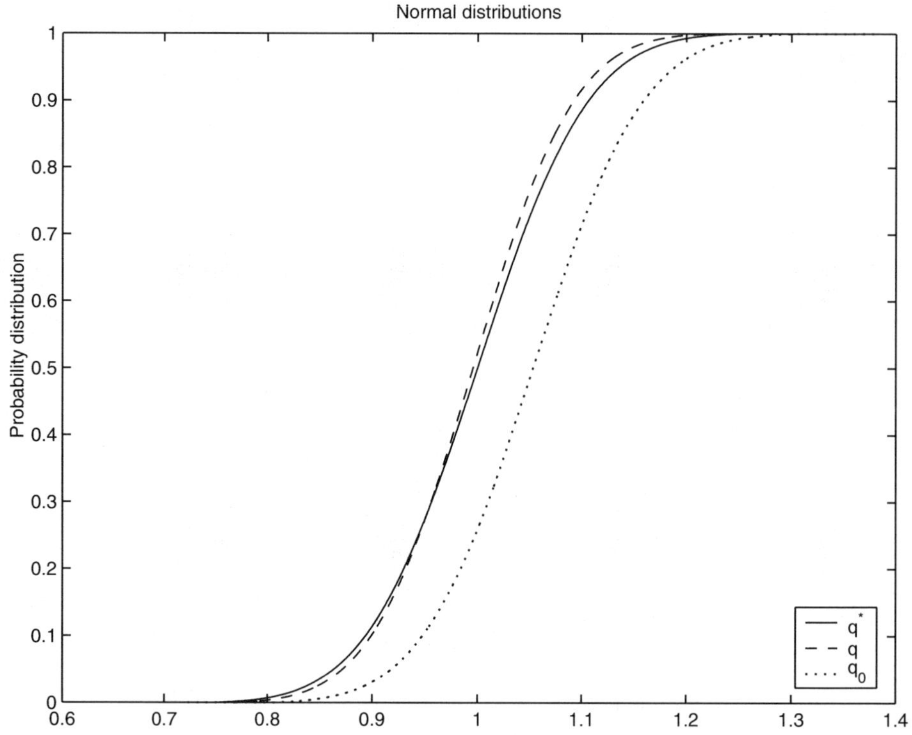

Figure 6.1. *Example solution for the parametric method applied to the TCE PBPK-hybrid model.*

Note that (6.29) can be rewritten as

$$J(P_M) = \sum_{i=1}^{N} \left| \sum_{j=1}^{M} x(t_i; q_j^M) p_j - \hat{d}_i \right|^2, \tag{6.30}$$

so that the minimization of (6.30) is equivalent to solving a constrained quadratic programming problem for $\{p_1, \ldots, p_M\}$ with constraints $p_j \geq 0$ and $\sum_{j=1}^{M} p_j = 1$.

Example results for the parametric and nonparametric methods are given in Figures 6.1 and 6.2, respectively. In each case, the observations used in the parameter estimation problems were generated by solving the TCE model (6.7)–(6.26) with a fixed parameter set q^*. In this case, the solution $x(t_i; q)$ is the spatial mean adipocyte concentration of TCE given the parameter $q = \mathcal{D}_B$. The probability distributions obtained by the estimation methods are presented in Figures 6.1 and 6.2. In Figure 6.1, the solid line represents the true distribution corresponding to q^*, q_0 denotes the initial iterate used in the optimization procedure and q is the estimated parameterization.

For the parametric case, the data-generating probability distribution we chose is a bitruncated normal distribution for q^* with mean $\mu^* = 1$, standard deviation $\sigma^* = 0.0833$, and support over the interval $[\mu^* - 3\sigma^*, \mu + 3\sigma^*]$. The objective function (6.28) was minimized over the set Π of bitruncated normal distributions with parameterizations (μ, σ) and with finite support in $[\mu - 3\sigma, \mu + 3\sigma]$. See [27] for complete details and additional examples.

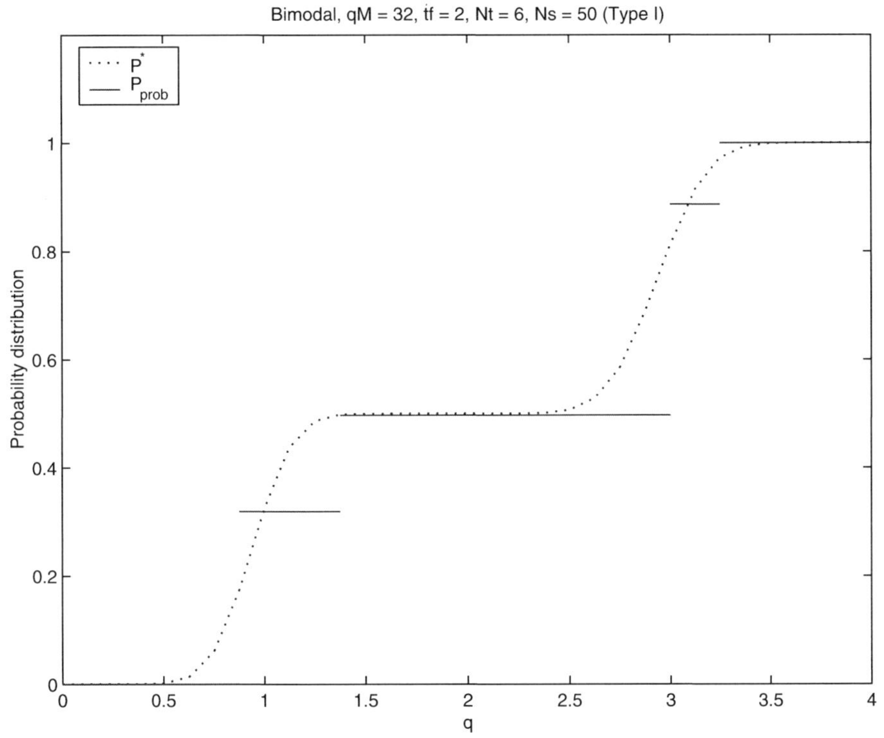

Figure 6.2. *Example solution for the nonparametric method applied to the TCE PBPK-hybrid model.*

For the nonparametric case, we used a bimodal gaussian distribution with means $\mu_1^* = 1$ and $\mu_2^* = 3$ and standard deviations $\sigma_1^* = 0.1667$ and $\sigma_2^* = 0.3333$. The objective function (6.30) was minimized using the quadratic programming routine `quadprog` in MATLAB®.[1] More details and examples for the nonparametric approach applied to the TCE model are given in [14].

6.2.2 Aggregate Dynamics

We turn next to the problems with aggregate dynamics (6.4) and OLS functional (6.5). For these problems one can also develop a general theoretical framework. We first outline the details for ODE systems such as population and SIR models.

Given the system dynamics

$$\frac{d\bar{x}}{dt} = g(t, \bar{x}(t), P), \qquad P \in \mathcal{P}(Q), \tag{6.31}$$

one first argues that $(t, \bar{x}, P) \to g(t, \bar{x}, P)$ is continuous from $[0, T] \times R^n \times \mathcal{P}(Q)$ to R^n, and locally Lipschitz in $\bar{x}$. Then by extension of standard continuous dependence on "parameters" results for ODEs, one obtains that $P \to \bar{x}(t; P)$ is continuous from $\mathcal{P}(Q)$ to R^n for each t. This again yields $P \to J(P) = \sum_i |\bar{x}(t_i; P) - \hat{d}_i|^2$ is continuous from $\mathcal{P}(Q)$ to R^1, where $\mathcal{P}(Q)$, with the Prohorov metric, is compact for Q compact.

[1]MATLAB is a registered trademark of The MathWorks, Inc. For MATLAB product information, please contact: The MathWorks, Inc., 3 Apple Hill Drive, Natick, MA 01760-2098 USA, 508-647-7000, Fax: 508-647-7101, info@mathworks.com, www.mathworks.com/

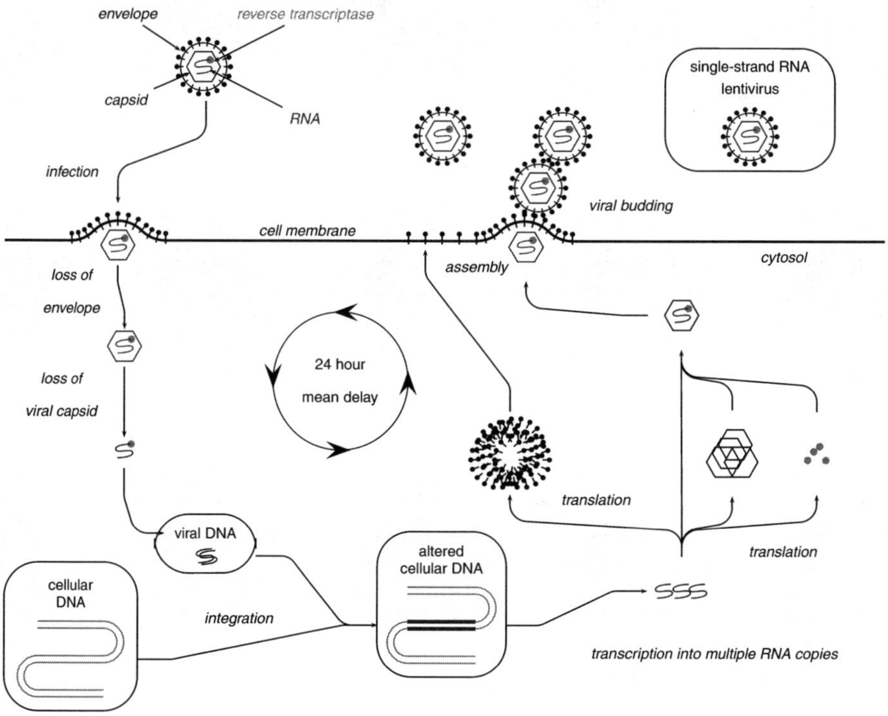

Figure 6.3. *HIV infection pathway.*

Then the general theory of Banks–Bihari [4] can be followed to obtain existence and stability for inverse problems (continuous dependence with respect to data of solutions of the inverse problem) as in Theorems 2 and 3 above. Moreover, an approximation theory as a basis for computational methods is obtained.

We illustrate the ideas in the situation where the underlying ODE system (6.31) is replaced by a nonlinear functional differential equation (FDE) system. This example arises in modeling progression of HIV for which a schematic of the cellular level infection pathway is given in Figure 6.3. This is explained more fully in [5, 17].

The model is a vector system for the variables $\bar{x}(t) = (V(t), A(t), C(t), T(t))$ given by

$$\dot{V}(t) = -cV(t) + n_A \int_{-r}^{0} A(t+\tau)dP_1(\tau) + n_C C(t) - pV(t)T(t), \tag{6.32}$$

$$\dot{A}(t) = (r_v - \delta_A - \delta X(t))A(t) - \gamma \int_{-r}^{0} A(t+\tau)dP_2(\tau) + pV(t)T(t), \tag{6.33}$$

$$\dot{C}(t) = (r_v - \delta_C - \delta X(t))C(t) + \gamma \int_{-r}^{0} A(t+\tau)dP_2(\tau), \tag{6.34}$$

$$\dot{T}(t) = (r_u - \delta_u - \delta X(t) - pV(t))(t) + S(t), \tag{6.35}$$

where $X = A + C + T$ and $V(t)$ is the *expected value* of the population count (number) of virus cells, $A(t)$ is the number of acutely infected cells, $C(t)$ is the *expected value* of the number of chronically infected cells, and $T(t)$ is the total number of target or uninfected cells, each at time t, respectively. The probability measures P_1, P_2 in the model arise because there are delays τ_1 and $\tau_1 + \tau_2$ from the time of acute cellular infection until a cell becomes productively infected and from the time of acute infection until chronic infection, respectively (see [5, Appendix A] or [17, Chapter 2] for a careful and detailed derivation). Biologically, these delay times must vary across the population and this variability is described by the PDFs P_1 and P_2 in the system (6.32)–(6.35). More specifically, the variables $V(t)$ and $C(t)$ have substructures (classes $V(t;\tau)$, $C(t;\tau)$ grouped according to their or their "mother's" delay times) which are averaged across the populations using the distributions P_1, P_2, respectively, so that

$$V(t) = \mathcal{E}[V(t;\tau)|P_1] = \int_{-r}^{0} V(t;\tau)dP_1(\tau),$$

$$C(t) = \mathcal{E}[C(t;\tau)|P_2] = \int_{-r}^{0} C(t;\tau)dP_2(\tau).$$

This yields the system (6.32)–(6.35) with vector valued measure-dependent ($P = (P_1, P_2)$) dynamics as formulated in (6.31) wherein the "state" variables are expected values of population counts. A careful consideration of the derivation of this system reveals that it does not arise from a *parameter-dependent* system for

$$x(t;q) = (V(t;q), A(t;q), C(t;q), T(t;q))$$

with parameters $q = (\tau_1, \tau_2)$, and thus the associated inverse problems for the estimation of P_1, P_2 are fundamentally different from those in the PBPK examples of section 6.2.1 above.

The dynamical system (6.32)–(6.35) for given P_1, P_2 is itself an infinite-dimensional state system (similar to a partial differential equation (PDE) system in this regard). To see this, we note that (6.32)–(6.35) can be written (see [5, 17] for details) in the form

$$\dot{\bar{x}}(t) = L(\bar{x}(t), \bar{x}_t) + f_1(\bar{x}(t)) + f_2(t), \quad t \geq 0, \tag{6.36}$$

where $\theta \to \bar{x}_t(\theta) \equiv \bar{x}(t+0)$, $-r \leq \theta \leq 0$, is a function from $[-r, 0]$ to R^4. This system requires initial data $(\bar{x}(0), \bar{x}_0)$ in the state space $\tilde{Z} = R^4 \times \mathcal{C}(-r, 0; R^4)$ which is readily recognized as being infinite dimensional. For such systems one needs an approximation theory and resulting computational methodology (e.g., finite element methods similar to those popular in PDE theory and implementation) even to carry out forward simulations (an integral part, of course, of most inverse problem methodologies). Fortunately, such a theory exists [15, 8, 9] in the context of abstract evolution equations

$$\dot{z}(t) = \mathcal{A}z(t) + (f_2(t), 0)$$

in a state space $Z = R^4 \times L_2(-r, 0; R^4)$, where

$$\mathcal{D}(\mathcal{A}) = \{(\phi(0), \phi) \in Z : \phi \in H^1(-r, 0; R^4)\}$$

and $\mathcal{A} : \mathcal{D}(\mathcal{A}) \subset Z \to Z$ is given by

$$\mathcal{A}(\phi(0), \phi) = \left(L(\phi(0), \phi) + f_1(\phi(0)), \frac{d}{d\theta}\phi \right)$$

for $\theta \to \phi(\theta)$ in $H^1(-r, 0; R^4)$.

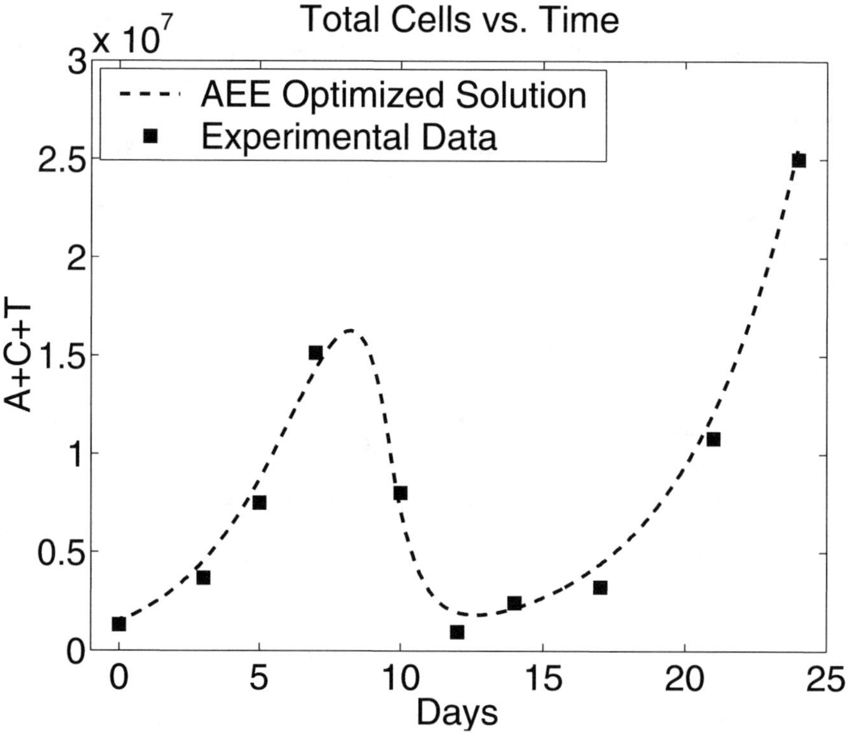

Figure 6.4. *Data versus best fit simulation.*

This theory can be used as a foundation to develop a theoretical and computational framework for inverse problems similar to that outlined for parameter-dependent systems such as the PBPK example in section 6.2.1. While the resulting wellposedness and method stability (see [17, Chapter 3]) statements are similar in spirit to the Banks–Bihari results given in section 6.2.1, the technical details are quite different and rely heavily on the FDE theory in [15, 8, 9]. Details are given in [5, 17].

The methodology outlined here (along with an ANOVA-type statistical methodology) was successfully used to analyze in vitro data [29] from the experiments of Dr. Michael Emerman of the Fred Hutchinson Cancer Research Center in Seattle. A comparison of the simulation of the model with minimizing $P^* = (P_1^*, P_2^*)$ obtained from the inverse problem (i.e., (6.5) with system (6.32)–(6.35)) to a set of Emerman's experimental data is depicted in Figure 6.4. We note that the measures P_1^*, P_2^* used for the simulation depicted here consisted of Dirac measures with single atoms at τ_1^* and $\tau_1^* + \tau_2^*$, respectively, where $\tau_1^* = 22.8$ hours and $\tau_2^* = 3.2$ hours.

6.3 Electromagnetic Imaging of Hidden Substances

In this section we summarize our efforts in modeling the use of electromagnetic pulsed signals to remotely extract information about geometric and chemical properties of substances. Our goal is to describe some existing methodologies developed for the propagation of microwaves in dielectric materials and related imaging techniques as well as to outline some directions in which this theory is currently being extended.

The interaction of very high frequency electromagnetic waves, X-rays, with materials has long been exploited for imaging purposes in medical diagnostics. Many novel techniques have been developed during the past several years to extend the capabilities of traditional X-ray methods. Moreover, waves at different frequency ranges of the electromagnetic spectrum have been utilized. A close inspection of the interaction of materials with electromagnetic radiation at different frequency ranges reveals different underlying mechanisms which need to be correctly captured in the appropriate models. At the same time, the diversity of this interaction makes possible a variety of applications from laser surgery to the detection of environmental contaminants. Some of these techniques have great potential to play an important role in the current efforts in providing a more secure environment from different forms of terrorist activities. As stated in section 6.1, interrogation of materials with electromagnetic waves could be useful in look-down surveillance, imaging of structures, identification of contaminants, airport security devices, and detection of hidden substances, explosives, chemicals, toxins, and bioagents.

The successful use of these techniques is wrought with many technical and theoretical challenges. While portable lasers and X-ray machines are widely available, other ranges of the EM spectrum are not as well represented. Terahertz signal generators and detectors are currently being developed and exhibit a great promise for providing novel imaging devices. Terahertz radiation has several advantages over traditional X-ray methods and is well-suited for imaging applications. T-rays have low photon energies and are nonionizing; thus they are thought to be safer than X-rays. Recently developed devices can generate very short (sub-ps) bursts of THz radiation consisting of only a few cycles of the electric field, yet spanning a broad bandwidth. THz waveforms passing through, or reflected from an object can be recorded in the time domain with very high signal-to-noise ratio. Many organic molecules show strong absorption and dispersion in this frequency range. These effects constitute the polarization mechanism of the molecules which has an influence on the electric field and the propagation of the electromagnetic wave inside the material. Since these transitions are characteristic to the particular molecules, detection of the temporal distortions produced thus yields information about the composition of the material in real time. For example, it is known that cancerous and benign tumors have different electromagnetic characteristics. Therefore an imaging device based on THz waves could not only give information about the structure of an object (geometrical properties) but also help in determining their composition and electromagnetic properties as well in a noninvasive way. As shown in [25], T-ray imaging can be useful both by sending a pulse through the material and detecting it on the other side or by sending a pulse toward the material and recording the reflections from the interface(s) (reflection imaging). This latter procedure is especially important when detectors cannot be placed on the other side of the object or when only slices of an object need to be evaluated. Potential applications range from medical and dental diagnostics to quality control in food processing, semiconductor and chip manufacturing and to the detection of hidden objects and substances in containers. It has been demonstrated in [32] that THz imaging can potentially be used to identify specific powders in mail.

The technical advances in generating electromagnetic radiation in different ranges of the spectrum and their emerging applications in both medical and general imaging fields call for better theoretical understanding and accurate models of the interaction of electromagnetic signals and various substances. In developing these models special attention has to be placed on the specific frequency range and intensity of the electromagnetic radiation, the type of the interrogating signal that is used and the type of material that it encounters. For example, in the high optical range one generally assumes a nonlinear relationship between the electric field and the polarization, and uses the slowly moving envelope assumption to derive the nonlinear Schrödinger equation for the propagation of wave-packets in a dielectric medium from Maxwell's equations. While the latter is a reasonable assumption for pulses that are "long" compared to a characteristic frequency, it may be inadequate to account for ultrashort pulses. In that case a different, full-wave derivation is necessary to capture transient effects.

In the microwave range of the electromagnetic spectrum one can assume that the relationship between the electric field and the polarization is linear for most materials. In the following we will summarize a model developed in [7] for the propagation of windowed microwave (3–100 GHz) pulses in a dielectric medium. In that work the basic question, which was answered in the affirmative, was whether a variational formulation of Maxwell's equations for a specific one-dimensional situation could successfully be used in the identification of geometric *and* dielectric properties of a material slab that is interrogated by microwave pulses from antenna sources.

6.3.1 Variational Approach for Microwave Pulse Propagation

In this one-dimensional model an infinite slab of material is placed in the interval $\Omega = [z_1, z_2]$ with faces parallel to the xy plane, as depicted in Figure 6.5. The interrogating signal is assumed to be a short planar electromagnetic pulse normally incident on the material and the electric field is polarized with oscillations in the xz plane only.

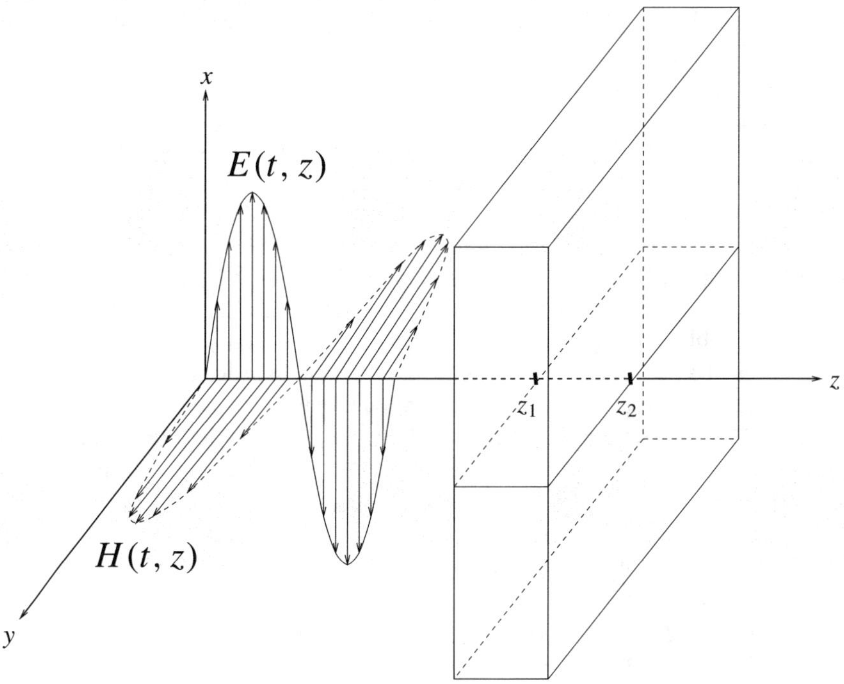

Figure 6.5. *Geometry of the physical problem.*

Thus the electric field is parallel to the $\hat{\imath}$ axis at all points in Ω_0 and the magnetic field $\vec{H}$ is parallel to the $\hat{\jmath}$ axis. Since the material properties are assumed to be homogeneous in the x and y variables, it can be shown that the propagating waves in Ω are also reduced to one nontrivial component [7]. This makes it possible to represent the fields in Ω and Ω_0 with the scalar functions $E(t, z)$ and $H(t, z)$. Under these assumptions, Maxwell's equations reduce to

$$\frac{\partial E}{\partial z} = -\mu_o \frac{\partial H}{\partial t}, \tag{6.37}$$

$$-\frac{\partial H}{\partial z} = \frac{\partial D}{\partial t} + \sigma E + J_s \tag{6.38}$$

for the scalar fields E and H. The magnetic field can be eliminated from the equations by taking the spatial derivative of (6.37) and the time derivative of (6.38) and using the equation for electric flux density $D = \epsilon E + P$, where $\epsilon = \epsilon_0(1 + (\epsilon_r - 1)I_\Omega)$, to obtain

$$\mu_0 \epsilon \ddot{E} + \mu_0 \ddot{P} + \mu_0 \sigma \dot{E} - E'' = -\mu_0 \dot{J}_s. \tag{6.39}$$

A general integral equation model can be employed to describe the behavior of the media's macroscopic electric polarization P:

$$P(t, x) = \int_0^t g(t - s, x) E(s, x) ds. \tag{6.40}$$

This constitutive law is given in terms of a susceptibility kernel g and expresses the fact that the material responds to the electric field in finite time. This formulation is sufficiently general to include microscopic polarization mechanisms such as dipole or orientational polarization as well as ionic and electronic polarization (see later) [3, 21]. We note that $P(0, x)$ is assumed to be 0. To allow for a component of the polarization which depends instantaneously on the electric field one can include a term $\epsilon_0 \chi E$ in D. Hence,

$$D = \epsilon_0(1 + \chi)E + P \tag{6.41}$$

$$= \epsilon_0 \epsilon_r E + P, \tag{6.42}$$

where $\epsilon_r = 1 + \chi \geq 1$ is a relative permittivity which can be treated as a spatially dependent parameter.

In this problem the location of the boundary at $z = z_1$ is assumed known, while the location of the original back boundary at $z = z_2$, i.e., the depth of the slab, is unknown. The unknown boundary creates computational difficulties in the inverse problem since changing domains would involve changing discretization grids in the usual finite element schemes. Thus the method of mappings [11, 10, 26] is applied to transform the problem to a known reference domain. The domain of the computation is defined to be the interval $\tilde{\Omega} = [0, 1]$. An absorbing boundary condition is placed at the $z = 0$ boundary of the interval to prevent the reflection of waves. This can be expressed by

$$\left[\frac{1}{c} \frac{\partial E}{\partial t} - \frac{\partial E}{\partial z} \right]_{z=0} = 0, \tag{6.43}$$

where $c^2 \equiv 1/\epsilon_0 \mu_0$. A supraconductive backing is placed on the slab at $z = z_2$. The boundary conditions on this supraconductive reflector (after mapping z_2 to the reference point $z = 1$) are given by $E(t, 1) = 0$. Substituting an expression for $\ddot{P}$ derived from (6.40), we obtain the strong form of the equation

$$\tilde{\epsilon}_r \ddot{E}(t, z) + \frac{1}{\epsilon_0} I_\Omega(z)(\sigma(z) + g(0, z))\dot{E}(t, z)$$

$$+\frac{1}{\epsilon_0} I_\Omega(z)\dot{g}(0,z)E(t,z) + \int_0^t I_\Omega(z)\frac{1}{\epsilon_0}\ddot{g}(t-s,z)E(s,z)ds \tag{6.44}$$
$$- c^2 E''(t,z) = -\frac{1}{\epsilon_0}\dot{J}_s(t,z),$$

where indicator functions I_Ω have been added to explicitly enforce the restriction of polarization and conductivity to the interior of the transformed medium $\Omega = [z_1, 1]$ and $\tilde{\epsilon}_r = \epsilon/\epsilon_0 = 1 + (\epsilon_r - 1)I_\Omega \geq 1$ throughout $[0, 1]$.

Due to the form of the interrogating inputs, the dielectrically discontinuous medium interfaces, and the possible lack of smoothness in mapping the original domain $\Omega_0 \bigcup \Omega = [0, z_2]$ to the reference domain $\tilde{\Omega} = [0, 1]$, one should not expect classical solutions to Maxwell's equations in strong form. Thus it is desirable to write the system equations in weak or variational form. Using the spaces $H = L_2(0, 1)$ and $V = H_R^1(0, 1) = \{\phi \in H^1(0,1) | \phi(1) = 0\}$ and the boundary conditions (6.43), equation (6.44) can be written in weak form as

$$\langle \tilde{\epsilon}_r \ddot{E}, \phi \rangle + \langle \gamma \dot{E}, \phi \rangle + \langle \beta E, \phi \rangle + \langle \int_0^t \alpha(t-s,\cdot)E(s,\cdot)ds, \phi \rangle$$
$$+ \langle c^2 E', \phi' \rangle + c\dot{E}(t,0)\phi(0) = \langle \mathcal{J}(t,\cdot), \phi \rangle \tag{6.45}$$

with initial conditions

$$E(0,z) = \Phi(z), \qquad \dot{E}(0,z) = \Psi(z),$$

where the coefficients are given by

$$\alpha(t,z) = \frac{1}{\epsilon_0} I_\Omega(z)\ddot{g}(t,z), \qquad \beta(z) = \frac{1}{\epsilon_0} I_\Omega(z)\dot{g}(0,z),$$
$$\gamma(z) = \frac{1}{\epsilon_0} I_\Omega(z)(\sigma(z) + g(0,z)), \qquad \mathcal{J}(t,z) = -\frac{1}{\epsilon_0}\dot{J}_s(t,z)$$

and $\langle \cdot, \cdot \rangle$ is the L^2 inner product. The functions α, β, and γ are dependent on parameters which must be identified. These functions are assumed to be in L^∞ but may lack any additional regularity.

Existence, uniqueness, and regularity of solutions are established in [7], and a comprehensive approximation framework is developed for the forward as well as the inverse problems. It is shown computationally that it is possible to simulate and identify Debye and Lorentz polarization mechanisms in media using first reflected pulses. The thickness of a layered slab using reflected signals from a supraconductive back boundary can also be accurately estimated. It is demonstrated computationally that this model captures transient effects and shows the formation of Brillouin precursors (see Figure 6.6) inside the material [7].

In summary, this approach is amenable to ultrashort input pulses and provides a complete theoretical and computational framework for the direct and the inverse problem in this one-dimensional model.

This work has been extended in different directions. A corresponding analysis with acoustic reflectors at the back of the slab of material and pressure dependent Maxwell system coefficients is developed in [2]. It is shown that instead of a supraconductive backing (which is not practical in many medical or remote imaging applications), an acoustic wave can be employed to reflect the electromagnetic signal. Moreover, these reflections can again be used to identify geometric and dielectric properties of the material.

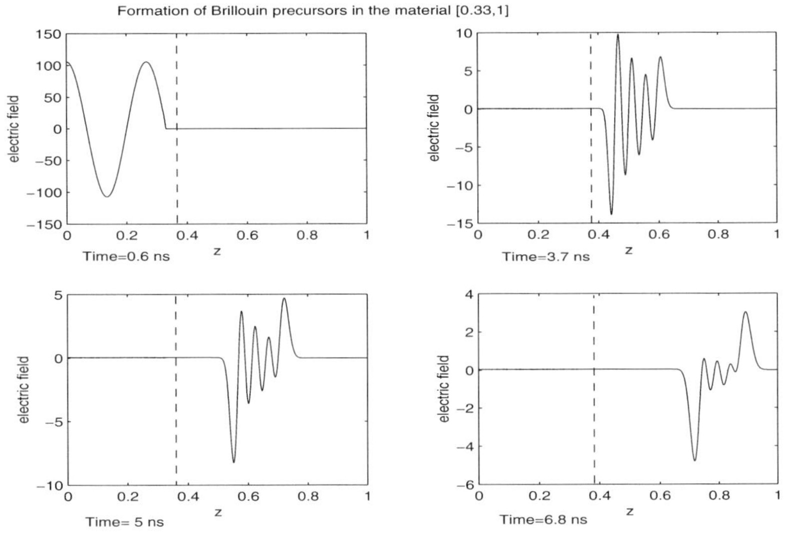

Figure 6.6. *Formation of Brillouin precursors using a linear Debye model.*

To develop and use a similar methodology for THz signals we must capture the response of materials to higher frequencies. Thus, we need to represent the absorption and dispersion properties of the material by accurately modeling the underlying polarization mechanisms. As interrogating frequencies increase, it is not unreasonable to expect that nonlinearities play a nontrivial role.

6.3.2 Polarization Mechanisms

Polarization, the general macroscopic response of a material to an electric field, is an important dielectric characteristic specific to a given material and hence is important to any interrogation methodology. It depends heavily on the molecular structure of the material.

Dielectric materials contain bound negative and positive charges that are not free to move as charges do in conductors. These charges are kept in place by atomic and molecular forces. When subjected to an external electric field, dipole moments are induced in the atoms and molecules. The electric polarization vector is defined as the dipole moment per unit volume. The mechanism by which these dipole moments are created is different in different materials, whether gases, liquids, or solids. Molecules of certain gases (e.g., oxygen) contain a symmetric pair of atoms in each molecule and thus have no inherent dipole moments. Such molecules are called nonpolar. In others, (e.g., water vapor) the center of gravity of the positive charge (in this case on the hydrogen atom) and the negative charge (on the oxygen) do not coincide, and the total charge distribution on the molecule has a dipole moment. These molecules are called polar.

First we consider nonpolar molecules. When an electric field is applied to the atoms of such molecules the electrons are forced in one direction, while the nucleus is forced in the opposite direction by the field. Thus there is a net displacement of the centers of charge, and a dipole moment is created. This displacement of the electron distribution is called electronic polarization. In a changing electric field the displacement of the center of charge of the electrons is usually modeled by a harmonic oscillator and this gives rise to the Lorentz model for electronic polarization:

$$\ddot{\vec{P}} + \frac{1}{\tau}\dot{\vec{P}} + \omega_0^2 \vec{P} = \epsilon_0 \omega_p^2 \vec{E},$$

where ϵ_0 is the dielectric constant, and ω_p is the so called plasma frequency given by $\omega_p = \sqrt{\epsilon_s - \epsilon_\infty}$, with ϵ_s and ϵ_∞ being the relative permittivities of the material in the limit of the static and very high frequencies, respectively.

The same mechanism can be observed in polar molecules. However, in addition to this effect, the electric field forces a portion of the originally randomly oriented internal dipoles to line up with the applied field, producing a net moment per unit volume. This is called dipole or orientational polarization and is described by the Debye model, which captures the relaxation of the molecules once the electric field is turned off:

$$\tau \dot{\vec{P}} + \vec{P} = \epsilon_0(\epsilon_s - \epsilon_\infty)\vec{E}. \tag{6.46}$$

It takes time for the molecules to line up because of their moment of inertia, so this mechanism becomes less pronounced if the material is subjected to very high frequencies. In that case the molecules simply cannot follow the changing electric field sufficiently fast and at some level appear to "freeze."

Polarization in denser materials, liquids and solids, is even more complicated. Here the electric field around each molecule is influenced not only by the external field but by the field of the neighboring molecules as well, giving rise to vibrational polarization (a form of electronic polarization). In solids that are made up of ionic crystals, e.g., NaCl, the positive and negative ions are displaced as a result of an applied field, which is called ionic polarization. In certain crystals there is a permanent internal polarization in the sense that each unit cell of the lattice has a permanent dipole moment. If the relative position of the lattice points change, e.g., by heating or stressing the material, external fields appear creating pyroelectricity and piezoelectricity, respectively. For an ideal dielectric, orientational polarization dominates for lower frequencies, giving way to vibrational and electronic polarization as the frequency increases. At very high frequencies (X-rays, gamma rays) there is almost no polarization since the material simply cannot "follow the wave" due to inertial effects.

In all of these models so far we assumed that the relationship between the applied electric field and the polarization is linear, given by, in general, an integral convolution. However, it is known that in the optical range this relationship becomes nonlinear (more so for noninfinitesimal fields), as evidenced by nonlinear optical effects like solitons, second harmonic generation, and self-focusing [31]. For some materials this transition starts to take place in the infrared range. For example, while for microwaves a linear model is appropriate (indeed, a Debye model provides a good fit for water), nonlinear effects, especially for noninfinitesimal amplitudes, need to be taken into account for higher frequency ranges. There is experimental evidence for small but significant departure from strict linearity at high values of the electric field [30, p. 245]. An example is the Kerr effect, in which insulating liquids, containing anisotropic molecules, become doubly refracting when subjected to very strong fields. As suggested in [30], this could be modeled by the constitutive relation $\vec{P} = \vec{E} + s|\vec{E}|^2\vec{E}$. However, we have already seen that inertial effects, i.e., the finite time response of the material, may be important so instead we will consider a Debye model where the electric field provides nonlinear forcing. For a centrosymmetric medium we might assume

$$\tau \dot{\vec{P}} + \vec{P} = \tilde{f}(\vec{E}),$$

where $\tilde{f}(\vec{E}) = c_1\vec{E} + c_2|\vec{E}|^2\vec{E}$ for $|\vec{E}| < M$ and 0 otherwise; i.e., $\tilde{f}$ is a saturated cubic nonlinearity. In integral form we obtain the relationship

$$\vec{P}(t,\vec{x}) = \int_0^t g(t-s,\vec{x})\tilde{f}(\vec{E}(s,\vec{x}))ds, \tag{6.47}$$

where $g(t,\vec{x}) = e^{\frac{-t}{\tau}}$. We note that a nonlinearly driven Lorentz model,

$$\ddot{\vec{P}} + \frac{1}{\tau}\dot{\vec{P}} + \omega_0^2\vec{P} = \epsilon_0\omega_p^2\tilde{f}(\vec{E}),$$

leads to a similar integral representation with kernel function $g(t,\vec{x}) = \frac{\epsilon_0\omega_p^2}{\nu_0}e^{-\frac{1}{2\tau}t}\sin(\nu_0 t)$, where $\nu_0 = \sqrt{\omega_0^2 - \frac{1}{4\tau^2}}$. As a first step, we considered a general model with nonlinear polarization in [13].

6.3.3 Variational Formulation of the Model with Nonlinear Polarization

We consider a polarization mechanism of the form (6.47) with $\tilde{f} = E + f(E)$ together with the one-dimensional model outlined above. As before, an infinite slab of material with supraconductive backing is interrogated by a normally incident polarized plane wave windowed pulse originating at an antenna source $z = 0$ in free space $\Omega_0 = [0, z_1]$. The slab of material in $\Omega = [z_1, z_2]$ is assumed to be homogeneous in the directions orthogonal to the direction z of propagation of the plane wave. As we have already noted, under these assumptions it is possible to represent the strength of the electric and magnetic fields in Ω and Ω_0 by the scalar functions $E(t,z)$ and $H(t,z)$, respectively. One can readily eliminate the magnetic field from the full Maxwell equations and substitute the assumed constitutive law for the polarization to arrive at the strong formulation of the problem with initial and boundary conditions similar to those in section 6.3.1:

$$\begin{aligned}
&\hat{\varepsilon}_r\ddot{E}(t,z) + \frac{1}{\varepsilon_0}I_\Omega(z)(\sigma(z) + g(0,z))\dot{E}(t,z)\\
&\quad + \frac{1}{\varepsilon_0}I_\Omega(z)\dot{g}(0,z)E(t,z) + \int_0^t \frac{1}{\varepsilon_0}I_\Omega(z)\ddot{g}(t-s,z)E(s,z)ds\\
&\quad + \frac{1}{\varepsilon_0}I_\Omega(z)\dot{g}(0,z)f(E(t,z)) + \int_0^t \frac{1}{\varepsilon_0}I_\Omega(z)\ddot{g}(t-s,z)f(E(s,z))ds\\
&\quad + \frac{1}{\varepsilon_0}I_\Omega(z)g(0,z)\frac{d}{dt}f(E(t,z)) - c^2E''(t,z)\\
&= -\frac{1}{\varepsilon_0}\dot{J}_s(t,z), \quad t > 0, \quad 0 < z < z_2,
\end{aligned} \tag{6.48}$$

$$\left[\frac{1}{c}\frac{\partial E}{\partial t} - \frac{\partial E}{\partial z}\right]_{z=0} = 0, \quad t > 0, \tag{6.49}$$

$$E(t,z_2) = 0, \quad t > 0, \tag{6.50}$$

$$E(0,z) = \Phi(z), \qquad \dot{E}(0,z) = \Psi(z), \quad 0 < z < z_2. \tag{6.51}$$

In the physical problem z_2 is assumed to be unknown, and it is desirable to estimate it from given data. Since the theoretical analysis is constructive in the sense that the numerical method we use to solve this problem (for both forward and inverse problems) follows the theoretical arguments, it is desirable to convert the problem to a fixed spatial domain, e.g., [0, 1], as explained above and in [7]. Thus we use the method of maps and subsequently formulate the variational problem as follows.

We let $H = L^2(0,1)$, $V = H^1_R(0,1) = \{\phi \in H^1(0,1) | \phi(1) = 0\}$ leading to the Gelfand triple [23, 33] $V \hookrightarrow H \hookrightarrow V^*$. We say that $E \in L^\infty(0,T;V)$ with $\dot{E} \in L^2(0,T;H)$, $\ddot{E} \in L^2(0,T;V^*)$, is a weak solution if it satisfies for every $\varphi \in V$

$$\begin{aligned}\langle \bar{\varepsilon}_r \ddot{E}, \varphi\rangle_{V^*,V} + \langle \gamma \dot{E}, \varphi\rangle + \langle \beta E, \varphi\rangle + \Big\langle \int_0^t \alpha(t-s,\cdot)E(s,\cdot)ds, \varphi\Big\rangle \\ + \langle \beta f(E), \varphi\rangle + \Big\langle \int_0^t \alpha(t-s,\cdot) f(E(s,\cdot))ds, \varphi\Big\rangle + \Big\langle \hat{\gamma}\frac{d}{dt} f(E), \varphi\Big\rangle \\ + \langle c^2 h' E', \varphi'\rangle + c\dot{E}(t,0)\varphi(0) = \langle \mathcal{J}(t,\cdot), \varphi\rangle_{V^*,V} \qquad (6.52)\end{aligned}$$

and

$$E(0,z) = \Phi(z), \qquad \dot{E}(0,z) = \Psi(z). \qquad (6.53)$$

Using a Galerkin-type approach and special considerations for the nonlinear terms we were able to show [13] that, under fairly general assumptions on the nonlinearity f, a unique global weak solution exists and depends continuously on initial data.

Thus the one-dimensional problem with nonlinearly forced dynamics for the polarization is well posed. This system can also be thought of as a type of approximation (using truncated Taylor expansions) to the nonlinear polarization dynamics:

$$\dot{\vec{P}} + f(\vec{P}) = k\vec{E} \qquad (6.54)$$

and

$$\ddot{\vec{P}} + \gamma\dot{\vec{P}} + f(\vec{P}) = k\vec{E}, \qquad (6.55)$$

which represent nonlinear Debye and Lorentz mechanisms and are suggested in [18]. Currently a study is underway to compare these different systems theoretically and computationally.

6.3.4 Extension to Higher Dimensions

To extend the above methodology to more realistic situations one needs to formulate the problems in higher (two or three) dimensions and demonstrate the applicability of the variational framework in that setting. The work on microwave interrogating signals has been extended to two dimensions computationally [6] for a diagonally anisotropic slab of material. The extensions to higher dimensions and higher frequencies are closely related and several new challenges arise.

Theoretically, the one-dimensional model formulated above depends on the tacit assumption that the polarization field $\vec{P}$ in the dielectric remains parallel to the electric field $\vec{E}$. Even then, the usual Maxwell equation $\nabla \cdot \vec{D} = 0$ along with the constitutive law $\vec{D} = \epsilon_0\epsilon_r\vec{E} + f_1(\vec{P})\vec{P}$ need not result in $\nabla \cdot \vec{E} = 0$. This is important in deriving the second order form of Maxwell's equation where the identity $\nabla \times \nabla \times \vec{E} = \nabla(\nabla \cdot \vec{E}) - \nabla^2\vec{E}$ results in the simple Laplacian only if $\nabla \cdot \vec{E} = 0$ or if one assumes this term is negligible as often done in nonlinear optics (see [18, pp. 54–60]). We believe that it may be important to consider the *full* system to capture the dynamics of the propagated electromagnetic signal.

Experimentally it is known that birefringence occurs in anisotropic dielectrics as a result of the different phase velocities for different directions of the electric field polarization. Birefringence can also occur without anisotropy when two different modes of the electric field are coupled nonlinearly. It is present in living organisms even at microwave frequencies, but its effect is small at 1–3 GHz. At frequencies higher than 10 GHz the effect cannot be neglected and anisotropy needs to be taken into account even if linear polarization dynamics are assumed. Anisotropic effects and the tensor nature of the dielectric constant is especially important for the detection of aerosols, suspended particles in fluids, and bacteria (e.g., anthrax) with membranes of complex geometries. At even higher frequencies where nonlinearities in the polarization dynamics become pronounced, it is expected that there are strong interactions between the nonlinear and anisotropic effects, so their correct modeling is crucial for the accurate representation of the propagation and reflection dynamics.

In the computational treatment of the two- or three-dimensional interrogation problem one encounters several difficulties. Naturally, the higher spatial dimension involves increased computational complexities, especially with nonlinear polarization dynamics. However, there are additional inherent challenges. As described in [6], the interrogating signals from a finite antenna produce oblique incident waves on a planar medium, and they must be treated in reflections as well. Thus one cannot use the uniformity assumption as in the one-dimensional model to reduce the problem to a finite computational domain. In this case the infinite spatial domain must be approximated by a finite computational domain with artificial boundaries. At these boundaries some type of boundary damping must be employed to remove unwanted numerical reflections. In [6] perfectly matched layers along with Enquist–Majda absorbing boundary conditions are used to successfully control these reflections. Another possibility that is currently being explored is to enlarge the computational domain so that reflections from the sample and from the artificial boundaries might be separated in time.

In summary, we believe that the variational framework for the interrogation problem is suitable for capturing important dynamic effects associated with the propagation of electromagnetic pulses in different materials. Although it is challenging both theoretically and computationally, it has a great potential for providing a firm foundation for novel imaging methods which can contribute to the current efforts for greater security against terrorist activities.

6.4 Concluding Remarks

In the preceding sections, we have discussed methodologies that we believe have high potential for use in homeland security. Specifically, the methods in section 6.2 are directly applicable in epidemic and disease transmission models discussed elsewhere in this volume. In these models one will usually have population level or aggregate data and the rates to be estimated (infectivity, death, recovery, etc.) can, in general, be expected to be distributed over the population.

Application of the imaging/interrogation techniques of section 6.3 are somewhat less obvious. For success in direct use, we must yet develop specific polarization models for biological and chemical agents (toxins, viruses, powders, etc.) that might be employed in deliberate attacks on populations.

The atmosphere of the real threat of terrorism at home and abroad has unfortunately initiated a new environment and urgency for scientific and technological research. While some in our community suggest [19] "for the most part we do not need new methods," our view is somewhat different. While it is true that we in the mathematical sciences community have techniques and approaches that may be extremely important in the new problems arising in the war on bioterrorism, as we enjoin this fight we will find much work to do to pursue our ideas in a relevant manner. It is not true that we have all the tools we need nor are those we do have in the needed form for immediate application. Our strong belief is that more will be required of mathematics and statistics than collecting of existing tools and applying them with relatively straightforward modifications.

The focus in this chapter is quite narrow and the actual effort on terrorism requires a multidisciplinary as well as interdisciplinary approach beyond that of this volume and beyond that which the community has embraced to date. There is a virtual catalogue of far ranging topics from the engineering, physical, mathematical, and biological sciences: data mining, network analysis, biomathematics, genomics, operations research (game theory, risk analysis, logistics), etc., which must be combined with the social and psychological sciences (individual and group behavior, e.g., fanaticism, cognition, etc.) in ways and on a scale unprecedented in the history of science. And this must be done with a new sense of urgency. For example, the development of agent-specific biosensors (discussed in some detail in another chapter of this book), sometimes in the context of "smart" materials, has for some time been a priority at several of our national labs; the needs have been heightened by events of the past several years.

Lest our view appear too pessimistic, we hasten to add that while we do not have ready "solutions" to questions and problems that perhaps are only now being precisely formulated, the mathematical and statistical sciences do have a rich history of model development with associated tools and techniques. This will undoubtedly provide a solid foundation that will prove extremely valuable in the pursuit of many specific problems related to terrorism in general and bioterrorism in particular. We are optimistic about the value we can bring to society in this essential effort.

Acknowledgments

The research reported here was supported in part by the U.S. Air Force Office of Scientific Research under grant AFOSR F49620-01-1-0026 and in part by the Joint DMS/NIGMS Initiative to Support Research in the Area of Mathematical Biology under grant 1R01GM67299-01.

The authors are grateful to Dr. Richard Albanese, Dr. Carlos Castillo-Chavez, and Dr. Marie Davidian for several informative discussions. Part of this chapter was completed while H.T.B. was a visitor to the Mittag Leffler Institute of the Royal Swedish Academy of Sciences, Djursholm, Sweden. Collaboration was also facilitated while all authors were visitors at the Statistical and Applied Mathematical Sciences Institute (SAMSI), Research Triangle Park, NC, which is funded by the National Science Foundation under grant DMS-0112069.

Bibliography

[1] R. A. ALBANESE, H. T. BANKS, M. V. EVANS, AND L. K. POTTER, *Physiologically based pharmacokinetic models for the transport of trichloroethylene in adipose tissue*, Bull. Math. Biol., 64 (2002), pp. 97–131.

[2] R. A. ALBANESE, H. T. BANKS, AND J. K. RAYE, *Non-destructive evaluation of materials using pulsed microwave interrogating signals and acoustic wave induced reflections*,

Inverse Problems, 18 (2002), pp. 1935–1958.

[3] C. A. BALANIS, *Advanced Engineering Electromagnetics*, Wiley, New York, 1989.

[4] H. T. BANKS AND K. L. BIHARI, *Modeling and estimating uncertainty in parameter estimation*, Inverse Problems, 17 (2001), pp. 1–17.

[5] H. T. BANKS, D. M. BORTZ, AND S. E. HOLTE, *Incorporation of variability into the mathematical modeling of viral delays in HIV infection dynamics*, Math. Biosci., 183 (2003), pp. 63–91.

[6] H. T. BANKS AND B. L. BROWNING, *Time Domain Electromagnetic Scattering Using Perfectly Matched Layers*, Technical Report CRSC-TR02-24, Center for Research in Scientific Computation, North Carolina State University, Raleigh, NC, 2002.

[7] H. T. BANKS, M. W. BUKSAS, AND T. LIN, *Electromagnetic Material Interrogation Using Conductive Interfaces and Acoustic Wavefronts*, Frontiers Appl. Math., SIAM, Philadelphia, 2000.

[8] H. T. BANKS AND J. A. BURNS, *Hereditary control problems: Numerical methods based on averaging approximations*, SIAM J. Control Optim., 16 (1978), pp. 169–208.

[9] H. T. BANKS AND F. KAPPEL, *Spline approximations for functional differential equations*, Journal of Differential Equations, 34 (1979), pp. 496–522.

[10] H. T. BANKS, F. KOJIMA, AND W. P. WINFREE, *Boundary estimation problems arising in thermal tomography*, Inverse Problems, 6 (1990), pp. 897–921.

[11] H. T. BANKS AND F. KOJIMA, *Boundary shape identification problems in two dimensional domains related to thermal testing of materials*, Quart. Appl. Math., 47 (1989), pp. 273–293.

[12] H. T. BANKS AND K. KUNISCH, *Estimation Techniques for Distributed Parameter Systems*, Birkhäuser, Boston, 1989.

[13] H. T. BANKS AND G. A. PINTER, *Maxwell-systems with nonlinear polarization*, Nonlinear Anal. Real World Appl., 4 (2003), pp. 483–501.

[14] H. T. BANKS AND L. K. POTTER, *Probabilistic Methods for Addressing Uncertainty and Variability in Biological Models: Application to a Toxicokinetic Model*, Technical Report CRSC-TR02-27, Center for Research in Scientific Computation, North Carolina State University, Raleigh, NC, 2002; Mathematical Biosci., submitted.

[15] H. T. BANKS, *Identification of nonlinear delay systems using spline methods*, in Nonlinear Phenomena in Mathematical Sciences, V. Lakshmikantham, ed., Academic Press, New York, 1982, pp. 47–55.

[16] P. BILLINGSLEY, *Convergence of Probability Measures*, Wiley, New York, 1968.

[17] D. M. BORTZ, *Modeling, Analysis and Estimation of an In Vitro HIV Infection Using Functional Differential Equations*, Ph.D. thesis, North Carolina State University, Raleigh, NC, 2002.

[18] R. W. BOYD, *Nonlinear Optics*, Academic Press, San Diego, 1992.

[19] C. Castillo-Chavez and F. S. Roberts, eds., *Report on DIMACS Working Group Meeting: Mathematical Sciences Methods for the Study of Deliberate Releases of Biological Agents and Their Consequences*, preliminary draft, DIMACS Center, Rutgers University, Piscataway, NJ, 2002.

[20] M. Davidian and D. Giltinan, *Nonlinear Models for Repeated Measurement Data*, Chapman and Hall, London, 1998.

[21] R. S. Elliot, *Electromagnetics: History, Theory and Applications*, IEEE Press, New York, 1993.

[22] B. G. Lindsay, *Mixture Models: Theory, Geometry and Applications*, NSF-CBMS Regional Conf. Ser. Probab. Statist. 5, Institute of Mathematical Statistics, Haywood, CA, 1995.

[23] J. L. Lions, *Optimal Control of Systems Governed by Partial Differential Equations*, Springer-Verlag, New York, 1971.

[24] M. A. Medinsky and C. D. Klaassen, *Toxicokinetics*, in Casarett and Doull's Toxicology: The Basic Science of Poisons, 5th ed., C. D. Klaassen, ed., McGraw–Hill, New York, 1996, pp. 187–198.

[25] D. M. Mittleman, M. Gupta, R. Neelamani, R. G. Baraniuk, J. V. Rudd, and M. Koch, *Recent advances in terahertz imaging*, Appl. Phys. B, 68 (1999), pp. 1085–1094.

[26] O. Pironneau, *Optimal Shape Design for Elliptic Systems*, Springer-Verlag, New York, 1983.

[27] L. K. Potter, *Physiologically Based Pharmacokinetic Models for the Systemic Transport of Trichloroethylene*, Ph.D. thesis, North Carolina State University, Raleigh, NC, 2001.

[28] M. S. Roberts and M. Rowland, *A dispersion model of hepatic elimination: 1. Formulation of the model and bolus considerations*, J. Pharmacokinetics and Biopharmaceutics, 14 (1986), pp. 227–260.

[29] M. E. Rogel, L. I. Wu, and M. Emerman, *The human immunodeficiency virus type 1 vpr gene prevents cell proliferation during chronic infection*, J. Virology, 69 (1995), pp. 882–888.

[30] B. K. P. Scaife, *Principles of Dielectrics*, Clarendon Press, Oxford, 1989.

[31] C. Sulem and P. L. Sulem, *The Nonlinear Schrodinger Equation, Self-Focusing and Wave Collapse*, Springer-Verlag, New York, 1999.

[32] S.Wang, B. Ferguson, C. Mannella, D. Gray, D. Abbott, and X. C. Zhang, *Powder detection using THz imaging*, in OSA Technical Digest, Postconference ed., Optical Society of America, Washington, DC, 2002, p. 132.

[33] J. Wloka, *Partial Differential Equations*, Cambridge University Press, Cambridge, UK, 1992.

Chapter 7

Models for the Transmission Dynamics of Fanatic Behaviors

Carlos Castillo-Chavez and Baojun Song†*

7.1 Introduction

Bioterrorism is driven by organized groups of individuals who recruit from highly fragile populations. Simple models are used to strengthen the common-sense view that the most effective counterterrorist approach consists of depriving fanatic groups of recruitment sources. Hence, approaches to infiltrate groups of converts in order to generate change are, in fact, unlikely to succeed. The overall objective of this chapter is to motivate and encourage the development of models for the spread of fanatic behaviors in susceptible (fragile) populations. The understanding of the transmission dynamics of such behaviors is likely to increase our knowledge of the mechanisms behind the evolution of cultural norms and values, and vice versa.

This chapter explicitly focuses on the generation of qualitative results that may be useful in the fight against fanaticism, the *force* behind the acts of terrorism. The focus is on the generation of qualitative results that may be useful in the generation of novel insights and confirmation of common-sense views. Our model and approach may also help identify pressure points or highlight strengths and redundancies of systems (landscapes) that facilitate the spread of fanatic behaviors. Our model is a *crude* attempt to model forces (such as peer pressure) and structures that facilitate the establishment of extreme behaviors. Extensions of the crude approach outlined in this chapter may help address several relevant questions including such questions as, What social landscapes support the existence of fanatic subcultures? How do cultural norms change from repeated acts of terrorism? What impact do contact structures (peer and group pressure) have on the spread of extreme ideologies? What is the role of "core" (ultra) fanatics on the spread and persistence of extreme behaviors? What is the role of competing fanatic ideologies on their joint dynamics?

It is worth stressing the fact that we make no attempt to define "fanatic" behavior or even what terrorism is. The overall situation is highly complex. Ideally one would be

*Department of Biological Statistics and Computational Biology, Cornell University, Ithaca, NY 14853-7801, and Center for Nonlinear Studies, Los Alamos National Laboratory, Los Alamos, NM 87545.
†Department of Mathematical Sciences, Montclair State University, Upper Montclair, NJ 07043.

interested in understanding the the processes or mechanisms that lead to insurgency (not defined here either) movements. Why some insurgency movements choose acts of terror against civilians while others do not is not an easy question to answer. What makes rational individuals join insurgency movements and, sometimes, fanatic activities? It is not the goal of this chapter to address these issues or even to pretend that they can be answered by mathematics. The sole objectives of this effort is to highlight the fact that group pressure leads to multiple steady states and to the type of dynamics that are not "typical" of classical contact epidemiological processes.

Although serious efforts to develop a framework to study the dynamics of human behavior at the population level have been developed [8, 4, 16, 12], there is no comprehensive mathematical framework or approach for the systematic study of the transmission of ideas. An effort, in the spirit of population biology "thinking," to model the spread of culture was developed by Boyd and Richerson [4]. Others like Lumsden and Wilson [16] and Feldman and Cavalli-Sforza [12] have also tried to develop a similar program using paradigms based on evolutionary biology. Here, we study the dynamics of the spread of extreme behaviors as some type of epidemiology contact processes. We are aware that our approach and the associated caricature model (as most sociological models) can be easily derailed or deconstructed. We hope that our efforts are not taken that lightly, as we believe that epidemiological models still represent a reasonable starting point for the study of the spread and growth of behaviors that are the engine behind most acts of terrorism.

The models in this chapter implicitly assume a fanatic hierarchy that implies that transmission is not an individual-to-individual process but rather part of a group effort. The hierarchy is determined by the level of individual commitment to the ideology. It is assumed that fanatic individuals are most effective at transmitting the ideology to vulnerable members of populations. Fortunately, not everybody is vulnerable. Consequently, the role of recruitment is essential. Here, recruitment is modeled after the work in [13, 14, 17]. Hence, in our model, individuals must be first recruited into the right environment (core group or vulnerable population), as it is assumed that only individuals who join the susceptible core population can be converted. The specifics of the model are outlined in the next section.

7.2 Model for the Transmission Dynamics of Extreme Ideologies in Vulnerable Populations

The description of our simple model requires the introduction of the following definitions and assumptions. The host population is divided into two subpopulations: the noncore (typically large) and the core or vulnerable subpopulation. The noncore population, defined as $G(t)$, includes everybody and it is the source of raw material, that is, the recruitment pool for the core. The core subpopulation is subdivided into three (hierarchical) classes: the naive (susceptible or vulnerable) subpopulation, which includes those members of the core who have not yet been converted and is defined as $S(t)$; the semifanatic subpopulation, $E(t)$, which includes those who have just converted as well as those who may not be fully committed; and the fanatic subpopulation, (t), which includes individuals who have completely internalized the extreme ideology. The model equations read as

$$\frac{dG}{dt} = \Lambda - \beta_1 G \frac{C}{T} + \gamma_1 S + \gamma_2 E + \gamma_3 F - \mu G, \tag{7.1}$$

$$\frac{dS}{dt} = \beta_1 G \frac{C}{T} - \beta_2 S \frac{E+F}{C} - \gamma_1 S - \mu S, \tag{7.2}$$

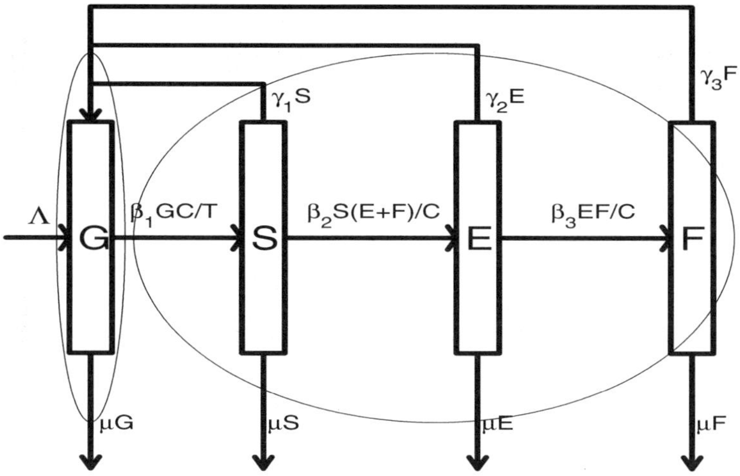

Figure 7.1. *Flow chart of model* (7.1)–(7.4).

$$\frac{dE}{dt} = \beta_2 S \frac{E+F}{C} - \beta_3 E \frac{F}{C} - \gamma_2 E - \mu E, \tag{7.3}$$

$$\frac{dF}{dt} = \beta_3 E \frac{F}{C} - \gamma_3 F - \mu F, \tag{7.4}$$

where the total population, T, and the core subpopulation, C, are

$$T = G + S + E + F$$

and

$$C = S + E + F.$$

Here

$$G(0) > 0, \qquad S(0) \geq 0, \qquad E(0) \geq 0, \qquad F(0) \geq 0.$$

Figure 7.1 shows the flow chart that provides a pictorial representation of the dynamics associated with model (7.1)–(7.4). The general population is replenished via an assumed constant birth rate, Λ. The core population recruits individuals from the general population through individual "contacts" between core and noncore individuals. A contact here has a broad meaning; for example, individuals can reach individuals who are physically located at a distant place via phone calls. This sort of contact distinguishes itself from regular contacts in epidemiology. The core population builds its membership from the general population and converts them at the rate $\beta_1 GC/T$, where β_1 measures the strength of the recruitment force associated with the members of the core population. Hence, β_1 is assumed to be proportional to the number of contacts per unit time as well as to the likelihood of success (recruitment). Migration from the noncore to the core ceases whenever there is no raw material ($G = 0$) or when the core no longer exists ($C = 0$). γ_i denotes the per-capita recovery rate for each subpopulation in the core. Hence, $1/\gamma_i$ is the average residence time for each subpopulation in the core ($i = 1$ corresponds to susceptible population, $i = 2$

semifanatics, and $i = 3$ fanatics). That is, it is assumed that the residence times in the core are exponentially distributed. Our implicit hierarchical assumptions require γ_3 to be small; thus the average residence time of fanatics is long $(1/\gamma_3 \gg 1)$. Semifanatics and fanatics are assumed to have frequent contacts with susceptible individuals in the core. The epidemiological approach expects that these deliberate contacts may result in the conversion of some susceptible individuals. Hence, we model the rate of conversion from the S-class to the E-class by $\beta_2 S(E+F)/C$. Similarly, the rate of conversion from the E-class to the F-class is modeled by the nonlinear term, $\beta_3 EF/C$. Finally, it is assumed (this is not critical) that everybody in the population has the same average natural mortality rate, μ.

The dynamics of the total population T (core and noncore subpopulations) are governed by the equation $dT/dt = \Lambda - \mu T$, that is, $\lim_{t\to\infty} T(t) = \Lambda/\mu$. Results on the relationship between the transient dynamics and their "natural" limiting systems [6, 18] are used to reduce the dimension of this model. Hence, we replace T with Λ/μ and $G(t)$ with $\Lambda/\mu - C(t)$ in the above system of differential equations for the state variables S, E, and F. The reduced system is "dynamically" equivalent. Hereafter, to shorten notation, we denote $\gamma_i + \mu$ by γ_i $(i = 1, 2, 3)$.

Rescaling the resulting equations by Λ/μ leads to the following equivalent system:

$$\frac{dS}{dt} = \beta_1(1-C)C - \beta_2 S\frac{E+F}{C} - \gamma_1 S, \tag{7.5}$$

$$\frac{dE}{dt} = \beta_2 S\frac{E+F}{C} - \beta_3 E\frac{F}{C} - \gamma_2 E, \tag{7.6}$$

$$\frac{dF}{dt} = \beta_3 E\frac{F}{C} - \gamma_3 F, \tag{7.7}$$

$$C = S + E + F.$$

The domain of interest is the simplex in $\mathbb{R}^3$ defined by the set

$$\Omega = \{(S, E, F) \in \mathbb{R}^3_+ : 0 \le S + E + F \le 1\}.$$

Features that distinguish this model from "somewhat" similar epidemiological models include the fact that the core population size is not constant as well as the fact that the recruitment into the core population is of logistic form $(\beta_1(1-C)C)$. In the following sections we explore the dynamics of model (7.5)–(7.7).

7.3 Global Thresholds

The qualitative features of model (7.5)–(7.7) are characterized by a sequence of thresholds which are based on the local dynamics. However, two global thresholds can also be identified. The first global threshold identifies necessary conditions for the formation of a (viable) core group.

7.3.1 Establishment of a Core Population

Theorem 1. *Assume* $\gamma_1 = \text{Min}\{\gamma_1, \gamma_2, \gamma_3\}$. *Whenever* $R_1 = \beta_1/\gamma_1 \le 1$, $\lim_{t\to\infty} C(t) = 0$.

Proof. Summing (7.5), (7.6), and (7.7), we obtain

$$\frac{dC}{dt} = \beta_1(1-C)C - \gamma_1 S - \gamma_2 E - \gamma_3 F$$

$$\leq \beta_1(1-C)C - \gamma_1 C \leq \beta_1 C - \gamma_1 C$$
$$= -\gamma_1 C(1-R_1).$$

It follows from the above inequality that $C(t)$ decays at least exponentially fast to zero. □

Theorem 1 identifies a precondition for the establishment of a core population. If $R_1 \leq 1$, the core population cannot become established. Since the general population never goes extinct, there is always a viable recruitment source for the core. However, the core can only be established if either the conversion rate β_1 is high enough or the residence time $1/\gamma_1$ is long enough, or both ($R_1 > 1$). Conversely, increasing resistance to vulnerability (that is, decreasing β_1), reducing the average core residence time $1/\gamma_1$, or both may result on the elimination of the core population.

7.3.2 Persistence of a Fanatic Ideology

A second global threshold controls the establishment of the fanatic population and, consequently, the persistence of the fanatic ideology. From (7.7) it follows that if $R_3 = \beta_3/\gamma_3 \leq 1$ then $\lim_{t\to\infty} F(t) = 0$. That is, R_3 gives sufficient conditions for the elimination of the fanatic population. The fact that the invariant superplane $F = 0$ is a global attractor, whenever $R_3 \leq 1$, can be used to reduce the dimension of model (7.5)–(7.7). In fact, using the limiting equation approach [6, 18], the model is reduced to the following (equivalent) two-dimensional system:

$$\frac{dS}{dt} = \beta_1(1-C)C - \beta_2 S\frac{E}{C} - \gamma_1 S, \tag{7.8}$$
$$\frac{dE}{dt} = \beta_2 S\frac{E}{C} - \gamma_2 E, \tag{7.9}$$
$$C = S + E.$$

Choosing the Dulac function $D(S, E) = 1/(SE)$, we check that the divergence of (7.8)–(7.9) is $-\beta_1(S^2 + E(1-E))/(S^2E) < 0$. Hence, there are no closed trajectories. It should be easy to carry out the analysis of this two-dimensional model, but this is not the case. The fact that the vector field is not analytic at the origin implies that linearization fails. Alternative analytical approaches can be used to handle this problem. In this work we split a complex singular point into several simple ones by the Briot–Bouquet transformation [19]. Using this technique and the Dulac function given above, we establish Theorem 2 which describes the *global* dynamics of (7.8)–(7.9). For its detailed proof we refer the reader to [1, 2].

Theorem 2. *Let* $\nu = \gamma_1/\gamma_2$ *and* $R_d = \nu R_1 R_2/(R_2 - 1 + \nu)$.

(i) *If* $R_1 < 1$ *and* $R_2 < 1$, *or* $1 < R_2$ *and* $R_d < 1$, *then* $(0, 0)$ *is the global attractor.*

(ii) *If* $1 < R_1$ *and* $R_2 < 1$, *then* $(1 - 1/R_1, 0)$ *is the global attractor.*

(iii) *If* $1 < R_2$ *and* $R_d > 1$, *then the positive equilibrium* (S^*, E^*) *with*

$$S^* = \frac{(R_d - 1)(R_2 - 1 + \nu)}{\nu R_1 R_2^2}, \tag{7.10}$$

$$E^* = \frac{(R_2 - 1)(R_d - 1)(R_2 - 1 + \nu)}{\nu R_1 R_2^2} \tag{7.11}$$

is the global attractor.

We use phase-parameter portraits (Figure 7.2) to describe Theorem 2. In the R_1R_2-plane, the curve $R_1 = (\nu - 1 + R_2)/(\nu R_2)$ (i.e., $R_d = 1$) is the boundary between regions III and IV in the left-hand figure and between regions II and IV in the right-hand figure in Figure 7.2. It has different position for a fixed level $\nu > 1$ and $\nu < 1$. From Theorem 2, $(1 - 1/R_1, 0)$ is the global attractor in region I; $(0, 0)$ is the global attractor in regions II and IV; (S^*, E^*) is the global attractor in region III. Apparently, the asymptotic behaviors of system (7.8)–(7.9) are the same whenever parameters fall in regions II and IV. However, it is interesting to note the appearance of a family of homoclinic trajectories when R_1 and R_2 fall in region IV (see Figure 7.3). Hence, the number of individuals in the core group will expand from a tiny initial size before the core population moves towards extinction; the core cannot become established.

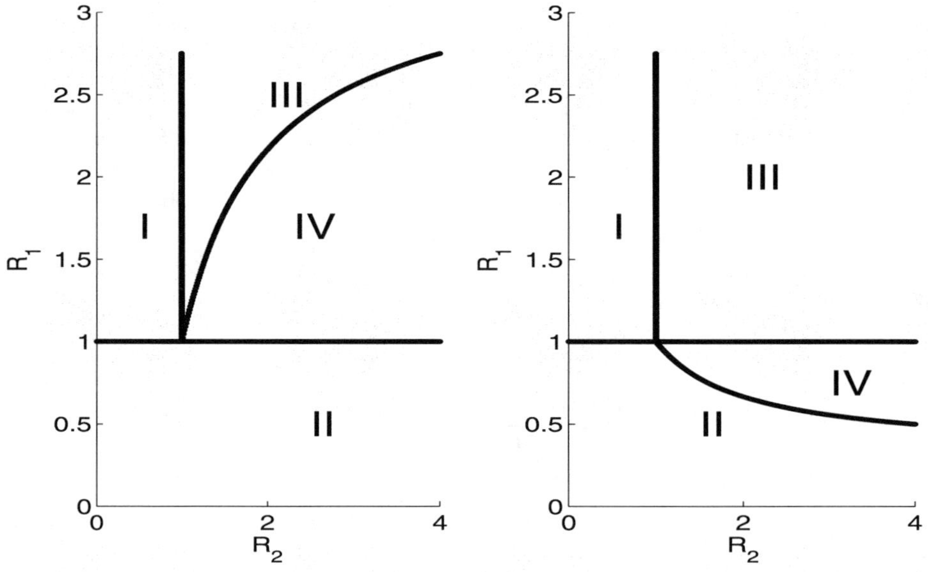

Figure 7.2. *Phase-parameter portraits for system* (7.8)–(7.9). $(1 - 1/R_1, 0)$ *is the global attractor in region* I; $(0, 0)$ *is the global attractor in regions* II *and* IV; (S^*, E^*) *is the global attractor in region* III. *The figure on the left corresponds to the case* $\nu < 1$, *while the figure on the right assumes that* $\nu > 1$. *The boundary between regions* III *and* IV *on the left and between regions* II *and* IV *on the right is given by the equation* $R_1 = (\nu - 1 + R_2)/(\nu R_2)$ (*i.e.,* $R_d = 1$).

Figures 7.3, 7.4, and 7.5 give three typical phase portraits.

7.4 Local Thresholds

Local thresholds and their role in the characterization of the model local dynamics are the subject of this section.

7.4.1 Boundary Equilibria

There are three boundary equilibria:

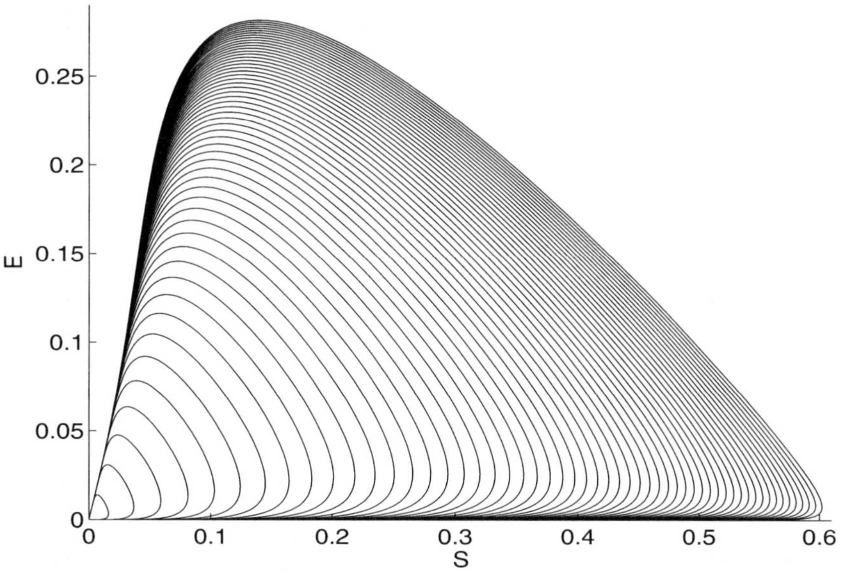

Figure 7.3. *Phase portrait. When $R_3 < 1$, the superplane $F = 0$ is a global attractor. Whenever R_2 and R_1 are in region* II *(Figure* 7.2*),* $(0, 0)$ *is the global attractor. In region* IV *all trajectories are homoclinic and approach* $(0, 0)$ *forward and backward.*

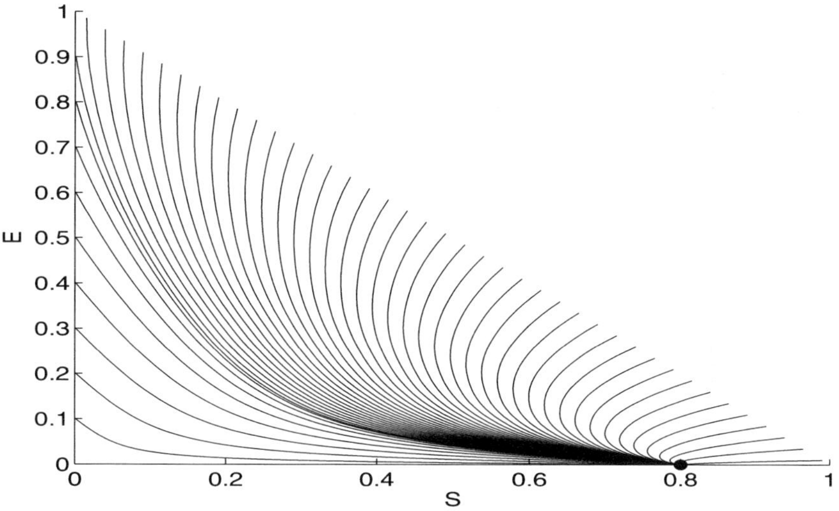

Figure 7.4. *Phase portrait. When $R_3 < 1$, the superplane $F = 0$ is a global attractor. When $R_2 < 1$ and $R_1 > 1$, $(1 - 1/R_1, 0)$ is the global attractor.*

$E_0 = (0, 0, 0)$ always viable;

$E_1 = (1 - R_1^{-1}, 0, 0)$ exists only if $R_1 > 1$;

$E_2 = \{S^*, E^*, 0\}$ exists if $R_2 = \frac{\beta_2}{\gamma_2} > 1$ and $R_d \geq 1$, where S^* and E^* are given by (7.10) and (7.11), respectively.

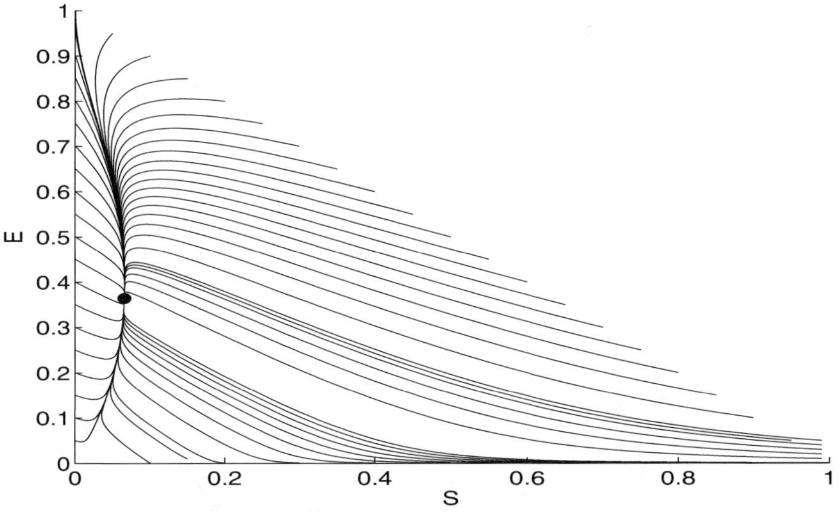

Figure 7.5. *Phase portrait. When $R_3 < 1$, the superplane $F = 0$ is the global attractor. If R_2 and R_1 are in region* III, *then (S^*, E^*) is the global attractor.*

Theorem 3. *E_1 is locally stable only if $R_1 > 1$ and $R_2 < 1$.*

The proof of this theorem is established from the linearization matrix of (7.5)–(7.7) around E_1:

$$J_{E_1} = \begin{pmatrix} \gamma_1(1 - R_1) & 2\gamma_1 - \beta_1 - \beta_2 & \beta_1 - \beta_2 \\ 0 & \gamma_2(R_2 - 1) & \beta_2 \\ 0 & 0 & -\gamma_3 \end{pmatrix}.$$

Theorem 4. *E_2 is stable if and only if $R_2 > 1$, $R_d > 1$, and $R_0 = (R_3 - 1)(R_2 - 1) < 1$. That is, whenever E_2 exists, it is stable, provided that $R_0 < 1$.*

Proof. Let $J_{E_2} = (f_{ij})_{3\times 3}$ be the communication matrix of the system (7.5)–(7.7) around E_2. The entries of J_{E_2} are $f_{31} = f_{32} = 0$ and

$$\begin{aligned}
f_{11} &= 2\gamma_2 + \frac{2}{R_2}(\gamma_1 - \gamma_2) - \beta_1 - \gamma_1 - \beta_2(1 - R_2^{-1})^2, \\
f_{12} &= 2\gamma_2 + \frac{2}{R_2}(\gamma_1 - \gamma_2) - \beta_1 - \beta_2 R_2^{-2}, \\
f_{13} &= 2\gamma_2 + \frac{2}{R_2}(\gamma_1 - \gamma_2) - \beta_1 - \beta_2 R_2^{-2}, \\
f_{21} &= \beta_2(1 - R_2^{-1})^2, \\
f_{22} &= \beta_2 R_2^{-2} - \gamma_2, \\
f_{23} &= \beta_2 R_2^{-2} - \beta_3(1 - R_2^{-1}), \\
f_{33} &= (R_0 - 1)\gamma_3 / R_2.
\end{aligned}$$

The stability of E_2 is determined by the upper left submatrix and the sign of f_{33}. Instead, whenever $R_0 < 1$, $f_{33} < 0$. Let $M = (f_{ij})_{2\times 2}$; then $|M| = \gamma_1\gamma_2(1 - R_2^{-1})(R_1 R_2 - 1 +$

$\nu(R_2 - 1)) > 0$ and $\text{Trace}(M) = (R_2^{-1} - 1)(\gamma_2(R_d - 1) + \gamma_1) - \gamma_2(\sqrt{R_2} - \frac{1}{R_2})^2 < 0$. Hence, Theorem 4 holds. □

7.4.2 Multiple Equilibria and the Turning Point

This subsection shows that our model can support multiple positive equilibria and introduces the *turning point* as a "replacement" concept for the basic reproductive number.

To avoid messy algebra, it is assumed in this subsection that $\gamma_1 = \gamma_2$. This assumption does not effect the results on the existence of multiple equilibria.

The equilibria of model (7.5)–(7.7) are the solutions of the following system:

$$0 = \beta_1(1 - C)C - \beta_2 S\frac{E + F}{C} - \gamma_1 S, \tag{7.12}$$

$$0 = \beta_2 S\frac{E + F}{C} - \beta_3 E\frac{F}{C} - \gamma_2 E, \tag{7.13}$$

$$0 = \beta_3 E\frac{F}{C} - \gamma_3 F. \tag{7.14}$$

Now, we proceed to look for nonzero equilibria that may provide information on the location (coordinates) of the turning point. The steps needed to compute such positive equilibria are outlined. Starting from (7.14), we express C and S in terms of E and F, that is,

$$\left.\begin{aligned} C &= R_3 E \\ \text{and}\quad S &= (R_3 - 1)E - F \end{aligned}\right\}. \tag{7.15}$$

Summation of (7.12), (7.13), and (7.14) leads to the equation

$$\beta_1(1 - C)C - \gamma_1 S - \gamma_2 E - \gamma_3 F = 0. \tag{7.16}$$

Substituting the expressions for C and S given in (7.15) into (7.16) and solving the resulting equation for F (noticing $\gamma_1 = \gamma_2$ and the definition of R_1) lead to

$$F(E) = \frac{\gamma_1 R_3 E}{\gamma_1 - \gamma_3}(1 - R_1 + R_1 R_3 E) := Ex, \tag{7.17}$$

where

$$x = \frac{\gamma_1 R_3}{\gamma_1 - \gamma_3}(1 - R_1 + R_1 R_3 E). \tag{7.18}$$

It follows from (7.15) and (7.17) that S can also be expressed in terms of E. In fact,

$$S(E) = (R_3 - 1 - x)E. \tag{7.19}$$

After substituting the expressions (7.15), (7.17), and (7.19) for C, F, and S, respectively, into (7.13) and cancelling a factor of E, we arrive at the following quadratic equation in x:

$$\frac{\beta_2}{R_3}x^2 - \gamma_3\left(\frac{\beta_2}{\beta_3}(R_3 - 2) - 1\right)x + (1 - R_0)\frac{\gamma_2}{R_3} = 0. \tag{7.20}$$

The goal is to understand how the solutions of (7.20) vary as β_3 changes. Here, we are interested in the existence of positive equilibria when $R_0 < 1$, because we have known that

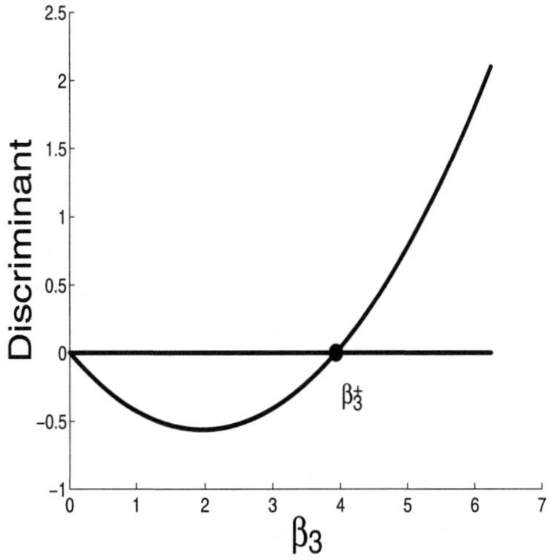

Figure 7.6. *Discriminant of* (7.20).

there is only one positive solution and the second solution is negative whenever $R_0 > 1$. In terms of β_3, $R_0 \leq 1$ is identical to $\beta_3 \leq \beta_3^* = \frac{\gamma_3 R_2}{R_2 - 1}$. $\beta_3 \leq \beta_3^*$ indicates that (7.20) has two roots with the same sign, if any. Define $\beta_3^b = \frac{2\beta_2\gamma_3}{\beta_2 - \gamma_3}$. If $\beta_3 = \beta_3^b$ the coefficient of the linear term in (7.20) is zero, which means that when $\beta_3 > \beta_3^b$ the coefficient is negative; otherwise it is nonnegative. In other words, if a solution of (7.20) exists when $\beta_3 > \beta_3^b$, then both solutions are positive, while if $\beta_3 < \beta_3^b$, then both solutions are negative. The crucial point is therefore to establish conditions for the existence of solutions to (7.20). Examination of the discriminant, although the algebra is rather messy, allows us to compute a threshold (Figure 7.6 shows how the discriminant changes with β_3) to guarantee the existence of the solutions to (7.20). The result is that (7.20) has real solutions if and only if $\beta_3 \geq \beta_3^+ = \frac{4\gamma_3\beta_2}{(\beta_2-\gamma_3)^2}(\gamma_2 - \gamma_3)$. To study the relationship among β_3^b, β_3^+, and β_3^*, we have the following lemmas.

Lemma 5. *If* $R_2 < 2 - \frac{\gamma_3}{\gamma_2}$ *and* $\beta_2 > \gamma_3$, *then* $\beta_3^b < \beta_3^+$.

Proof. Rewrite β_3^+ as $\beta_3^+ = \frac{4\gamma_3\beta_2}{(\beta_2-\gamma_3)^2}(\gamma_2 - \gamma_3) = \frac{2\gamma_3\beta_2}{(\beta_2-\gamma_3)}\frac{2(\gamma_2-\gamma_3)}{(\beta_2-\gamma_3)} = \beta_3^b \frac{2(\gamma_2-\gamma_3}{(\beta_2-\gamma_3)}$. Hence, it is equivalent to show $\frac{2(\gamma_2-\gamma_3}{(\beta_2-\gamma_3)} > 1$. In fact, $R_2 < 2 - \frac{\gamma_3}{\gamma_2}$ can be rewritten as $\beta_2 - \gamma_2 < \gamma_2 - \gamma_3$, or $\frac{\gamma_2-\gamma_3}{\beta_2-\gamma_2} > 1$. The desired result follows from the last inequality. Indeed, we have shown $2\beta_3^b < \beta_3^+$. □

Lemma 6. $\beta_3^+ \leq \beta_3^*$.

Proof. It is equivalent to prove $(4(\gamma_2 - \gamma_3))/(\beta_2 - \gamma_3)^2 \leq 1/(\beta_2 - \gamma_2)$, which can be rewritten as $4(\gamma_2 - \beta_2 + \beta_2 - \gamma_3)(\beta_2 - \gamma_2) \leq (\beta_2 - \gamma_3)^2$. The last inequality is identical to $-(2(\beta_2 - \gamma_2) - (\beta_2 - \gamma_3))^2 \leq 0$. Thus the desired result follows. □

Lemma 7. *Assume that* $R_2 < 2 - \gamma_3/\gamma_2$ *and that* $\beta_2 > \gamma_3$; *then if* $\beta_3^+ \leq \beta_3 \leq \beta_3^*$, *there are* *two positive equilibria.*

Proof. First we show that (7.20) has two positive solutions. $\beta_3 \geq \beta_3^+$ gives the existence of real solutions of (7.20); $\beta_3 < \beta_3^*$ implies that the two solutions have the same sign; finally $\beta_3^+ \leq \beta_3$ also implies that the two solutions are positive since $\beta_3^b \leq \beta_3^+$.

Now, it must be proved that the corresponding components E and S are positive. Let x^+ and x^- be the two positive solutions of (7.20), that is,

$$x^{\pm} = \frac{\beta_3}{2}\left(\frac{1}{\gamma_3} - \frac{1}{\beta_2}\right) - 1 \pm \sqrt{\frac{\beta_3^2}{2^2}\left(\frac{1}{\gamma_3} - \frac{1}{\beta_2}\right)^2 + \frac{\beta_3}{\beta_2}\left(1 - \frac{\gamma_2}{\gamma_3}\right)}. \qquad (7.21)$$

Corresponding to x^+ and x^-, we have two equilibria, (S^+, E^+, F^+) and (S^-, E^-, F^-), respectively. It follows from (7.18) that $E^{\pm} > 0$ because $R_1 > 1$. Noticing that $\gamma_2 > \gamma_3$, from (7.21), we obtain that $x^+ < R_3 - 1$ which gives $S^+ > 0$ from (7.19). Hence, $S^- > 0$ since $x^- < x^+$.

The final step is to verify $S^+ + E^+ + F^+ \leq 1$ and $S^- + E^- + F^- \leq 1$, which guarantees that the equilibria are located in the domain of interest. Here, we only give the proof of that $S^+ + E^+ + F^+ \leq 1$. The same procedure can be used to show $S^- + E^- + F^- \leq 1$. To this end, we solve for E from (7.18) and get

$$E^+ = \frac{(\gamma_2 - \gamma_3)x^+/R_3 - (1 - R_1)}{R_1 R_3}.$$

Rewrite $S^+ + E^+ + F^+$ as

$$\begin{aligned} S^+ + E^+ + F^+ &= (R_3 - 1 - x^+)E^+ + E^+ + x^+E^+ \\ &= R_3E^+ = \frac{1}{R_1}\left(\frac{\gamma_2 - \gamma_3}{R_3 - 1}x^+ - 1\right) + 1. \end{aligned}$$

It suffices to prove that $(\gamma_2 - \gamma_3)x^+/(R_3 - 1) - 1 < 0$, which can be derived from the fact that $x^+ < R_3 - 1$. This ends the proof of Lemma 7. □

Note that from (7.21) the two branches of the solutions meet at $\beta_3 = \beta_3^+$ (the expression under the square root is equal to zero). Figure 7.7 plots each component of the positive equilibria (endemic equilibria) versus β_3. It can be seen from this figure that two branches (solid lines and dashed lines) meet when $\beta = \beta_3^+$. Hence, β_3^+ will define the turning point, named β_3^T, that is,

$$\beta_3^T = \frac{4\gamma_3\beta_2}{(\beta_2 - \gamma_3)^2}(\gamma_2 - \gamma_3). \qquad (7.22)$$

In conclusion, the backward bifurcation starts from $\beta_3 = \beta_3^*$ and turns around at $\beta_3 = \beta_3^T$. Whenever $0 < \beta_3 < \beta_3^T$, there is no positive endemic equilibrium, so that either E_1 or E_2 is an attractor, depending on the values of R_2, R_d, and/or R_1. Once the core group has become established ($R_1 > 1$), the turning point does not depend on the S class. The magnitude of β_3^T is determined only by the fanatic populations.

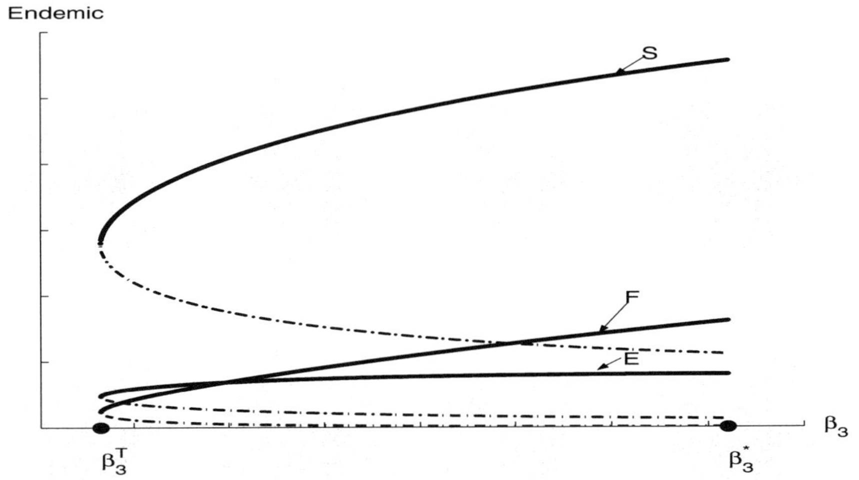

Figure 7.7. *Backward bifurcation and turning point.*

7.4.3 Bifurcation Approach

Section 7.4.2 established the existence of multiple equilibria. We now characterize their stability using a theory that is based on general center manifold theory. This theory is applied to study the bifurcation direction at $R_0 = 1$. Consider a general system of ODEs with a parameter ϕ:

$$\frac{dx}{dt} = f(x, \phi), \qquad f : \mathbb{R}^n \times \mathbb{R} \to \mathbb{R}^n, \quad \text{and} \quad f \in \mathbb{C}^2(\mathbb{R}^n \times \mathbb{R}). \tag{7.23}$$

Without losing generality, we assume that 0 is an equilibrium for system (7.23) for all values of ϕ, that is,

$$f(0, \phi) \equiv 0 \quad \text{for all } \phi. \tag{7.24}$$

Theorem 8 below provides a criteria for determining if a bifurcation is backward (subcritical). The generality of this theorem and its proof can be found in [7, 9, 10, 17].

Theorem 8. *Assume the following:*

(1) $D_x f(0,0) = \left(\frac{\partial f_i}{\partial x_j}(0,0)\right)$ *is the linearization matrix of system* (7.23) *around the equilibrium 0 with ϕ evaluated at 0. Zero is a simple eigenvalue of $D_x f(0,0)$, and all other eigenvalues have negative real parts.*

(2) *Matrix $D_x f(0,0)$ has a right null vector (the dominant right eigenvector of $D_x f(0,0)$) w and a left null vector v.*

Let f_k be the kth component of f and

$$a = \sum_{k,i,j=1}^{n} v_k w_i w_j \frac{\partial^2 f_k}{\partial x_i \partial x_j}(0,0), \tag{7.25}$$

$$b = \sum_{k,i=1}^{n} v_k w_i \frac{\partial^2 f_k}{\partial x_i \partial \phi}(0, 0). \tag{7.26}$$

If both a and b are positive, equilibrium 0 is asymptotical stable and there exist unstable positive equilibria when $\phi < 0$ and $|\phi| \ll 1$. That is, the direction of the bifurcation at $\phi = 0$ is backward.

Applying Theorem 8 to our model leads us to the following conclusion.

Theorem 9. *If*

(i) $R_1 > 1$ *and* $R_2 > 1$,

(ii) $R_1 + R_2^{-1} < 2$,

(iii) $R_2 < 2 - \frac{\gamma_3}{\gamma_1}$,

then system (7.5)–(7.7) *undergoes a backward bifurcation at* $R_0 = (R_3 - 1)(R_2 - 1) = 1$.

Proof. We choose β_3 as the bifurcation parameter; that is, $\beta_3 = \beta_3^* = (\gamma_3 R_2)/(R_2 - 1)$ corresponds to $R_0 = 1$. We have computed the linearization matrix of system (7.5)–(7.7) around E_2 in section 7.4.2. Now we apply the additional condition $R_0 = 1$ to it. J_{E_2} is the linearization matrix of system (7.5)–(7.7) around E_2 when $R_0 = 1$. In fact, $f_{3j} = 0$ for $j = 1, 2, 3$, and

$$f_{11} = \gamma_1 \left(1 - R_1 - \frac{(R_2 - 1)^2}{R_2}\right) < 0, \qquad f_{12} = \gamma_1 (2 - R_1 - R_2^{-1}) > 0,$$

$$f_{13} = \gamma_1 (2 - R_1 - R_2^{-1}) > 0, \qquad f_{21} = \gamma_1 R_2 (1 - R_2^{-1})^2 > 0,$$

$$f_{22} = \gamma_1 (R_2^{-1} - 1) < 0, \qquad f_{23} = \gamma_1 \left(R_2^{-1} - \frac{\gamma_3}{\gamma_2}\right) > 0,$$

where $f_{23} > 0$ is obtained from the fact that $(\gamma_2 - \gamma_3)^2 > 0$ because the inequality $R_2 < 2 - \gamma_3/\gamma_2$ can be rewritten as $R_2^{-1} > (\gamma_3/\gamma_2)(\gamma_2^2)(2\gamma_2\gamma_3 - \gamma_3^2)$. $\lambda_1 = 0$ is a simple eigenvalue of J_{E_2}. The other two eigenvalues have negative real parts because $|M| > 0$ and $\text{Trace}(M) < 0$ as it was shown in Theorem 9. The right dominant eigenvalues of J_{E_2} are the nonzero solutions of the system

$$\begin{pmatrix} f_{11} & f_{12} & f_{13} \\ f_{21} & f_{22} & f_{23} \end{pmatrix} \begin{bmatrix} w_1 \\ w_2 \\ w_3 \end{bmatrix} = 0. \tag{7.27}$$

Choosing w_3 to be the free variable, letting $w_3 = 1$, and solving for w_1 and w_2 from (7.27), we obtain

$$\begin{pmatrix} f_{11} & f_{12} \\ f_{21} & f_{22} \end{pmatrix} \begin{bmatrix} w_1 \\ w_2 \end{bmatrix} = \begin{bmatrix} -f_{13} \\ -f_{23} \end{bmatrix}. \tag{7.28}$$

It follows from $f_{11} < 0$, $f_{22} < 0$, $f_{12} > 0$, and $f_{21} > 0$ that the matrix $-\begin{pmatrix} f_{11} & f_{12} \\ f_{21} & f_{22} \end{pmatrix}$ is an M-matrix. Recalling $f_{13} > 0$ and $f_{23} > 0$, it follows from the properties of M-matrices that the solution of the linear system (7.28) is positive [3]. We therefore obtain a positive right eigenvector $w = [w_1 \quad w_2 \quad 1]'$ for the matrix J_{E_2}. One can check that $v = [0 \quad 0 \quad 1]$ is a left dominant eigenvector for J_{E_2}. The application of Theorem 8 requires the computation of two quantities a and b using w, v, and the second derivatives of the vector field. The results are

$$\begin{aligned} a &= \sum_{i,j=1}^{3} w_i w_j \frac{\partial^2 f_3}{\partial x_i \partial x_j} \\ &= 2w_1 w_3 \frac{\partial^2 f_3}{\partial S \partial F} + 2w_2 w_3 \frac{\partial^2 f_3}{\partial E \partial F} + w_3 w_3 \frac{\partial^2 f_3}{\partial F \partial F} \\ &= \gamma_3 \frac{R_1(R_2-1)}{R_1-1}\left(-2w_1 + \frac{1}{R_2-1}(2w_2+1)\right), \\ b &= 2\sum_{i=1}^{3} w_i \frac{\partial^2 f_3}{\partial x_i \partial \beta_3} = 2w_3 \frac{\partial^2 f_3}{\partial F \partial \beta_3} = 2\gamma_3 > 0. \end{aligned}$$

We have to go further by showing $a > 0$. It is enough to show $-2w_1 + 2w_2/(R_2 - 1) > 0$. We know $|M| = \gamma_2^2(R_2 - 1)(R_1 - 1) > 0$. Hence, we can explicitly write the solution to system (7.28):

$$\begin{aligned} w_1 &= \frac{1}{\gamma_2^2(R_2-1)(R_1-1)}(f_{12}f_{23} - f_{13}f_{22}), \\ w_2 &= \frac{1}{\gamma_2^2(R_2-1)(R_1-1)}(f_{13}f_{21} - f_{11}f_{23}). \end{aligned}$$

Hence,

$$\begin{aligned} &-2w_1 + \frac{2w_2}{R_2-1} \\ &= \frac{2}{\gamma_2^2(R_2-1)(R_1-1)}\left(\left(\frac{f_{21}}{R_2-1} + f_{22}\right) f_{13} - f_{23}\left(\frac{f_{11}}{R_2-1} + f_{12}\right)\right). \end{aligned}$$

It can be verified that $f_{21}/(R_2-1)+f_{22} = 0$ and $f_{11}/(R_2-1)+f_{12} = \gamma_1(1-R_1)R_2/(R_2-1)$. Therefore, the fact that $f_{23} > 0$ leads to the following inequality:

$$\begin{aligned} &\frac{1}{R_2-1} - 2w_1 + \frac{2w_2}{R_2-1} \\ &= \frac{1}{R_2-1} + f_{23}\frac{2}{\gamma_2^2(R_2-1)(R_1-1)}\gamma_1(R_1-1)\frac{R_2}{R_2-1} > 0. \qquad \square \end{aligned}$$

7.4.4 The Role of the Threshold R_d

The appearance of the equilibrium E_2 depends on the magnitude of R_d. When $R_d < 1$, there are only two "legal" equilibria, E_0 and E_1. However, E_1 is unstable. A possible attractor could be E_0. We show numerically that E_0 is actually an attractor (the vector field is nonanalytic at the origin). The results are established here by showing that all trajectories approach the origin, as can be seen in Figure 7.8. The core population can be eradicated by intervention at two levels: by reducing recruitment into the core population (making $R_1 < 1$) or by managing to make $R_d < 1$ (when $R_1 > 1$). In the special case when $\gamma_1 = \gamma_2$, R_1 and R_d are identical. Hence, R_d is a measure of recruitment from the general population (same as R_1). Hence, it is natural to expect that $R_d < 1$ would lead to the extinction of the core.

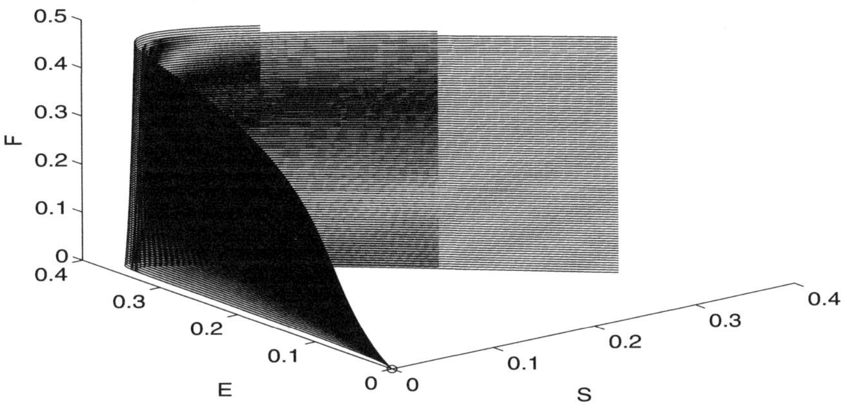

Figure 7.8. *The origin is an attractor when $R_d < 1$. All trajectories end up at the origin.*

7.5 Conclusions and Discussion

Table 7.1 summarizes the results of the mathematical analysis. From this summary, we see that the most effective approach for the eradication of an extreme ideology comes from sufficient efforts to limit recruitment into the core group. The control threshold is $R_1 = \beta_1/\gamma_1$. One way to bring this quantity below 1 is to reduce the value of β_1. This reduction corresponds to an increase in the resistance of the general population to "advances" from the core.

Table 7.1. *Summary of analytical results.*

Definitions	Thresholds	Attractors
$R_1 = \frac{\beta_1}{\gamma_1}$	$R_1 < 1$	$\{0, 0, 0\}$
$R_2 = \frac{\beta_2}{\gamma_2}$	$R_2 < 1 < R_1$	$\{1 - \frac{1}{R_1}, 0, 0\}$
$R_d = \frac{\nu R_1 R_2}{(R_2 - 1 + \nu)}$	$R_d < 1$	$\{0, 0, 0\}$
	$R_3 < 1$, $R_2 < 1$, and $R_1 < 1$	$\{0, 0, 0\}$
$R_3 = \frac{\beta_3}{\gamma_3}$	$R_3 < 1$ and $R_2 < 1 < R_1$	$\{1 - \frac{1}{R_1}, 0, 0\}$
	$R_3 < 1 < R_2$ and $R_d > 1$	$\{S^*, E^*, 0\}$
$R_0 =$	$R_0 < 1$	$\{0, 0, 0\}$ or $\{S^+, E^+, F^+\}$
$(R_2 - 1)(R_3 - 1)$	$R_0 > 1$	$\{S^+, E^+, F^+\}$

The elimination of the fanatic population is critically important. $R_3 = \beta_3/\gamma_3 < 1$ implies that the fanatic population will crash regardless of its size. Since the value of γ_3 is actually a tiny number or, equivalently, since the residence time $1/\gamma_1$ is long, then it is quite unlikely that R_3 could be made less than one. The existence of a homoclinic bifurcation (even when $R_3 < 1$) implies that, even though the core population is on its way to extinction, it can still experience, grow, and expand, in finite time, before it begins to decay.

The appearance of a backward bifurcation at $R_0 = 1$ implies that the elimination of the fanatic population with extreme ideology is very difficult. It is easy to see that a small number of individuals in the fanatic class (founder members) may successfully invade the general population. In this situation, it becomes extremely difficult to eliminate an established fanatic population because the threshold R_0 alone does not determine the fate of such a population (the coordinates of the turning point are the key).

The results in this chapter should not be taken as a prediction of reality. The development of ideologies is an extremely complex social process. At best, this model is only a useful metaphor that highlights the potential differences between "typical" contact and "conversion" contact social processes. As pointed out in the introduction, "extreme" fanatic behaviors are a particular (not always present) manifestation of insurgency movements (such as those in Bosnia, Guatemala, El Salvador, etc.). Popular support for such movements, given the cost to individuals and families, is not well understood. It could be easily argued that these movements are often important forces of change. Why do some movements use terror against civilians? Why would rational people join these groups? Here, we have introduced a simple model that at some level corroborates common sense, that is, the view that it is nearly impossible to eliminate extreme groups using within group intervention strategies. On the other hand, limiting the recruitment capabilities is effective (of course, we provide no explicit ways of implementing such efforts). Extensions of this model could be used to test alternative intervention strategies such as the use of competing (preselected) ideologies and the introduction of a menu of "multiple" (ideological) opportunities.

Finally, the issue of data to validate model predictions is a commonly asked question. The model predictions here are clear (and simplistic): The spread of extreme ideologies depends on initial conditions (multiple attractors), the spread (when it takes place) is extremely fast (groups become established on a fast time scale), within-group education efforts (intervention) are ineffective, recruitment is the most sensitive point of the structure, and resource reduction may reduce the size of the group but is unlikely to lead to its extinction. How can they be validated? Certainly, observations on the current situation in Afghanistan seem to fit the profile, but this is not validation. The collection of data to validate these predictions is not likely to generate "quality" data. However, the use of alternative models, such as individual (agent)–based models, may help us test the relevance of these predictions under reasonable accepted assumptions of individual behavior.

Acknowledgments

The work in this project arose from efforts to enhance the role of mathematics in homeland security issues via a workshop at DIMACS in the spring of 2002. We thank Fred Roberts, Martin Meltzer, Marcello Pagano, Karl Hadeler, Simon A. Levin, Ed Kaplan, Mac Hyman, Gerardo Chowell, Zhilan Feng, and Ira Schwartz for valuable discussions that indirectly impacted this work. In addition, we would like to acknowledge the input and suggestions provided by Carlos W. Castillo-Garsow, Miriam Nuno, and Erika Camacho. This work has been partially supported by NSF, NSA, and Sloan Foundation grants.

Bibliography

[1] F. Berezovskaya, G. Karev, and R. Arditi, *Parametric analysis of the ratio-dependent predator-prey model*, J. Math. Biol., 43 (2001), pp. 221–246

[2] F. Berezovsky, G. Karev, B. Song, and C. Castillo-Chavez, *Simple Epidemic Models with Surprising Dynamics*, manuscript, 2002.

[3] A. Berman and R. J. Plemmons, *Nonnegative Matrices in the Mathematical Sciences*, Academic Press, New York, 1970.

[4] R. Boyd and P. Richerson, *Culture and the Evolutionary Process*, University of Chicago Press, Chicago, London, 1976.

[5] C. Castillo-Chavez and F. Roberts, *Report on DIMACS Working Group Meeting: Mathematical Sciences Methods for the Study of the Deliberate Releases of Biological Agents and Their Consequences*, DIMACS Center, Rutgers University, Piscataway, NJ, 2002.

[6] C. Castillo-Chavez and R. H. Thieme, *Asymptotically autonomous epidemic models*, in Mathematical Population Dynamics: Analysis of Heterogeneity, O. Arimo, D. E. Axelrod, and M. Kimmel, eds., Wuerz Publishing, Winnipeg, MB, Canada, 1995, pp. 33–50.

[7] C. Castillo-Chavez and B. Song, *An overview of dynamical models of tuberculosis*, in Summer School on Mathematical Biology, A. Margheri, C. Rebelo, and F. Zanolin, eds., Centro Internacional de Matemática, Lisbon, accepted, 2002.

[8] D. Dennett, *Darwin's Dangerous Idea*, Simon and Schuster, New York, London, Toronto, Sydney, Tokyo, Singapore, 1995.

[9] J. Dushoff, W. Huang, and C. Castillo-Chavez, *Backward bifurcations and catastrophe in simple models of fatal diseases*, J. Math. Biol., 36 (1998), pp. 227–248.

[10] P. van den Driessche and J. Watmough, *Reproductive numbers and sub-threshold endemic equilibria for compartment models of disease transmission*, Math. Biosci., 180 (2002), pp. 183–201.

[11] L. L. Cavalli-Sforza and M. W. Feldman, *Cultural Transmission and Evolution: A Quantitative Approach*, Princeton University Press, Princeton, NJ, 1981.

[12] M. W. Feldman and L. L. Cavalli-Sforza, *On the theory of evolution under genetic and cultural transmission with application to the lactose absorption problem*, in Mathematical Evolutionary Theory, M. W. Feldman, ed., Princeton University Press, Princeton, NJ, 1989, pp. 145–173.

[13] K. P. Hadeler and C. Castillo-Chavez, *A core group model for disease transmission*, J. Math. Biosci., 128 (1995), pp. 41–55.

[14] H. W. Hethcote and J. A. Yorke, *Gonorrhea Transmission Dynamics and Control*, Lecture Notes in Biomath 56, Springer-Verlag, Berlin, 1984.

[15] W. Huang, K. Cooke, and C. Castillo-Chavez, *Stability and bifurcation for a multiple group model for the dynamics of HIV/AIDS transmission*, SIAM J. Appl. Math., 52 (1989), pp. 835–854.

[16] C. LUMSDEN AND E. O. WILSON, *Genes, Mind and Culture: The Coevolutionary Process*, Havard University Press, Cambridge, MA, 1981.

[17] B. SONG, *Dynamical Epidemical Models and Their Applications*, Ph.D. thesis, Cornell University, Ithaca, NY, 2002.

[18] R. H. THIEME, *Asymptotically autonomous differential equations in the plane*, Rocky Mountain J. Math., 24 (1994), pp. 351–380.

[19] Y. YE, *Theory of Limit Cycles*, Trans. Math. Monogr. 66, AMS, Providence, RI, 1986.

Chapter 8

An Epidemic Model with Virtual Mass Transportation: The Case of Smallpox in a Large City

Carlos Castillo-Chavez, Baojun Song,† and Juan Zhang‡*

8.1 Deliberate Releases of Biological Agents

Smallpox was eradicated in 1980, but two *official* repositories were left in the world: one at the Centers for Disease Control and Prevention (CDC) in Atlanta, US and the second at the State Research Center of Virology and Biotechnology in Koltsovo, Novosibirsk, Russia. US concerns about the actual number or location of sources of smallpox are due, in part, to the fact that the Soviet government began growing and stockpiling large quantities of smallpox, specially adapted for use in bombs and missiles, right after the elimination of the virus from human populations [54]. Hence, concerns about our nation's readiness in the event of the deliberate release of a biological agent, like smallpox, have been discussed prior to the tragic events of September 11, 2001. The response, at least at some level, as the probability of such an event seems to have increased. For example, there were about 50 million smallpox vaccine doses stockpiled worldwide before September 11. Now there are about 300 million in the US alone. The necessity for such a measure is obvious because smallpox kills about 30% of infected (unvaccinated) persons. The absence of specific treatment, combined with the loss of the world's population herd immunity, makes smallpox a possible weapon of terror [25].

Of course, smallpox is not the only concern. Past human epidemics and pandemics of influenza, combined with its high genetic variability, its potential genetic manipulation (engineering of new strains), and its past lethal impact on human and animal populations, bring influenza viruses to the forefront of candidates for deliberate releases [8, 9, 33, 36, 15, 14, 41, 46].

The First National Symposium on Medical and Public Health Response to Bioterrorism, held on February 16–17, 1999 in Arlington, Virginia, was motivated, in part, by the 1995 sarin gas attack by a Japanese religious cult, which actually attempted to aerosolize anthrax in the Tokyo area; the establishment of Russia's bioweapons research program; and the emi-

*Department of Biological Statistics and Computational Biology, Cornell University, Ithaca, NY 14853-7801, and Center for Nonlinear Studies, Los Alamos National Laboratory, Los Alamos, NM 87545..

†Department of Mathematical Sciences, Montclair State University, Upper Montclair, NJ 07043.

‡Department of Mathematics, Xi'an Jiaotong University, Xi'an, 710049, People's Republic of China.

gration of a large number of scientists after the crash of former Soviet Union. At this meeting, D. A. Henderson raised the following questions [26]: "What is the possible aftermath of an act of biological terrorism? Which biological threats warrant the most concern?"

Smallpox vaccination programs ceased more than 25 years ago, and consequently, the world's population lacks herd immunity [25]. History shows that a lack or reduced herd immunity can have serious consequences. A frightening example of the impact of a fatal disease in a population with no recent history of exposure to such a disease is easily detected from the data on malaria deaths after funding for control measures that had eliminated malaria for a decade was canceled (United Nations' Garki project). The devastating impact of smallpox and other communicable diseases on the Incas in Peru and the Aztecs in Mexico nearly five centuries ago has been well documented [54, 28, 25, 16].

Bioterrorism is now defined as the deliberate release of biological agents and their consequences [10]. This is not a novel activity. Smallpox was used as a weapon by British forces during the French and Indian Wars via the distribution of smallpox-infected blankets before cowpox vaccination against smallpox took practice in 1796 [43]. Biological weapons have been recently utilized by Iraq in its wars with Iran and local Kurds. The distribution of anthrax spores through the US mail in 2001 has shown that the type and number of channels for bioterrorist activities is so extensive that efforts to address the source of such attacks (individuals) must become a priority (see the contribution in this volume by Castillo-Chavez and Song).

In this chapter, we focus on release targets that include city subway systems or airport hubs. A deliberate release in New York City's subway system, which carries 4.3 million people weekly, could have catastrophic consequences.

8.2 Smallpox

Because we apply our framework to the situation of a deliberate release of smallpox in a large city, we review some of its epidemiology. Smallpox is a viral communicable disease that is passed from person to person either via the inhalation of air droplets or from aerosols expelled from the oropharynx of infected persons or by direct physical contact with infectious individuals. Transmission can occur whenever an individual is within a seven-foot radius of the infectious person; however, long-distance airborne transmission is also possible [31, 5, 44]. After exposure, infected individuals experience an incubation (latent/exposed) period which lasts between 7 and 17 days; the average is between 12 and 14 days. During the incubation period, individuals do not show symptoms nor do they feel sick. Twelve to fourteen days after infection, those infected typically become febrile, have severe aching pains and high fever (101–104 degress Fahrenheit), and often must stay in bed (prodrome phase). Infected persons are most contagious 2 to 3 days after the prodrome state; the period of infectiousness lasts about 4 days. Afterwards, a rash develops over the face and spreads to the extremities. This rash soon becomes vesicular and, later, pustular. The patient remains febrile throughout the evolution of the rash and customarily experiences considerable pain as the pustules grow and expand. Gradually, scabs form and eventually separate, leaving pitted scars. Death usually occurs during the second week [28].

8.3 Vaccination Strategies

There is no proven treatment for smallpox, but it can be prevented through vaccination. Receiving the vaccine before exposure or within 3–4 days after exposure will protect 97% of vaccinated individuals. Acquired immunity could last as long as 5 years. Vaccination throughout the US ceased 30 years ago, but it has been selectively restarted by the Bush administration (to medical emergency personnel and the military). Hence, most of the population is highly susceptible to a smallpox invasion.

During the last phases associated with the smallpox-eradication campaign headed by the World Health Organization, ring vaccination was the preferred (successful) policy. The CDC still seems to recommend this strategy in the event of a smallpox outbreak. This strategy, also referred to as traced vaccination (TV) [32], focuses on providing vaccinations exclusively to individuals who have had direct contact with those who have become infected. On October 25, 2002, following the US Senate Armed Services Committee's Subcommittee on Emerging Threats and Capabilities examination of the Dark Winter [51] scenarios and their implications for the government's response in the event of a bioterrorist attack on the US, Randy Larsen, director of the ANSER Institute for Homeland Security, asserted that vaccinating everyone (against smallpox) would not be the best option, given the high number of people in the US population with compromised immune systems (who likely would have strong adverse reactions to the vaccine). Larsen encouraged the stockpiling of vaccines and the training of medical professionals [48]. The paper *Smallpox: Right Topic, Wrong Debate* by Larsen supports the use of TV for small-scale attack and mass vaccination (MV) for large-scale attack [35]. The obvious question is, how do we establish a priori the scale of an attack? After computing the total deaths and cases resulting from the use of either MV or TV before and right after a deliberate release, E. Kaplan [32] concluded, with his model, that MV is the best approach. A. Kemper and colleagues (as well as Kaplan and collaborators) considered in their model the impact of serious adverse vaccine effects, including deaths. Their model favors MV [34]. In the same issue of the journal *Effective Clinical Practice* in which Kemper's article was published, J. Modlin responds [38], questioning the adverse reaction figures and number of deaths used by Kemper and collaborators (see "A Mass Smallpox Vaccination Campaign: Reasonable or Irresponsible?"). Meltzer and colleagues [37] construct a Markov chain model that incorporated both vaccination and quarantine. From his results, Meltzer concludes that it is enough to stockpile 40 million doses of smallpox, a figure that is consistent with those recommended by CDC and the Working Group on Civilian Biodefense in 1999 [25]. However, the US has stockpiled 286 million doses, including 155 million doses bought from a British firm [17]. Was this a reasonable decision? Meltzer's model assumes that the basic reproductive number for smallpox is approximately 3 (a value that is consistent with previous estimates [1]). Estimates of the basic reproductive number using data from particular European countries (in the 1960s and 1970s) support values as large as 10 and even 20 [25]. R. Gani and S. Leach argue that the value of this important number is likely to be between 3.5 and 6 [24].

The estimation of optimal vaccination strategies depend on many factors, including an accurate estimate of the value of the basic reproductive number. A sensible approach is to develop planning policies based on worst-case scenarios. The question of how to define or model worst-case scenario is challenging. In this volume, an effort to define this concept (and the challenges associated with its definition) are carried out by Chowell and Castillo-Chavez (Chapter 2).

8.4 Model Equations and Parameters

Smallpox and anthrax have been identified as some of the most likely biological agents to be used in a deliberate release [40]. This view is supported by the facts that they are easily aerosolized and generate high case fatality rates. The earliest mathematical work on smallpox is attributed to Daniel Bernoulli [3]. His main objective was to calculate the adjusted life table if smallpox were to be eliminated as a cause of death [20]. A series of models have been developed to explore the consequences of the use of smallpox as a biological agent [32, 37, 27, 26].

Smallpox is a deadly threat to civilian populations if used as a biological weapon because of its high case-fatality rate, the absence of specific treatment, and its targeting of highly susceptible and mobile subpopulations. Potential release targets include city mass transportation systems, airport hubs or terminals. For example, New York City (NYC), with a population of about 8 million that includes 4.3 million subway users during weekdays alone, is a likely target for a deliberate release.

The main purpose of this chapter is to expand the framework that we have developed for the spread of tuberculosis on a mass transportation system [7] to include the possibility of smallpox transmission. The focus is on disease dynamics and its control in a city with a widely used mass transportation system, like a subway.

The city is divided into N neighborhoods. Within each neighborhood the population is subdivided into subway users (SU) and non–subway users (NSU). It is assumed that SU have contacts with SU and NSU in their own neighborhoods. SU may also have contacts with SU from different neighborhoods when they share a subway ride. Contacts between SU individuals from different neighborhoods outside the subway are considered negligible (as a first-order approximation) for the purpose of disease transmission. It is also assumed that NSU have most of their contacts (that could lead to disease transmission) only within their own neighborhood.

If a fixed number of infected individuals is introduced in the subway, then the first cases of infection would occur in the SU population. Newly infected individuals will then take the virus back to their own neighborhoods, generating infections in the NSU and SU populations. Once the attack is recognized and smallpox is detected, vaccination starts.

Within each neighborhood, individuals fall into one of four classes according to their epidemiological status. S_i, E_i, I_i, and R_i denote the numbers of NSU in neighborhood i who are susceptible, exposed, infectious, and recovered, respectively; W_i, X_i, Y_i, and Z_i denote the corresponding epidemiological classes for SU individuals. The total population sizes of the two groups are $Q_i = S_i + E_i + I_i + R_i$ and $T_i = W_i + X_i + Y_i + Z_i$. The constants a_i and b_i denote the per-capita contact rates of NSU and SU in neighborhood i. In addition, $\omega_i = \rho_i/(\sigma_i + \rho_i)$ and $\tau_i = \sigma_i/(\sigma_i + \rho_i)$, where ρ_i and σ_i denote the rates at which the SUs get on and off the subway, respectively, represent the fractions of "contact" of time that an SU individual spends on or off the subway, respectively. If individuals mix according to the proportional mixing scheme [6, 2, 39, 30] then the mixing probabilities are given by the following formulas:

(1) $P_{a_i a_i} = \tilde{P}_{a_i} = \frac{a_i Q_i}{a_i Q_i + b_i \tau_i T_i}$ is the mixing probability between NSU from the same neighborhood i.

(2) $P_{a_i b_i} = \tilde{P}_{b_i} = \frac{b_i \tau_i T_i}{a_i Q_i + b_i \tau_i T_i}$ is the mixing probability of NSU and SU from the same neighborhood i.

(3) $P_{b_i a_i} = \bar{P}_{a_i} = \frac{a_i Q_i}{a_i Q_i + b_i \tau_i T_i} \tau_i$ is the mixing probability of SU and NSU from the same neighborhood i.

(4) $P_{b_i b_i} = \bar{P}_{b_i} = \frac{b_i \tau_i T_i}{a_i Q_i + b_i \tau_i T_i} \tau_i$ is the mixing probability between SU from the same neighborhood i.

(5) $P_{b_i b_j} = \bar{P}_{b_j^i} = \frac{b_j \omega_j T_j}{\sum_{k=1}^{N} b_k \omega_k T_k} \omega_i$ is the mixing probability between SU from neighborhoods i and j.

Table 8.1. *Definitions of parameters. i refers to the index of a neighborhood.*

Parameters	Definitions
Λ_i	recruitment rate of SU
A_i	recruitment rate of NSU
μ	natural mortality rate
d	mortality rate due to smallpox
q_i	per capita vaccination rate
l_1, l_2	vaccination efficacy in susceptible and exposed populations
ϕ	progression rate from latent to infectious
α	recovery rate
σ_i	the rate at which an SU leaves the subway
ρ_i	the rate at which an SU gets in the subway
a_i	average number of contacts of NSU per unit of time
b_i	average number of contacts of SU per unit of time
β_i	transmission rate per contact
$\frac{1}{\rho_i}$	the average time spent on the subway
$\frac{\sigma_i}{\sigma_i+\rho_i}$	the proportion of time spent off the subway (SU)
$\frac{\rho_i}{\sigma_i+\rho_i}$	the proportion of time spent on the subway (SU)

(6) $P_{a_i a_j} = 0$ means NSU from neighborhoods i and j do not have contacts assuming $i \neq j$.

(7) $P_{a_i b_j} = 0$ means NSU from neighborhood i and SU from neighborhood j have no contacts assuming $i \neq j$.

For each neighborhood, the following two "conditional probability" identities hold:

$$\tilde{P}_{a_i} + \tilde{P}_{b_i} = 1, \quad i = 1, 2, \ldots, N, \tag{8.1}$$

$$\bar{P}_{a_i} + \bar{P}_{b_i} + \sum_{j=1}^{N} \bar{P}_{b_j} = \tau_i + \omega_i = 1, \quad i = 1, 2, \ldots, N. \tag{8.2}$$

The model equations are given below (parameters are defined in Table 8.1):

$$\frac{dW_i}{dt} = \Lambda_i - V_i(t) - (\mu + q_i l_1) W_i, \tag{8.3}$$

$$\frac{dX_i}{dt} = V_i(t) - (\mu + \phi + q_i l_2) X_i, \tag{8.4}$$

$$\frac{dY_i}{dt} = \phi X_i - (\mu + \alpha + d) Y_i, \tag{8.5}$$

$$\frac{dZ_i}{dt} = \alpha Y_i - \mu Z_i + q_i l_1 W_i + q_i l_2 X_i, \tag{8.6}$$

$$\frac{dS_i}{dt} = A_i - B_i(t) - (\mu + q_i l_1) S_i, \tag{8.7}$$

$$\frac{dE_i}{dt} = B_i(t) - (\mu + \phi + q_i l_2)E_i, \tag{8.8}$$
$$\frac{dI_i}{dt} = \phi E_i - (\mu + \alpha + d)I_i, \tag{8.9}$$
$$\frac{dR_i}{dt} = \alpha I_i - \mu R_i + q_i l_1 S_i + q_i l_2 E_i, \quad i = 1, \ldots, N, \tag{8.10}$$

where the (force of) infection rate for NSU is

$$B_i(t) = \beta_i a_i S_i \left[\tilde{P}_{a_i} \frac{I_i}{T_i \tau_i + Q_i} + \tilde{P}_{b_i} \frac{Y_i \tau_i}{T_i \tau_i + Q_i} \right] \tag{8.11}$$

and the (force of) infection rate for SU is

$$V_i(t) = \beta_i b_i W_i \left[\bar{P}_{a_i} \frac{I_i}{T_i \tau_i + Q_i} + \bar{P}_{b_i} \frac{Y_i \tau_i}{T_i \tau_i + Q_i} + \sum_{j=1}^{N} \bar{P}_{b_j^i} \frac{Y_j \omega_j}{T_j \omega_j} \right] \tag{8.12}$$

with

$$Q_i(t) = S_i(t) + E_i(t) + I_i(t) + R_i(t),$$
$$T_i(t) = W_i(t) + X_i(t) + Y_i(t) + Z_i(t).$$

Incidence rates (force of infection) are modeled more or less in the "usual" way, but the addition of multiple neighborhoods and the classification (SU and NSU) complicate the expressions (see (8.12)). In addition, it is worth pointing out that we are ignoring the possible immigration of individuals in the E, I, or R classes.

Proportionate mixing [6, 2, 39, 30] has been used. In proportionate mixing, the probability that a type 1 individual has a contact with a type 2 individual, given that the type 1 individual has had a contact, is equal to the *weighted* proportion of type 2 individuals' activity in the total population (that is, it is independent of type 1 individuals). Here, SU do not spend their entire time in their "home" neighborhood, and consequently, some modifications are required via the appropriate budgeting of SU contacts. For example, $P_{a_i a_i} = \tilde{P}_{a_i} = \frac{a_i Q_i}{a_i Q_i + b_i \tau_i T_i}$ gives the mixing probability between NSU from the same neighborhood i. The numerator $a_i Q_i$ is the average activity of NSU, while the denominator $a_i Q_i + b_i \tau_i T_i$ gives the average total activities in neighborhood i (SU within neighborhood activity has to be weighted by the additional factor τ_i).

8.5 The Basic Reproductive Number

Here, we consider the case where there are two "neighborhoods" or subpopulations. The first includes regular city residents while the second is made of temporary residents, that is, tourists. The computation of the basic reproductive number for $N = 2$ follows below.

At the disease-free steady state, the total population size for the ith neighborhood is $\frac{\Lambda_i + A_i}{\mu}$, of which $T_i^0 = \frac{\Lambda_i}{\mu}$ are SU while $Q_i^0 = \frac{A_i}{\mu}$ are NSU. The disease-free equilibrium of system (8.3)–(8.10) is $(w_i, 0, 0, z_i, s_i, 0, 0, r_i)$ with

$$w_i = T_i^0 \frac{\mu}{\mu + q_i l_1}, \qquad z_i = T_i^0 \frac{q_i l_1}{\mu + q_i l_1},$$
$$s_i = Q_i^0 \frac{\mu}{\mu + q_i l_1}, \qquad r_i = Q_i^0 \frac{q_i l_1}{\mu + q_i l_1},$$

and, as it is customary, the local asymptotic stability of this equilibrium is determined by the basic reproductive number.

The disease-free equilibrium is the demographic steady state with the following mixing probabilities (the superscript 0 represents the demographic steady state):

$$\bar{P}^0_{b^1_1} = \frac{b_1\omega_1\Lambda_1}{\sum_{j=1}^N b_j\omega_j\Lambda_j}\omega_1, \tag{8.13}$$

$$\bar{P}^0_{b^1_2} = \frac{b_2\omega_2\Lambda_2}{\sum_{j=1}^N b_j\omega_j\Lambda_j}\omega_1, \tag{8.14}$$

$$\bar{P}^0_{b^2_1} = \frac{b_1\omega_1\Lambda_1}{\sum_{j=1}^N b_j\omega_j\Lambda_j}\omega_2, \tag{8.15}$$

$$\bar{P}^0_{b^2_2} = \frac{b_2\omega_2\Lambda_2}{\sum_{j=1}^N b_j\omega_j\Lambda_j}\omega_2, \tag{8.16}$$

$$\tilde{P}^0_{a_i} = \frac{a_i A_i}{a_i A_i + b_i\tau_i\Lambda_i}, \tag{8.17}$$

$$\tilde{P}^0_{b_i} = \frac{b_i\tau_i\Lambda_i}{a_i A_i + b_i\tau_i\Lambda_i}, \tag{8.18}$$

$$\bar{P}^0_{a_i} = \tilde{P}^0_{a_i}\tau_i, \tag{8.19}$$

$$\bar{P}^0_{b_i} = \tilde{P}^0_{b_i}\tau_i. \tag{8.20}$$

Following the next generation operator approach [11, 18, 19], we linearize the equations of $X_1, Y_1, E_1, I_1, X_2, Y_2, E_2, I_2$ around the disease-free equilibrium. The Jacobian matrix has dimension of 8 by 8, and it is denoted by J.

Table 8.2. *Nonzero entries of the Jacobian matrix J.*

$J_{11} = -(\mu + \phi + q_1 l_2)$	$J_{12} = \beta_1 b_1 (\bar{P}^0_{b_1} \frac{\tau_1 w_1}{Q^0_1 + \tau_1 T^0_1} + \bar{P}^0_{b^1_1} \frac{w_1}{T^0_1})$
$J_{14} = \beta_1 b_1 \bar{P}^0_{a_1} \frac{w_1}{Q^0_1 + \tau_1 T^0_1}$	$J_{21} = J_{43} = J_{65} = J_{87} = \phi$
$J_{16} = \beta_1 b_1 \bar{P}^0_{b^1_2} \frac{w_1}{T^0_2}$	$J_{22} = J_{44} = J_{66} = J_{88} = -(\mu + d + \alpha)$
$J_{32} = \beta_1 a_1 \tilde{P}^0_{b_1} \frac{\tau_1 s_1}{Q^0_1 + \tau_1 T^0_1}$	$J_{33} = -(\mu + \phi + q_1 l_2)$
$J_{34} = \beta_1 a_1 \tilde{P}^0_{a_1} \frac{s_1}{Q^0_1 + \tau_1 T^0_1}$	$J_{52} = \beta_2 b_2 \bar{P}^0_{b^2_1} \frac{w_2}{T^0_1}$
$J_{55} = -(\mu + \phi + q_2 l_2)$	$J_{56} = \beta_2 b_2 \left(\bar{P}^0_{b_2} \frac{\tau_2 w_2}{Q^0_2 + \tau_2 T^0_2} + \bar{P}^0_{b^2_2} \frac{w_2}{T^0_2}\right)$
$J_{58} = \beta_2 b_2 \bar{P}^0_{a_2} \frac{w_2}{Q^0_2 + \tau_2 T^0_2}$	$J_{76} = \beta_2 a_2 \tilde{P}^0_{b_2} \frac{\tau_2 s_2}{Q^0_2 + \tau_2 T^0_2}$
$J_{77} = -(\mu + \phi + q_2 l_2)$	$J_{78} = \beta_2 a_2 \tilde{P}^0_{a_2} \frac{s_2}{Q^0_2 + \tau_2 T^0_2}$

The nonzero entries in J are listed in Table 8.2. J has the form of $\begin{pmatrix} A & B \\ C & D \end{pmatrix}$, where

$$A = \begin{pmatrix} J_{11} & J_{12} & 0 & J_{14} \\ J_{21} & J_{22} & 0 & 0 \\ 0 & J_{32} & J_{33} & J_{34} \\ 0 & 0 & J_{43} & J_{44} \end{pmatrix}, \qquad B = \begin{pmatrix} 0 & J_{16} & 0 & 0 \\ 0 & 0 & 0 & 0 \\ 0 & 0 & 0 & 0 \\ 0 & 0 & 0 & 0 \end{pmatrix},$$

$$C = \begin{pmatrix} 0 & J_{52} & 0 & 0 \\ 0 & 0 & 0 & 0 \\ 0 & 0 & 0 & 0 \\ 0 & 0 & 0 & 0 \end{pmatrix}, \qquad D = \begin{pmatrix} J_{55} & J_{56} & 0 & J_{58} \\ J_{65} & J_{66} & 0 & 0 \\ 0 & J_{76} & J_{77} & J_{78} \\ 0 & 0 & J_{87} & J_{88} \end{pmatrix}.$$

We write J as $J = M - M_d$, where $M_d = \operatorname{diag}\{-J_{ii}\}$ and $M = J + M_d$. The dominant eigenvalue of MM_d^{-1} gives the basic reproductive number. Since MM_d^{-1} has the same form as J, we use Laplace's formula to expand the characteristic equation $\det(MM_d^{-1} - \lambda I) = 0$. If $f(\lambda)$ denotes the characteristic equation, then $f(\lambda) = f_A(\lambda)f_D(\lambda) + f_B(\lambda)f_C(\lambda)$, where

$$f_A(\lambda) = \begin{vmatrix} -\lambda & -\frac{J_{12}}{J_{22}} & 0 & -\frac{J_{14}}{J_{44}} \\ -\frac{J_{21}}{J_{11}} & -\lambda & 0 & 0 \\ 0 & -\frac{J_{32}}{J_{22}} & -\lambda & -\frac{J_{34}}{J_{44}} \\ 0 & 0 & -\frac{J_{43}}{J_{33}} & -\lambda \end{vmatrix}, \qquad f_D(\lambda) = \begin{vmatrix} -\lambda & -\frac{J_{56}}{J_{66}} & 0 & -\frac{J_{58}}{J_{88}} \\ -\frac{J_{65}}{J_{55}} & -\lambda & 0 & 0 \\ 0 & -\frac{J_{76}}{J_{66}} & -\lambda & -\frac{J_{78}}{J_{88}} \\ 0 & 0 & -\frac{J_{87}}{J_{77}} & -\lambda \end{vmatrix},$$

$$f_C(\lambda) = \begin{vmatrix} -\frac{J_{21}}{J_{11}} & -\lambda & 0 & 0 \\ 0 & -\frac{J_{32}}{J_{22}} & -\lambda & -\frac{J_{34}}{J_{44}} \\ 0 & 0 & -\frac{J_{43}}{J_{33}} & -\lambda \\ 0 & -\frac{J_{52}}{J_{22}} & 0 & 0 \end{vmatrix}, \qquad f_B(\lambda) = \begin{vmatrix} 0 & -\frac{J_{16}}{J_{66}} & 0 & 0 \\ -\frac{J_{65}}{J_{55}} & -\lambda & 0 & 0 \\ 0 & -\frac{J_{76}}{J_{66}} & -\lambda & -\frac{J_{78}}{J_{88}} \\ 0 & 0 & -\frac{J_{87}}{J_{77}} & -\lambda \end{vmatrix}.$$

We rewrite $f_A(\lambda)$ as

$$f_A(\lambda) = \lambda^2(\lambda^2 - R_{1,1}^{n,n,L_1}) - (R_{1,1}^{u,u,L_1} + R_{1,1}^{u,u,L_2})(\lambda^2 - R_{1,1}^{n,n,L_1}) - R_{1,1}^{n,u,L_1}R_{1,1}^{u,n,L_1},$$

where

$$R_{1,1}^{n,n,L_1} = \frac{1}{\mu + d + \alpha}\frac{\phi}{\mu + \phi + q_1 l_2}\beta_1 a_1 \tilde{P}_{a_1}^0 \frac{s_1}{Q_1^0 + \tau_1 T_1^0}.$$

$R_{1,1}^{n,n,L_1}$ are the new cases generated from NSU by NSU within their neighborhoods;

$$R_{1,1}^{u,u,L_1} = \frac{1}{\mu + d + \alpha}\frac{\phi}{\mu + \phi + q_1 l_2}\beta_1 b_1 \tilde{P}_{b_1}^0 \frac{w_1 \tau_1}{Q_1^0 + \tau_1 T_1^0}$$

are the new cases generated from SU by SU within their neighborhoods;

$$R_{1,1}^{u,u,L_2} = \frac{1}{\mu + d + \alpha}\frac{\phi}{\mu + \phi + q_1 l_2}\beta_1 b_1 \tilde{P}_{b_1^1}^0 \frac{w_1}{T_1^0}$$

are the new cases generated from SU by SU when they meet on the subway;

$$R_{1,1}^{n,u,L_1} = \frac{1}{\mu + d + \alpha}\frac{\phi}{\mu + \phi + q_1 l_2}\beta_1 a_1 \tilde{P}_{b_1}^0 \frac{\tau_1 s_1}{Q_1^0 + \tau_1 T_1^0}$$

and

$$R_{1,1}^{u,n,L_1} = \frac{1}{\mu + d + \alpha}\frac{\phi}{\mu + \phi + q_1 l_2}\beta_1 b_1 \tilde{P}_{a_1}^0 \frac{w_1}{Q_1^0 + \tau_1 T_1^0},$$

which does not have a clear interpretation. However, the last two expressions will eventually disappear (see (8.34)–(8.36)). The superscript L_1 means that contact takes place off the subway; L_2 means that contact takes place on the subway.

In order to facilitate our understanding, we let $R_{i,j}^{\alpha,\beta,\gamma}$ denote the number of expected new cases of different type of individuals at the corresponding locations. Here α and β denote the susceptible and infectious types, respectively; i indexes the neighborhood for susceptible and j for infectious individuals; and γ represents the place where α-individuals contact β-individuals. Hence, for example, $R_{1,1}^{n,n,L_1}$ gives the expected number of new cases of NSU generated by NSU within the first neighborhood; $R_{2,2}^{n,u,L_1}$ gives the expected number of new cases of NSU generated by SU within the second neighborhood; $R_{2,2}^{u,n,L_1}$ gives the expected number of new cases of SU generated by NSU within the second neighborhood; $R_{1,2}^{u,u,L_2}$ denotes the expected number of new cases of SU from neighborhood 1 generated by SU from neighborhood 2 when they meet on the subway. All relevant $R_{i,j}^{\alpha,\beta,\gamma}$ are listed below:

$$R_{1,1}^{n,n,L_1} = \frac{1}{\mu+d+\alpha}\frac{\phi}{\mu+\phi+q_1l_2}\beta_1 a_1 \tilde{P}_{a_1}^0 \frac{s_1}{Q_1^0+\tau_1 T_1^0}, \tag{8.21}$$

$$R_{1,1}^{u,u,L_1} = \frac{1}{\mu+d+\alpha}\frac{\phi}{\mu+\phi+q_1l_2}\beta_1 b_1 \bar{P}_{b_1}^0 \frac{w_1\tau_1}{Q_1^0+\tau_1 T_1^0}, \tag{8.22}$$

$$R_{1,1}^{u,u,L_2} = \frac{1}{\mu+d+\alpha}\frac{\phi}{\mu+\phi+q_1l_2}\beta_1 b_1 \bar{P}_{b_1^1}^0 \frac{w_1}{T_1^0}, \tag{8.23}$$

$$R_{1,1}^{n,u,L_1} = \frac{1}{\mu+d+\alpha}\frac{\phi}{\mu+\phi+q_1l_2}\beta_1 a_1 \tilde{P}_{b_1}^0 \frac{\tau_1 s_1}{Q_1^0+\tau_1 T_1^0}, \tag{8.24}$$

$$R_{1,1}^{u,n,L_1} = \frac{1}{\mu+d+\alpha}\frac{\phi}{\mu+\phi+q_1l_2}\beta_1 b_1 \bar{P}_{a_1}^0 \frac{w_1}{Q_1^0+\tau_1 T_1^0}, \tag{8.25}$$

$$R_{2,2}^{n,n,L_1} = \frac{1}{\mu+d+\alpha}\frac{\phi}{\mu+\phi+q_2l_2}\beta_2 a_2 \tilde{P}_{a_2}^0 \frac{s_2}{Q_2^0+\tau_2 T_2^0}, \tag{8.26}$$

$$R_{2,2}^{u,u,L_1} = \frac{1}{\mu+d+\alpha}\frac{\phi}{\mu+\phi+q_2l_2}\beta_2 b_2 \bar{P}_{b_2}^0 \frac{w_2\tau_2}{Q_2^0+\tau_2 T_2^0}, \tag{8.27}$$

$$R_{2,2}^{u,u,L_2} = \frac{1}{\mu+d+\alpha}\frac{\phi}{\mu+\phi+q_2l_2}\beta_2 b_2 \bar{P}_{b_2^2}^0 \frac{w_2}{T_2^0}, \tag{8.28}$$

$$R_{2,2}^{n,u,L_1} = \frac{1}{\mu+d+\alpha}\frac{\phi}{\mu+\phi+q_2l_2}\beta_2 a_2 \tilde{P}_{b_2}^0 \frac{\tau_2 s_2}{Q_2^0+\tau_2 T_2^0}, \tag{8.29}$$

$$R_{2,2}^{u,n,L_1} = \frac{1}{\mu+d+\alpha}\frac{\phi}{\mu+\phi+q_2l_2}\beta_2 b_2 \bar{P}_{a_2}^0 \frac{w_2}{Q_2^0+\tau_2 T_2^0}, \tag{8.30}$$

$$R_{2,1,}^{u,u,L_2} = \frac{1}{\mu+d+\alpha}\frac{\phi}{\mu+\phi+q_2l_2}\beta_2 b_2 \bar{P}_{b_2^1}^0 \frac{w_2}{T_2^0}, \tag{8.31}$$

$$R_{1,2}^{u,u,L_2} = \frac{1}{\mu+d+\alpha}\frac{\phi}{\mu+\phi+q_1l_2}\beta_1 b_1 \bar{P}_{b_1^2}^0 \frac{w_1}{T_1^0}. \tag{8.32}$$

Using the above definitions we rewrite $f_D(\lambda)$, $f_B(\lambda)$, and $f_C(\lambda)$ as

$$\begin{aligned}
f_D(\lambda) &= \lambda^2(\lambda^2 - R_{2,2}^{n,n,L_1}) - (R_{2,2}^{u,u,L_1} + R_{2,2}^{u,u,L_2})(\lambda^2 - R_{2,2}^{n,n,L_1}) - R_{2,2}^{n,u,L_1} R_{2,2}^{u,n,L_1}, \\
f_B(\lambda) &= -\frac{\beta_1 b_1 w_1}{\beta_2 b_2 w_2} R_{2,1}^{u,u,L_2}(\lambda^2 - R_{2,2}^{n,n,L_1}), \\
f_C(\lambda) &= \frac{\beta_2 b_2 w_2}{\beta_1 b_1 w_1} R_{1,2}^{u,u,L_2}(\lambda^2 - R_{1,1}^{n,n,L_1}).
\end{aligned}$$

Hence,

$$\begin{aligned} f(\lambda) &= f_A(\lambda) f_D(\lambda) + f_B(\lambda) f_C(\lambda) \\ &= (\lambda^2(\lambda^2 - R_{1,1}^{n,n,L_1}) - (R_{1,1}^{u,u,L_1} + R_{1,1}^{u,u,L_2})(\lambda^2 - R_{1,1}^{n,n,L_1}) - R_{1,1}^{n,u,L_1} R_{1,1}^{u,n,L_1}) \\ &\quad (\lambda^2(\lambda^2 - R_{2,2}^{n,n,L_1}) - (R_{2,2}^{u,u,L_1} + R_{2,2}^{u,u,L_2})(\lambda^2 - R_{2,2}^{n,n,L_1}) - R_{2,2}^{n,u,L_1} R_{2,2}^{u,n,L_1}) \\ &\quad - (R_{1,2}^{u,u,L_2}(\lambda^2 - R_{1,1}^{n,n,L_1}))(R_{2,1}^{u,u,L_2}(\lambda^2 - R_{2,2}^{n,n,L_1})). \end{aligned} \tag{8.33}$$

We claim that

$$R_{1,1}^{u,u,L_1} R_{1,1}^{n,n,L_1} - R_{1,1}^{n,u,L_1} R_{1,1}^{u,n,L_1} = 0, \tag{8.34}$$

$$R_{2,2}^{u,u,L_1} R_{2,2}^{n,n,L_1} - R_{2,2}^{n,u,L_1} R_{2,2}^{u,n,L_1} = 0, \tag{8.35}$$

$$R_{1,1}^{u,u,L_2} R_{2,2}^{u,u,L_2} - R_{1,2}^{u,u,L_2} R_{2,1}^{u,u,L_2} = 0. \tag{8.36}$$

In fact, substituting (8.21), (8.23), (8.24), and (8.25) into (8.34), we obtain

$$\begin{aligned} &R_{1,1}^{u,u,L_1} R_{1,1}^{n,n,L_1} - R_{1,1}^{n,u,L_1} R_{1,1}^{u,n,L_1} \\ &= \frac{1}{(\mu+d+\alpha)^2} \frac{\beta_1 a_1 \phi}{\mu+\phi+q_1 l_2} \frac{\beta_1 b_1 \phi}{\mu+\phi+q_1 l_2} \frac{w_1 s_1}{(Q_1^0+\tau_1 T_1^0)^2} (\tilde{P}_{a_1}^0 \tau_1 \bar{P}_{b_1}^0 - \tilde{P}_{b_1}^0 \tau_1 \bar{P}_{a_1}^0). \end{aligned}$$

Expressions (8.19) and (8.20) imply that $\tilde{P}_{a_1}^0 \tau_1 \bar{P}_{b_1}^0 - \tilde{P}_{b_1}^0 \tau_1 \bar{P}_{a_1}^0 = 0$. Similar arguments show $R_{2,2}^{u,u,L_1} R_{2,2}^{n,n,L_1} - R_{2,2}^{n,u,L_1} R_{2,2}^{u,n,L_1} = 0$. Substituting (8.23), (8.28), (8.32), and (8.31) into (8.36) leads to

$$\begin{aligned} &R_{1,1}^{u,u,L_2} R_{2,2}^{u,u,L_2} - R_{1,2}^{u,u,L_2} R_{2,1}^{u,u,L_2} \\ &= \frac{\beta_1 b_1 \beta_2 b_2}{(\mu+d+\alpha)^2} \frac{\phi}{\mu+\phi+q_1 l_2} \frac{\phi}{\mu+\phi+q_2 l_2} \frac{w_1 w_2}{T_1^0 T_2^0} (\bar{P}_{b_1^1}^0 \bar{P}_{b_2^2}^0 - \bar{P}_{b_2^1}^0 \bar{P}_{b_1^2}^0). \end{aligned}$$

It follows from (8.13), (8.14), (8.15), and (8.16) that (8.36) holds.

Actually, (8.36) and the vanishing of the constant term in (8.33) are equivalent since the constant term is $R_{1,1}^{n,n,L_1} R_{2,2}^{n,n,L_1}(R_{1,1}^{u,u,L_2} R_{2,2}^{u,u,L_2} - R_{1,2}^{u,u,L_2} R_{2,1}^{u,u,L_2})$. Hence, $\lambda = 0$ is a root of the characteristic equation. Using (8.34)–(8.35) and the fact that 0 is a root lead us to the following cubic equation for $x = \lambda^2$:

$$\begin{aligned} &x^3 - \sum_{i=1}^{2} (R_{i,i}^{n,n,L_1} + R_{i,i}^{u,u,L_1} + R_{i,i}^{u,u,L_2}) x^2 \\ &+ \left(\sum_{i=1}^{2} (R_{i,i}^{n,n,L_1} R_{i,i}^{u,u,L_2}) + \prod_{i=1}^{2} (R_{i,i}^{n,n,L_1} + R_{i,i}^{u,u,L_1} + R_{i,i}^{u,u,L_2}) - R_{1,2}^{u,u,L_2} R_{2,1}^{u,u,L_2} \right) x \\ &+ R_{1,2}^{u,u,L_2} R_{2,1}^{u,u,L_2} (R_{1,1}^{n,n,L_1} + R_{2,2}^{n,n,L_1}) - R_{2,2}^{n,n,L_1} R_{2,2}^{u,u,L_2} (R_{1,1}^{n,n,L_1} + R_{1,1}^{u,u,L_1} + R_{1,1}^{u,u,L_2}) \\ &- R_{1,1}^{n,n,L_1} R_{1,1}^{u,u,L_2} (R_{2,2}^{n,n,L_1} + R_{2,2}^{u,u,L_1} + R_{2,2}^{u,u,L_2}), \end{aligned}$$

or

$$x^3 - ax^2 + bx + c = 0, \tag{8.37}$$

where

$$a = \sum_{i=1}^{2}(R_{i,i}^{n,n,L_1} + R_{i,i}^{u,u,L_1} + R_{i,i}^{u,u,L_2}),$$

$$b = \sum_{i=1}^{2}(R_{i,i}^{n,n,L_1} R_{i,i}^{u,u,L_2}) + \prod_{i=1}^{2}(R_{i,i}^{n,n,L_1} + R_{i,i}^{u,u,L_1} + R_{i,i}^{u,u,L_2}) - R_{1,2}^{u,u,L_2} R_{2,1}^{u,u,L_2},$$

$$\begin{aligned} c = {} & R_{1,2}^{u,u,L_2} R_{2,1}^{u,u,L_2}(R_{1,1}^{n,n,L_1} + R_{2,2}^{n,n,L_1}) - R_{2,2}^{n,n,L_1} R_{2,2}^{u,u,L_2}(R_{1,1}^{n,n,L_1} + R_{1,1}^{u,u,L_1} + R_{1,1}^{u,u,L_2}) \\ & - R_{1,1}^{n,n,L_1} R_{1,1}^{u,u,L_2}(R_{2,2}^{n,n,L_1} + R_{2,2}^{u,u,L_1} + R_{2,2}^{u,u,L_2}). \end{aligned}$$

The dominant root of the cubic equation (8.37) gives the basic reproductive number [47]. That is,

$$\begin{aligned} \mathcal{R}_0 = {} & \frac{a}{3} + \sqrt[3]{\frac{2a^3 - 9ab - 27c}{54} + \sqrt{\left(\frac{3b - a^2}{9}\right)^3 + \left(\frac{2a^3 - 9ab - 27c}{54}\right)^2}} \\ & + \sqrt[3]{\frac{2a^3 - 9ab - 27c}{54} - \sqrt{\left(\frac{3b - a^2}{9}\right)^3 + \left(\frac{2a^3 - 9ab - 27c}{54}\right)^2}}. \end{aligned} \tag{8.38}$$

Since contacts between individuals are not independent, the average appearing in the formula in (8.38) is not given by a simple arithmetic or geometric average. A look at some special cases may be therefore helpful in understanding this *important* formula.

Case 1. Letting $\omega_2 = 0$ implies that $R_{1,2}^{u,u,L_2} = R_{2,2}^{u,u,L_2} = R_{2,1}^{u,u,L_2} = 0$. This assumption reduces (8.37) to

$$(x - (R_{2,2}^{n,n,L_1} + R_{2,2}^{u,u,L_1}))(x^2 - (R_{1,1}^{n,n,L_1} + R_{1,1}^{u,u,L_1} + R_{1,1}^{u,u,L_2})x + R_{1,1}^{n,n,L_1} R_{1,1}^{u,u,L_2}) = 0,$$

which has the following positive roots:

$$\begin{aligned} \lambda_1 = {} & \frac{1}{2}(R_{1,1}^{n,n,L_1} + R_{1,1}^{u,u,L_1} + R_{1,1}^{u,u,L_2}) \\ & + \frac{1}{2}\sqrt{(R_{1,1}^{n,n,L_1} + R_{1,1}^{u,u,L_1} + R_{1,1}^{u,u,L_2})^2 - 4R_{1,1}^{n,n,L_1} R_{1,1}^{u,u,L_2}}, \\ \lambda_2 = {} & R_{2,2}^{n,n,L_1} + R_{2,2}^{u,u,L_1}. \end{aligned}$$

Hence, $\mathcal{R}_0 = \max\{\lambda_1, \lambda_2\}$, while $\omega_1 = \omega_2 = 0$ gives

$$\mathcal{R}_0 = \max\{R_{1,1}^{n,n,L_1} + R_{1,1}^{u,u,L_1}, R_{2,2}^{n,n,L_1} + R_{2,2}^{u,u,L_1}\}.$$

Case 2. Only one neighborhood is included (N=1). The basic reproductive number is the dominant root of $f_A(\lambda) = 0$, that is,

$$\begin{aligned} \mathcal{R}_0 = {} & \frac{R_{1,1}^{n,n,L_1} + R_{1,1}^{u,u,L_1} + R_{1,1}^{u,u,L_2}}{2} \\ & + \frac{1}{2}\sqrt{(R_{1,1}^{n,n,L_1} + R_{1,1}^{u,u,L_1} + R_{1,1}^{u,u,L_2})^2 - 4R_{1,1}^{n,n,L_1} R_{1,1}^{u,u,L_2}}. \end{aligned}$$

Case 3. In the extreme case where N=1 and ω_1=0, the basic reproductive number reduces to $\mathcal{R}_0 = \frac{1}{2}(R_{1,1}^{n,n,L_1} + R_{1,1}^{u,u,L_1})$.

8.6 Parameter Values

We apply our model to a city that has some of the characteristics of NYC. First, we consider two "neighborhoods" or subpopulations; the first is composed of long-term residents (neighborhood 1) while the second is made up of tourists (neighborhood 2). The resident population of NYC is composed of about 8 million individuals, while the tourist population needs to be estimated. In 1999 there were 36.7 million tourists: 30.1 domestic and 6.6 international visitors [50]. Furthermore, it is known that 18.4 million visitors stayed overnight. On the other hand, 18.3 million were classified as daytrippers. In order to carry out "specific" simulations, we *assume* that international visitors stayed in NYC an average of 5 days, while domestic overnight visitors stayed only for 2 days. Hence, average number of visitors per day is $(18.3 + 11.8 \times 2 + 6.6 \times 5)/365 = 0.2$ million. That is, on a "typical" day there may be 0.2 million tourists in NYC. The implementation of our model uses days as the unit of time.

Rough estimations of most parameters are made based on the "general" knowledge of smallpox parameters and relevant "demographic" data of NYC. For instance, it is known that 30% of infections are likely to result in deaths (typically during the second week following infection). It is also known that once a person has been vaccinated, even if the vaccine was applied 30 years ago, she/he is more likely to survive a smallpox infection. Since about half of Americans were vaccinated about 30 years ago, we adjust the average case-fatality rate to be 0.15 (an exponential distribution is assumed for the times of death due to smallpox infection). We therefore estimate $d \approx -\ln(0.85)/14 = 0.0116$. Estimates for the rest of the parameters are listed in Table 8.3. Most parameter values were taken from those cited or used in [32, 37, 23, 24].

Table 8.3. *Estimation of parameters.*

μ	d	l_1	l_2	ϕ	α	β	a_1	a_2	b_1	b_2
0.033	0.0116	0.97	0.3	0.086	0.086	0.5	5	15	10	30

The average contact rate for NSU is estimated to be around 5. This estimate is based on estimates of the total effective contact number of 50 for the entire period of infection [23]. It is assumed that SU are more active than NSU (in terms of contacts that may lead to infection). Here, their contact rate is assumed to be twice that of NSU. The average daily contact rates for the second neighborhood (tourists) is assumed to be greater than those of residents. For the purposes of illustration, it is assumed that the contact rates of tourists are three times as large as those of residents. The demographics Λ_i, A_i, and μ are chosen to ensure that the resident and tourist populations are 8 million and 0.2 million, respectively. The selections of $\tau_1 = 0.6$ and $\tau_2 = 0.1$ come from our assumption that tourists are assumed to spend more time on the subway than residents do. It is also assumed that residents spend most of their time within their own neighborhood (that is, in noncrowded areas). It follows from our selection of parameters that the basic reproductive number (in the absence of vaccination) is 5.8 (see Figure 8.1; that is, $R_0(0, 0) = 5.8$).

Naturally, we are interested in the impact of the two key parameters, q_1 and q_2, the vaccination rates for the resident and tourist populations, respectively. They are control parameters. Hence, we would like to find out their "best" or "optimal" values, that is, the values of q_1 and q_2 that minimize disease prevalence and total deaths.

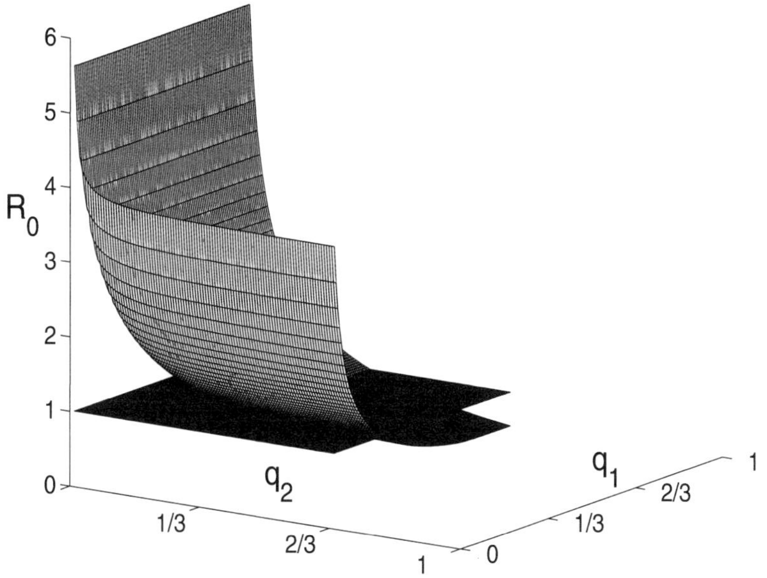

Figure 8.1. *Plot of R_0 versus q_1 and q_2.*

The initial conditions are set as follows: it is assumed that smallpox is released in the subway. Hence, initially $Y_1(0)$ and $Y_2(0)$ are positive. We let $Y_1(0) = 70$ and $Y_2(0) = 30$ (that is, we assume that there are 100 persons infected initially on the subway). The initial values of W_1, S_1, W_2, and S_2 are chosen to satisfy $W_1(0) + S_1(0) = 8,000,000$ and $W_2(0) + S_2(0) = 200,000$. The rest of initial values are set to zero, that is, $I_1(0) = I_2(0) = X_1(0) = X_2(0) = Z_1(0) = Z_2(0) = R_1(0) = R_2(0) = 0$.

8.7 Results and Discussions

Preliminary simulations support mass vaccination (MV) as the "best" strategy for NYC. The last smallpox outbreak in NYC (1947) is believed to have been started by 8 cases. (Six million people were vaccinated (almost everyone) [23].) Our initial introduction of 100 infections is from the assumptions of a deliberate release. Hence, model simulations agree with R. Larsen's recommendations in the event of a large-scale outbreak [35].

We looked at the impact of varying q_1 and q_2 on the basic reproductive number, R_0. Figure 8.1 gives a plot of $R_0(q_1, q_2)$ as a function of q_1 and q_2. The regions where $R_0(q_1, q_2) < 1$, $R_0(q_1, q_2) = 1$ and $R_0(q_1, q_2) > 1$ are marked in Figure 8.2 (the contour of $R_0(q_1, q_2) = 1$ from Figure 8.1). These graphs show that the vaccination rate for the residence population needs to be greater than 0.46. The tourist population is less important (for a city outbreak) than the resident population (its vaccination rate could be as low as 0.26). However, it is worth noticing that the tourist population cannot be ignored. For example, a vaccination rate for the tourist population that is less than 0.24 generates a basic reproductive number that is greater than 1. Furthermore, the fact that the world is highly connected suggests that a major outbreak in a city like NYC may generate a policy of MV for the whole country. Smallpox vaccination campaigns cannot ignore the global impact of nonresident populations.

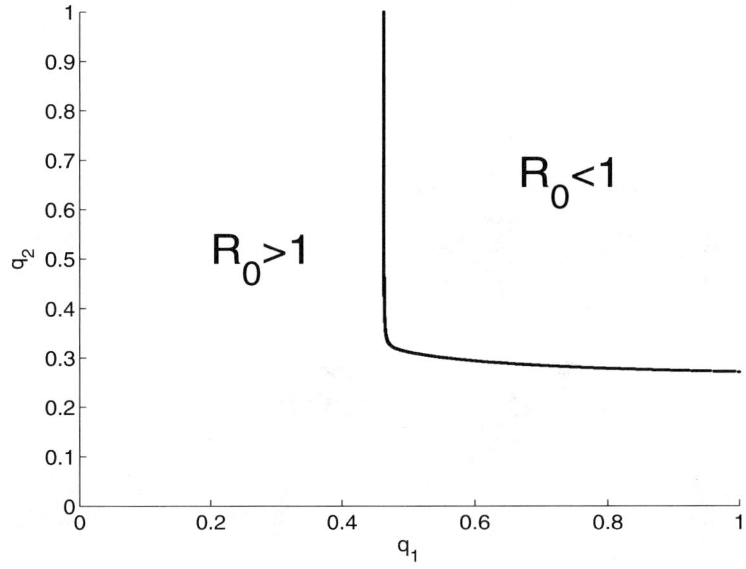

Figure 8.2. *Boundary curve of* $R_0(q_1, q_2) = 1$ *on the* q_1q_2*-plane.*

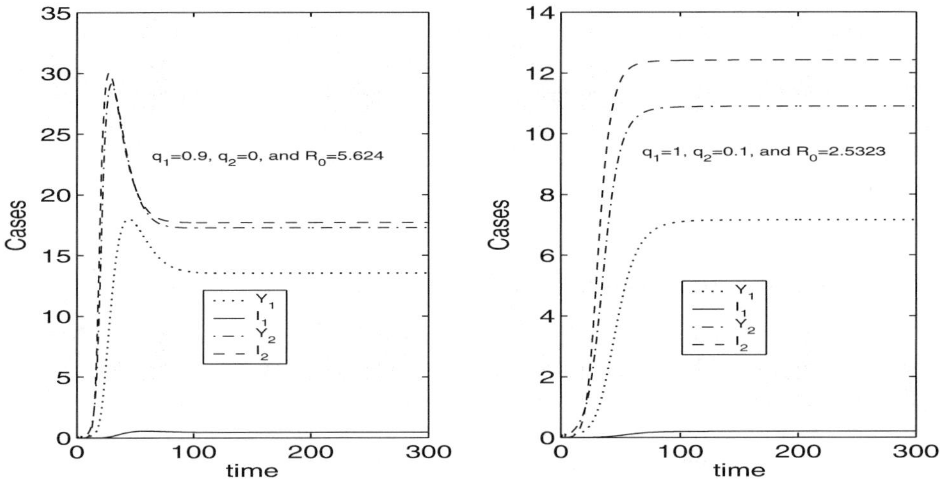

Figure 8.3. *The number of cases for* $q_1 = .9$ *and* $q_2 = 0$ *(left) and* $q_1 = 0.9$ *and* $q_2 = 0.1$ *(right). The disease becomes endemic. The units on the y-axis are* 1000.

In order to stress the importance of the last remark, we observe from Figure 8.1 that when q_2 is close to zero, the basic reproductive number is above 5. That smallpox will not be controlled if tourists (less than 2.5% of the entire population) are not included in the vaccination policy (see Figure 8.3). Note that vaccinating 90% of the residents without vaccinating visitors still gives a very large basic reproductive number, 5.624 (see the left graph in Figure 8.3). Mathematically, the disease becomes endemic, with most of the prevalence

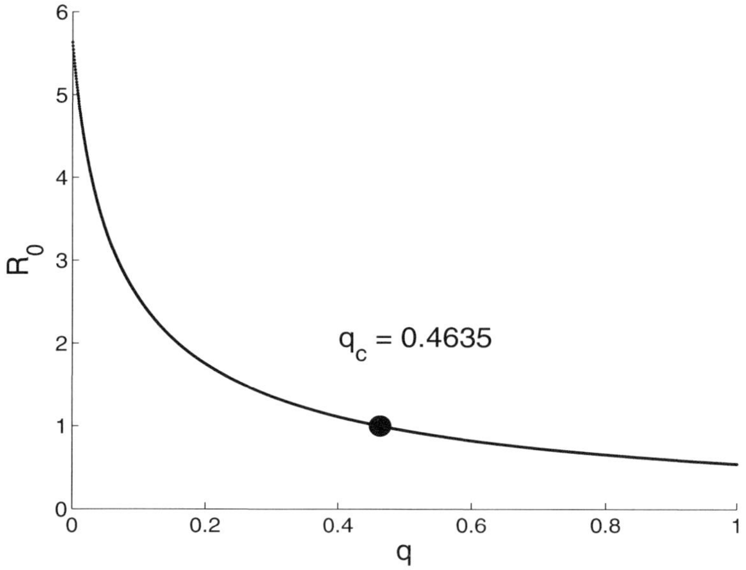

Figure 8.4. *Critical vaccination rate is* $q_c = 0.4635$.

being supported by the tourist population. The extreme case where all residents and 10% of tourist population are vaccinated is still not enough (see the right graph in Figure 8.3).

We now look at the special case where $q_1 = q_2$, that is, when the vaccination campaign does not differentiate between residents and tourists. In this case, a "single" critical value q_c (defined as the value where $\mathcal{R}_0(q_c) = 1$ when $q < q_c$, $\mathcal{R}_0(q) > 1$, and $\mathcal{R}_0(q) < 1$ when $q > q_c$) for the vaccination rate is $q_c = 0.4635$, as can be seen in Figure 8.4. This value is slightly greater than the critical value for an "isolated" resident population (no tourists).

The above results on critical vaccination rate values are consistent with those found in the literature on herd immunity and smallpox. Estimates for the values of herd immunity for smallpox as given in [1, 23] are on the order of 80%. The critical value (q_c) corresponds to this herd immunity level. This can be seen from the "back of the envelope computation" on a population in a steady state. If we keep this (q_c) vaccination rate for 3 consecutive days, then we achieve about 80% herd immunity since $q_c(1+(1-q_c)+(1-q_c)^2) = 0.8456$. The following similar computations are carried out in order to explore vaccine response readiness in the case of a smallpox attack. A vaccination rate $q = q_c = 0.4635$ creates the heaviest workload in the first day of its implementation, as $8.2 \times 0.4635 = 3.8$ million persons must be vaccinated. Use of 1947 NYC MV data [23] and assuming that one registered nurse could serve, on average, 200 individuals per day lead to a first-day requirement of about 20,000 nurses.

The selection of q_1 and q_2 values that make $R_0 < 1$ does control the disease eventually. However, a higher number of deaths would be expected if R_0 is just a little bit lower than a unit (only a little bit lower than the critical curve in Figure 8.2).

For example, $q_1 = q_2 = .5$ leads to the death of 156 infected with smallpox. Table 8.4 lists the total number of deaths due to smallpox infection for various combinations of q_1 and q_2.

Delays in implementation naturally lead to different scenarios. Ideally, everybody

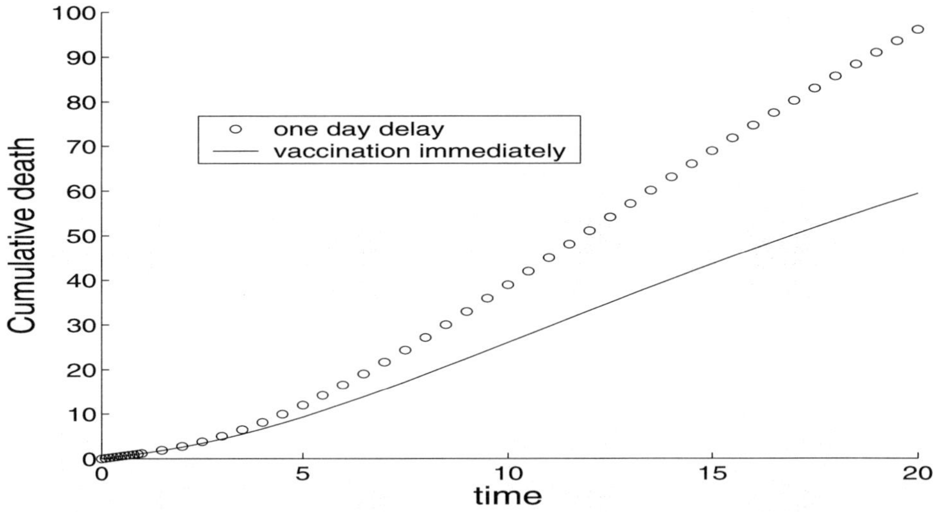

Figure 8.5. *Comparison in the cumulative number of deaths when vaccination begins a day later ($q_1 = q_2 = 0.8$). Twenty more civilians would die in a smallpox attack.*

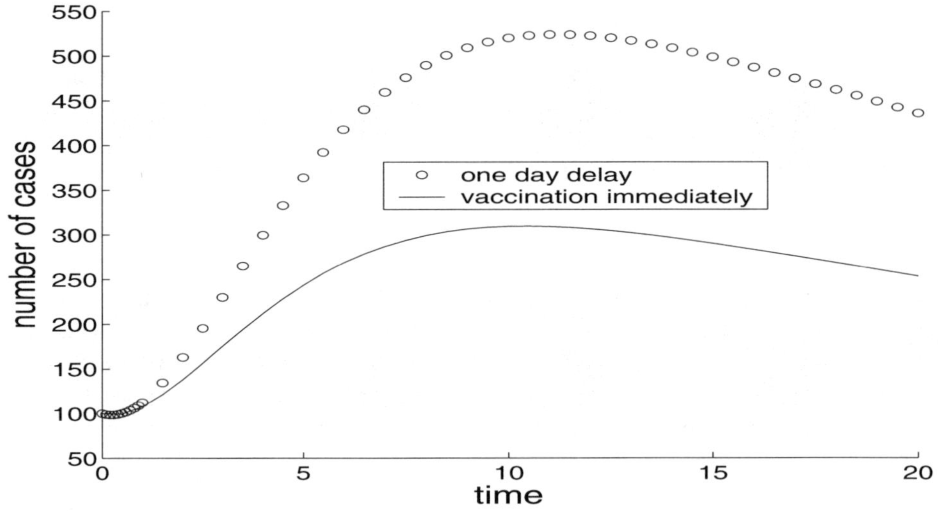

Figure 8.6. *Comparison in the cumulative number of deaths when vaccination begins one day later ($q_1 = q_2 = 0.8$).*

would be vaccinated as soon as the first infected is detected (and such an index case is immediately detected). If so, then 27 people would die from this attack under MV.

In other words, response time plays a critical role if the goal is to reduce the total number of deaths. Figures 8.5 and 8.6 compare the total number of cases and deaths in situations where there is an immediate response and a one-day delay.

As can be seen from theses two graphs, a one-day delay in the vaccine implementation

Table 8.4. *Total death due to smallpox infections. Rows correspond to q_1 and columns with q_2.*

	1.0	.95	.90	.85	.80	.75	.70	.65	.60	.55	.50	.45	.40
1.0	27	29	31	33	36	40	45	51	60	72	90	119	173
.95	27	29	31	34	37	41	46	52	61	73	91	121	178
.90	28	30	32	34	38	41	46	53	62	74	93	124	182
.85	28	30	32	35	38	42	47	54	63	76	96	128	188
.80	29	31	33	36	39	43	49	56	65	78	99	132	196
.75	30	32	34	37	41	45	50	57	67	81	102	137	205
.70	31	33	36	39	42	47	52	60	70	84	107	144	216
.65	33	35	37	40	44	49	55	63	73	89	113	153	232
.60	35	37	40	43	47	52	58	67	78	95	121	165	254
.55	38	40	43	47	51	57	64	73	86	104	133	183	288
.50	45	48	51	55	60	66	74	85	100	122	156	218	353

results in 26 additional deaths, and 100 more cases would result even if 80% of the total population is vaccinated ($q_1 = q_2 = 0.8$). The situation would be worse if there is a two-day delay in the vaccine implementation (66 more deaths). There would be 131 additional deaths if vaccination were to start 3 days later. Tables A.5–A.12 in the appendix include the computations for various delay situations.

Figure 8.7 shows the appearance of a second wave about 8 days after the initial release of smallpox in the public transportation system despite the vaccination policy. Figure 8.8 offers a comparison between the cumulative number of deaths for two particular vaccination combinations, $q_1 = q_2 = 0.8$ and $q_1 = q_2 = 0.5$.

The explicit quantitative predictions derived from our simulation results naturally depend on the selected parameters. However, the results on the criticality of the average response time and the importance of the tourist population are quite general. Furthermore, it is clear that delays in response to an attack (vaccination implementation) and exclusive focus on resident populations may result in serious consequences. Consequently, the development of effective surveillance systems is perhaps the most important issue. In summary, having enough vaccines stockpiles, while necessary, is not sufficient. The best policy must include the *timely* application of mass vaccination. Such a policy requires a fast mobilization plan that includes a "SWAT"-type medical team.

8.8 Conclusion

In this chapter, an epidemiological model for the transmission dynamics and control of smallpox with multiple levels of mixing is introduced. Mathematically, a complete analysis has not been carried out, except for the rather long and essential computation of the basic reproductive number, which we are able to carry out in the case of two "neighborhoods." Knowledge of R_0 is critical in the development of intervention strategies that reduce the likelihood or limit the impact of outbreaks. The model supports endemic equilibria, but the analysis is not obvious. The existence of such equilibria is not an irrelevant issue. Epidemic models that support multiple equilibria tend to support fast outbreaks. More importantly for the purposes of this chapter, the relevance of R_0 in this case has to be seriously questioned (see Chapter 7).

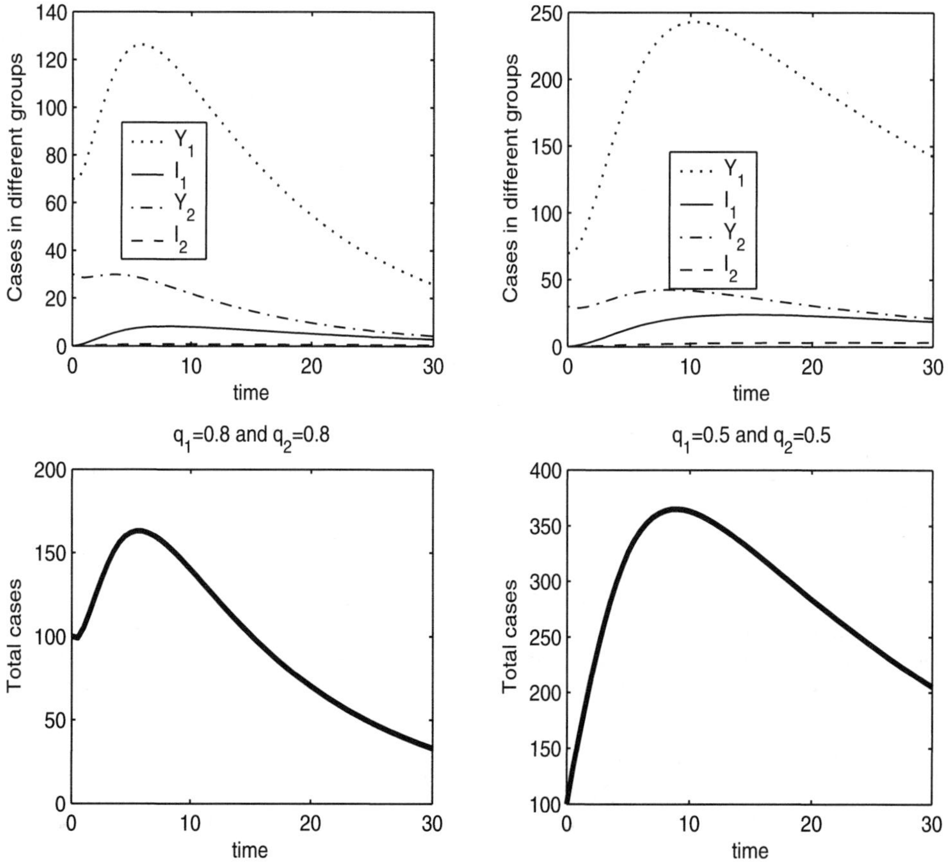

Figure 8.7. *The upper two graphs show the distribution of cases for SU and NSU in the resident and tourist populations. The bottom two graphs collect total cases in the entire population.*

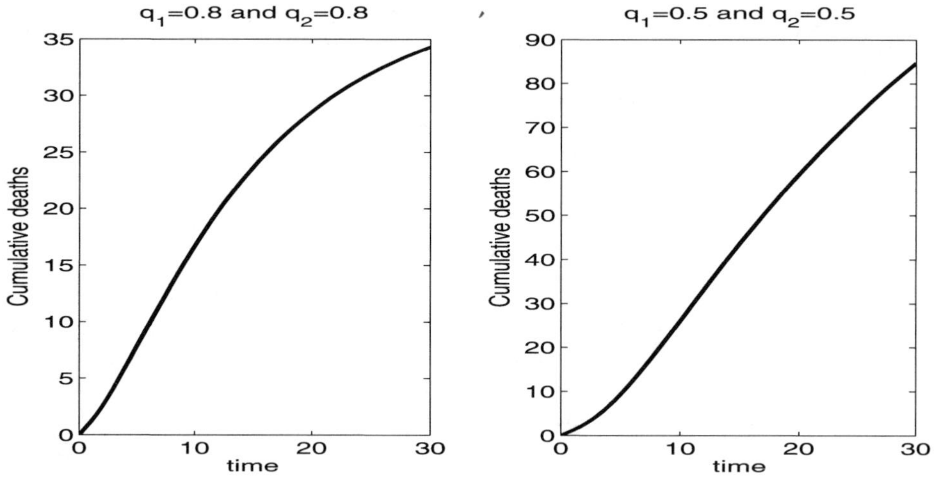

Figure 8.8. *Cumulative deaths for* $q_1 = q_2 = 0.8$ *(left) and* $q_1 = q_2 = 0.5$ *(right).*

Appendix

Table A.5. *Cumulative cases from 56 combinations of vaccination levels in the resident population when vaccinations are given after a one-day delay.*

	.3	.4	.5	.6	.7	.8	.9	1.0
.4	3652	3005	2680	2488	2362	2272	2205	2155
.5	2236	1846	1654	1542	1469	1418	1380	1351
.6	1544	1282	1155	1082	1034	1001	977	958
.7	1154	964	874	822	789	765	749	735
.8	911	767	698	660	635	619	606	596
.9	747	635	582	551	532	519	509	502
1.0	633	541	498	474	459	449	441	436

Table A.6. *Cumulative cases from 56 combinations of vaccination levels in the tourist population when vaccinations are given after a one-day delay.*

	.3	.4	.5	.6	.7	.8	.9	1.0
.4	935	467	295	212	165	136	116	102
.5	836	415	263	190	149	124	107	94
.6	776	385	245	178	141	118	101	90
.7	739	366	234	171	135	114	98	87
.8	711	353	226	165	131	111	96	86
.9	691	342	220	162	129	109	94	84
1.0	677	335	216	159	127	107	93	84

Table A.7. *Cumulative number of deaths from 56 combinations of vaccination levels in the resident population when vaccinations are given after a one-day delay.*

	.3	.4	.5	.6	.7	.8	.9	1.0
.4	243	208	191	180	172	166	162	159
.5	155	135	123	117	112	109	107	104
.6	110	96	89	85	82	80	78	76
.7	84	74	69	65	63	61	60	60
.8	67	60	56	53	52	51	50	49
.9	56	50	47	45	44	43	42	42
1.0	48	43	41	39	38	38	37	37

Table A.8. *Cumulative number of deaths from 56 combinations of vaccination levels in the tourist population when vaccinations are given after a one-day delay.*

	.3	.4	.5	.6	.7	.8	.9	1.0
.4	60	32	22	16	13	10	9	8
.5	55	29	20	15	12	10	8	8
.6	50	27	19	14	12	9	8	7
.7	48	26	18	14	11	9	8	7
.8	47	26	18	13	11	9	8	7
.9	45	25	18	13	11	9	8	7
1.0	45	24	17	13	11	8	8	7

Table A.9. *Cumulative number of cases from 56 combinations of vaccination levels in the tourist population when vaccinations are given after a two-day delay.*

	.3	.4	.5	.6	.7	.8	.9	1.0
4	1530	755	473	338	262	214	182	159
.5	1373	672	422	303	237	195	167	147
.6	1281	625	393	284	222	185	159	140
.7	1220	595	375	272	213	178	153	136
.8	1177	573	362	263	207	173	150	133
.9	1146	558	352	256	203	170	147	131
1.0	1121	545	345	252	200	168	145	129

Table A.10. *Cumulative number of cases from 56 combinations of vaccination levels in the resident population when vaccinations are given after a two-day delay.*

	.3	.4	.5	.6	.7	.8	.9	1.0
.4	5924	4897	4382	4076	3874	3731	3624	3542
.5	3650	3030	2725	2547	2430	2348	2287	2241
.6	2537	2119	1917	1801	1725	1672	1633	1603
.7	1907	1605	1461	1378	1325	1289	1261	1240
.8	1514	1284	1176	1115	1075	1048	1028	1012
.9	1250	1069	985	937	906	886	870	858
1.0	1064	917	849	810	787	769	757	748

Acknowledgments

C. C. C. wants to acknowledge the scientific exchanges that he had with the late Angel Capurro as they relate to this work. A version of this project was given to summer REU

Table A.11. *Cumulative number of deaths corresponding to 56 combinations of vaccination levels in the population if vaccinations are given after a two-day delay.*

	.3	.4	.5	.6	.7	.8	.9	1.0
.4	97	52	34	26	20	17	15	13
.5	89	47	31	24	18	16	14	12
.6	84	45	30	22	18	15	13	12
.7	79	43	29	22	17	15	13	12
.8	77	41	28	21	17	14	13	12
.9	75	40	27	21	16	14	13	11
1.0	73	40	27	20	16	14	12	11

Table A.12. *Cumulative number of deaths corresponding to 56 combinations of vaccination levels in a resident population if vaccinations are given after a two-day delay.*

	.3	.4	.5	.6	.7	.8	.9	1.0
.4	395	341	313	295	283	274	268	262
.5	254	221	204	193	186	181	176	174
.6	182	160	148	141	136	133	130	128
.7	140	123	115	110	107	104	103	101
.8	113	101	94	91	88	86	85	84
.9	95	84	79	76	75	73	72	71
1.0	81	73	69	67	65	64	63	63

students (Karen Rios-Soto, Emilia Sanchez-Huerta, and Guarionex Silva) working under our supervision. We thank Emilia, Karena, and Guarionex for their review of the literature, for asking lots of questions, and for attempting to tackle a difficult model. We also thank Erika Camacho, Carlos W. Castillo-Garsow, David Murillo, and Fabio Sanchez for their valuable comments. This work was partially supported from funds by NSA, NSF, and the Sloan Foundation.

Bibliography

[1] R. Anderson and R. May, *Infectious Disease of Humans*, Oxford University Press, Oxford, New York, Toronto, 1991.

[2] A. D. Barbour, *Macdonal's model and the transmission of bilharzia*, Trans. Roy. Trop. Med. Hygiene, 72 (1978), pp. 6–15.

[3] D. Bernoulli, *Essai d'une nouvelle analyse de la mortalité causée par la petite vérole, et des avantages de l'inoculation pour la prévenir*, Histoires et Mémoires de l'Académie Royale des Sciences de Paris, (1766), pp. 1–45 (in French); in Smallpox Inoculation: An Eighteenth Century Mathematical Controversy, L. Bradley, translator and commentator,

Adult Education Department, University of Nottingham, Nottingham, UK, 1971 (in English).

[4] F. BRAUER AND C. CASTILLO-CHAVEZ, *Mathematical Models in Population Biology and Epidemiology*, Springer-Verlag, New York, Berlin, Heidelberg, 2001.

[5] W. J. BROAD AND J. MILLER, *Report provides new details of Soviet smallpox accident*, The New York Times, June 15, 2002, section A, p. 1.

[6] S. BUSENBERG AND C. CASTILLO-CHAVEZ, *A general solution of the problem of mixing sub-populations, and its application to risk-and age-structured epidemic models for the spread of AIDS*, IMA J. Math. Appl. Med. Biol., 8 (1991), pp. 1–29.

[7] A. F. CAPURRO, C. CASTILLO-CHÁVEZ, J. VELASCO-HERNÁNDEZ, AND M. L. ZELLNER, *El transporte público y la dinámica de la tuberculosis a nivel poblacional*, in Aportaciones Matemáticas, Serie Comunicaciones 22, Sociedad Matemática Mexicana, México, DF, Mexico, 1998.

[8] C. CASTILLO-CHAVEZ, H. W. HETHCOTE, V. ANDREASON, S. A. LEVIN, AND W. LIU, *Cross-immunity in the dynamics of homogeneous and heterogeneous populations*, in Mathematical ecology. Proceedings, Autumn Course Research Seminars, Trieste 1986, L. Gross, T. G. Hallam, and S. A. Levin, eds., World Scientific, Singapore, 1988, pp. 303–316.

[9] C. CASTILLO-CHAVEZ, H. W. HETHCOTE, V. ANDREASON, S. A. LEVIN, AND W. LIU, *Epidemiological models with age structure, proportionate mixing, and cross-immunity*, J. Math. Biol., 27 (1989), pp. 233–258.

[10] C. CASTILLO-CHAVEZ AND F. ROBERTS, *Report on DIMACS Working Group Meeting: Mathematical Sciences Methods for the Study of the Deliberate Releases of Biological Agents and Their Consequences*, DIMACS Center, Rutgers University, Piscataway, NJ, 2002.

[11] C. CASTILLO-CHAVEZ, Z. FENG, AND W. HUANG, *On the computation of R_0 and its role on global stability*, in Mathematical Approaches for Emerging and Reemerging Infectious Diseases: An Introduction, IMA Vol. Math. Appl. 125, C. Castillo-Chavez, S. Blower, P. van den Driessche, D. Kirschner, and A. A. Yakubu, eds., Springer-Verlag, New York, 2002, pp. 229–250.

[12] C. CASTILLO-CHAVEZ, S. BLOWER, P. VAN DEN DRIESSCHE, D. KIRSCHNER, AND A. A. YAKUBU, EDS., *Mathematical Approaches for Emerging and Reemerging Infectious Diseases: An Introduction*, IMA Vol. Math. Appl. 125, Springer-Verlag, New York, 2002.

[13] C. CASTILLO-CHAVEZ, S. BLOWER, P. VAN DEN DRIESSCHE, D. KIRSCHNER, AND A. A. YAKUBU, EDS., *Mathematical Approaches for Emerging and Reemerging Infectious Diseases: Models, Methods and Theory*, IMA Vol. Math. Appl. 126, Springer-Verlag, New York, 2002.

[14] CENTERS FOR DISEASE CONTROL, *Summary of notifiable diseases, United States*, 1998, Morbidity and Mortality Weekly Rep., 47 (1999), pp. 78–83.

[15] CENTERS FOR DISEASE CONTROL, *Update: Influenza activity–United States,* 1999–2000 *season*, Morbidity and Mortality Weekly Rep., 49 (2000), pp. 53–57.

[16] CENTERS FOR DISEASE CONTROL, *Smallpox Information for the General Public*, Centers for Disease Control, Atlanta; available online from http://www.cdc.gov/nip/smallpox/public.htm; modified June 20, 2002; accessed July 2002.

[17] E. CHECK, *Need for vaccine stocks questioned*, Nature, 414 (2001), p. 677.

[18] O. DICKMANN, K. DIETZ, AND J. A. P. HEESTERBEEK, *On the definition and the computation of the basic reproductive ration R_0 in models for infectious diseases in heterogeneous populations*, J. Math. Biol., 28 (1990), pp. 365–382.

[19] O. DICKMANN AND J. A. P. HEESTERBEEK, *Mathematical Epidemiology of Infectious Diseases: Model Building, Analysis and Interpretation*, Wiley, Chichester, New York, Weihheim, Brisbane, Sigapore, Toronto, 2000.

[20] K. DIETZ AND J. P. A. HEESTERBEEK, *Bernoulli was ahead of modern epidemiology* Nature, 408 (2000), pp. 513–514.

[21] S. DUNHAM, *Mass transit defends itself against terrorism*, J. Homeland Security, March 2002; available online from http://www.homelandsecurity.org/journal/Articles/displayarticle.asp?article=47.

[22] C. J. DUNCAN, S. R. DUNCAN, AND S. SCOTT, *Oscillatory dynamics of smallpox and the impact of vaccination*, J. Theoret. Biol., 183 (1996), pp. 336–343.

[23] F. FENNER, D. A. HENDERSON, I. ARITA, Z. JEZEK, AND I. D. LADNYI, *Smallpox and Its Eradication*, World Health Organization, Geneva, 1988.

[24] R. GANI AND S. LEACH, *Transmission potential of smallpox in contemporary populations*, Nature, 415 (2001), pp. 748–751.

[25] D. A. HENDERSON, T. V. INGLESBY, J. G. BARTLETT, M. S. ASCHER, E. EITZEN, P. B. JAHRLING, J. HAUER, M. LAYTON, J. MCDADE, M. T. OSTERHOLM, T. O'TOOLE, G. PARKER, T. PERL, P. K. RUSSELL, AND K. TONAT, *Smallpox as a biological weapon*, J. Amer. Med. Assoc., 281 (1999), pp. 2127–2137; available online from http://jama.ama-assn.org/cgi/content/full/281/22/2127.

[26] D. A. HENDERSON, *About the first national symposium on medical and public health response to bioterrorism*, Emerging Infectious Diseases, 5 (1999), p. 491.

[27] D. A. HENDERSON, *The looming threat of bioterrorism*, Science, 283 (1999), pp. 1279–1282.

[28] D. A. HENDERSON, *Smallpox: Clinical and epidemiologic features*, Emerging Infectious Diseases, 5 (1999), pp. 537–539.

[29] H. W. HETHCOTE, *The mathematics of infectious diseases*, SIAM Rev., 42 (1999), pp. 599–653.

[30] J. A. JACQUEZ, C. P. SIMON, J. KOOPMAN, L. SATTENPIEL, AND T. PERRY, *Modeling and analyzing HIV transmission: Effect of contact patterns*, Math. Biosci., 92 (1988), pp. 119–199.

[31] L. H. KAHN, *Smallpox transmission risks: How bad?*, Science, 297 (2002), p. 50.

[32] E. H. Kaplan, D. L. Craft, and L. M. Wein, *Emergency response to a smallpox attack: The case for mass vaccination*, Proc. Natl. Acad. Sci. USA, 99 (2002), pp. 10935–10940.

[33] M. M. Kaplan and R. G. Webster, *The epidemiology of influenza*, Sci. Amer., 237 (1977), pp. 88–105.

[34] A. R. Kemper, M. M. Davis, and G. L. Freed, *Expected adverse events in a mass smallpox vaccination campaign*, Effective Clinical Practice, 5 (2002), pp. 84–90.

[35] R. Larsen, *Smallpox: Right topic, wrong debate*, J. Homeland Security, July, 2002; available online from http://www.homelandsecurity.org/HLSCommentary/20020725 .htm; accessed January 31, 2003.

[36] I. M. Longini, M. E. Halloran, A. Nizam, M. Wolff, P. M. Mendelman, P. E. Fast, and R. B. Belshe, *Estimation of the efficacy of life, attenuated influenza vaccine from a two-year, multi-center vaccine trial: Implications for influenza epidemic control*, Vaccine, 18 (2000), pp. 1902–1909.

[37] M. I. Meltzer, I. Damon, J. W. LeDuc, and J. D. Millar, *Modeling potential responses to smallpox as a bioterrorist weapon*, Emerging Infectious Diseases, 7 (2001), pp. 959–968.

[38] J. F. Modlin, *A mass smallpox vaccination campaign: Reasonable or irresponsible?*, Effective Clinical Practice, 5 (2002), pp. 98–99.

[39] A. Nold, *Heterogeneity in disease-transmission modeling*, Math. Biosci., 52 (1980), pp. 227–240.

[40] L. D. Rotz, A. S. Khan, S. R. Lillibridge, S. M. Ostroff, and J. M. Hughes, *Public health assessment of potential biological terrorism agents*, Emerging Infectious Diseases, 8 (2002), pp. 225–229.

[41] R. Snacken, A. P. Kendal, L. R. Haaheim, and J. M. Wood, *The next influenza pandemic: Lessons from Hong Kong*, 1997, Emerging Infectious Diseases, 5 (1999), pp. 195–203.

[42] B. Song, *Dynamical Epidemical Models and Their Applications*, Ph.D. thesis, Department of Biological Statistics and Computational Biology, Cornell University, Ithaca, NY, 2002.

[43] E. W. Stearn and A. E. Stearn, *The Effect of Smallpox on the Destiny of the Amerindian*, Bruce Humphries, Boston, 1945.

[44] P. F. Wehrle, J. Posch, K. H. Richter, and D. A. Henderson, *An airborne outbreak of smallpox in a German hospital and its significance with respect to other recent outbreaks in Europe*, Bull. World Health Org., 43 (1970), pp. 669–679.

[45] I. Weinstein, *An outbreak of smallpox in New York City*, Amer. J. Public Health, 37 (1947), p. 376.

[46] R. G. Webster, *Influenza: An emerging disease*, Emerging Infectious Diseases, 4 (1998), pp. 436–441; available online from http://www.cdc.gov/ncidod/eid/vol4no3/ webster.htm.

[47] E. W. WEISSTEIN, *CRC Concise Encylopedia of Mathematics*, CRC Press, Boca Raton, London, New York, Washington, DC, 1999, pp. 362–364.

[48] CENTER FOR STRATEGIC AND INTERNATIONAL STUDIES, *Dark Winter*, Center for Strategic and International Studies, Washington, DC; available online from http://www.csis.org/hill/darkwinter.htm; accessed January 31, 2002.

[49] CENTERS FOR DISEASE CONTROL, *CDC Interim Smallpox Response Plan and Guidelines (Draft* 2.0), Centers for Disease Control, Atlanta, 2001.

[50] ARCHIVES OF THE MAYOR'S PRESS OFFICE, *New York City Experiences Record Number of Visitors in 1999 with Nearly 37 Million Tourists Visiting New York City*, Release 291-00, Archives of the Mayor's Press Office, New York, August 7, 2000; available online from http://home.nyc.gov/html/om/html/2000b/pr291-00.html.

[51] ANSER INSTITUTE FOR HOMELAND SECURITY, *Dark Winter*, ANSER Institute for Homeland Security, Arlington, VA; available online from http://www.homelandsecurity.org/darkwinter/index.cfm; accessed January 31, 2002.

[52] NYC & COMPANY, *Visitor Statistics*, NYC & Company, New York; available online from http://www.nycvisit.com/content/index.cfm?pagePkey=598; updated 2002.

[53] CENTERS FOR DISEASE CONTROL, *Ebola Hemorrhagic Fever*, Special Pathogens Branch, Centers for Disease Control, Atlanta; available online from http://www.cdc.gov/ncidod/dvrd/spb/mnpages/dispages/ebola.htm; accessed July 2002.

[54] WEBMD, *What Is Smallpox?*, WebMD, Elmwood Park, NJ, 2001; available online from http://content.health.msn.com/content/pages/5/4058_261; accessed July 2002.

Chapter 9

The Role of Migration and Contact Distributions in Epidemic Spread

*K. P. Hadeler**

9.1 Introduction

There are two processes which govern the spread of an infectious disease within a population: the transmission of the disease from infected to susceptibles and the removal of infected, either by death or by transition to an immune state. If the infection rate is large and the removal rate is small, then there is an outbreak characterized by a rapid increase of the number of infected. Unless new susceptibles are introduced by loss of immunity or by demographic renewal, the outbreak eventually comes to a halt for lack of susceptibles. Then the outbreak leaves a final number of susceptibles who never get infected.

This general pattern can be modified by taking into account inhomogeneous populations such as populations structured with respect to age and gender, intermediate host populations and transmission by vectors, distinction between micro- and macroparasites, and, notably, distribution in space.

In mathematical models these features enter via additional variables and equations or by introduction of suitable quantities such as averages and moments; see the guidebook by Diekmann and Heesterbeek [6] for the basics of epidemic modeling. Also, with respect to space, one can take averages, but usually one is interested in the actual distribution in space and the speed of spatial spreading, particularly in the speed of epidemic fronts.

For a spatial epidemic the basic reproduction number is only one characteristic feature. It determines whether an outbreak can or will occur. Another feature is the speed of the epidemic front, which describes how the epidemic moves into a territory which is initially void of infected cases. It may seem even more important for the following two reasons.

First of all, it is a typical feature of simple reaction migration systems in *unbounded* domains that limit sets are formed by nonlinear waves or fronts traveling with constant speed. Hence, in epidemic models we can expect that epidemic fronts, in particular those with minimal speed, yield the essential features of spatial spread. Second, once we get the explicit dependence of the minimal speed on the model parameters, we can gain some insight

*Biomathematics, University of Tübingen, Auf der Morgenstelle 10, 72076 Tübingen, Germany.

into what determines the speed of propagation.

Since the 1960s, if not earlier, there have been models for the spread of an epidemic in space which are based either on contact distributions [18] or on processes describing movements of individuals in space [24]. These two approaches lead to two different classes of models which have, however, similar qualitative behavior. In particular, both classes of models produce traveling epidemic fronts for which the speed depends in a plausible manner on the parameters of the basic model (infection rate, recovery rate) and characteristic parameters of the spatial process (width of the contact distribution, diffusion coefficient). Mostly, a comparison of the two approaches has not been made; sometimes both mechanisms, i.e., contact and movement, have been combined in one complex model. On the other hand, one can postulate close connections between both approaches since contacts between individuals located at distinct space positions can be accomplished only by individuals moving in space. The goal of this chapter is to exhibit the connections between both models by appropriate scaling of migration and infection rates in a deterministic setting. For scaling of epidemic models in a statistical mechanics context, in particular for the case where migration rates dominate the infection and recovery rates, we refer the reader to [28, 29].

We start from the simplest case of the Kermack–McKendrick model (the SIR model) for susceptibles u and infected v:

$$\begin{aligned} \dot{u} &= -\beta u v, \\ \dot{v} &= \beta u v - \alpha v. \end{aligned} \tag{9.1}$$

The parameter $\beta > 0$ is the infection rate and $\alpha > 0$ is the recovery rate. The infection process is modeled by a mass action kinetics uv. Mass action kinetics is meaningful in the case of constant population size, i.e., in the absence of demographic changes (on the chosen time scale). In a more general model, one would assume an infection term uv/P, where P is the total population size, and perhaps a nonlinear dependence on v. Thus, in (9.1) the population size has been normalized to $P = 1$.

If the "basic reproduction number" $R_0 = \beta/\alpha$ exceeds 1, then a small number of infected ($v > 0$, $v \approx 0$) in an almost totally susceptible population ($u = 1 - v$) causes an outbreak. The variable v increases rapidly and then returns to zero, and the variable u decreases to a value $u_\infty > 0$, the number of remaining susceptibles ($u_\infty < \alpha/\beta$ is the lower solution of the equation $u - (\alpha/\beta)\log u = 1$).

Kendall [18] has extended this "local" model to a spatial model by introducing a contact distribution which tells how an infected at position y acts upon a susceptible at position x. This distribution is assumed to be translation invariant, symmetric, nonnegative, normalized to total mass 1; i.e., it is given by a kernel $k(x)$ satisfying $k(-x) = k(x)$, $k \geq 0$, $k * 1 = 1$. We further assume that k has a second moment. Then the contact model has the form

$$\begin{aligned} u_t(t,x) &= -\beta\, u(t,x) \int k(x-y) v(t,y) dy, \\ v_t(t,x) &= \beta\, u(t,x) \int k(x-y) v(t,y) dy - \alpha v(t,x). \end{aligned} \tag{9.2}$$

In order to keep a simple notation, we use the same letter k for the kernel and for the convolution operator, and we write kv instead of $k * v$ for the application of the operator to the function v. Then the system (9.2) becomes

$$\begin{aligned} u_t &= -\beta\, u\, k v, \\ v_t &= \beta\, u\, k v - \alpha v. \end{aligned} \tag{9.3}$$

On the other hand, one can assume that moving infectives are the main cause for spreading infections as in [24, 17, 8, 15, 16]. Then the system assumes the form of a migration model

$$\begin{aligned} u_t &= -\beta u v + \delta_0 K_0 u, \\ v_t &= \beta u v - \alpha v + \delta K v, \end{aligned} \tag{9.4}$$

where K_0 and K are either integral operators of the general form $Kv = k * v - v$, with $k * 1 = 1$, or diffusion operators, e.g., $Kv = \Delta v$. Here $\delta > 0$ plays the role of a diffusion coefficient.

For a general migration process governed by an operator $\delta(k * v - v)$ one can define a diffusion approximation. It can be formally obtained by a Taylor expansion in the convolution integral,

$$k * v \approx v + \sigma^2 v_{xx}, \tag{9.5}$$

where σ^2 is the second moment (which is assumed to exist)

$$\sigma^2 = \frac{1}{2} \int_{-\infty}^{\infty} y^2 k(y) dy. \tag{9.6}$$

Kendall [18] replaced the convolution term in the contact model (9.3) by a diffusion term as in (9.5) and arrived at a contact model

$$\begin{aligned} u_t &= -\beta\, u(v + \sigma^2 v_{xx}), \\ v_t &= \beta\, u(v + \sigma^2 v_{xx}) - \alpha v. \end{aligned} \tag{9.7}$$

The system (9.7) is a contact model although it contains a second-order derivative. Atkinson and Reuter [2] and Aronson [1] treated the full intregral equation. Diekmann [5] formulated a renewal equation for the incidence (the number of newly infected, $-u_t$ in our notation, also equal to the change in *accumulated* infected; see [20]) based on a "reproduction and dispersal kernel" (RD kernel) $A(\tau, x, \xi)$ which gives the infectivity of an individual presented to a susceptible at position x by an infected at position ξ with infective age τ. Most spatial epidemic models can be subsumed to this general framework. For the contact model (9.3) one gets $A(\tau, x, \xi) = \beta e^{-\alpha\tau} k(x - \xi)$; for the reaction-diffusion model (9.4) with $K_0 = 0$ the function $A(\tau, x, \xi)$ can be expressed in terms of the solution operator (the "heat kernel") of the linear equation $z_t = Kz$. Radcliffe and Rass [25] considered situations with several types of infected.

9.2 Sedentary and Migrating Infected

The concept of a sedentary and a migrating state has been used in ecological models. Lewis and Schmitz [19] studied an extension of the Fisher equation (or diffusive logistic equation) where a stationary (sedentary) and a mobile (migrating) state are coupled by Poisson processes. They determined the wave speed for this system in one important case. The system of Cook [4] on a dispersing and a nondispersing subpopulation coupled by the recruitment process shows similar general features. Mark Lewis and the author study the qualitative behavior of a Fisher equation with a sedentary state in a forthcoming paper [12]. Hillen [14] studies transport equations with resting phases systematically and derives reaction-advection-diffusion equations as limiting cases. Sattenspiel and Dietz [26] have designed an epidemic model (and applied to concrete situations) with residents and travelers in discrete space. The population resides in a number of villages. At a given moment, the model counts the number of residents of the ith village who are actually there and those who are visiting the jth village, for $j \neq i$.

We present a general but simple model for an infectious disease in continuous space with sedentary and migrating infected from which a migration model and a contact model are derived as special or limiting cases. We assume that there are susceptibles, u, and infected of two kinds, which we call sedentary, w, and migrating, z. Infected can switch between the sedentary and the migrating state according to Poisson processes with rates γ_1 and γ_2, respectively. The total number of infected is $v = w+z$. It must be pointed out that our model is not an epidemic subgroup model in the usual sense, e.g., [13, 10, 11], where subgroups interact only via infection. Here the switching rates do not depend on the state or level of infection; i.e., we model the situation where infected are not aware of their infection or do not care. However, sedentary and migrating infected differ in the level of infectivity, β_1 versus β_2, in most practical situations not for entirely disease-related reasons but because of social and sanitary conditions.

Introducing a sedentary phase leads to an epidemic model with two levels of mixing, as in [3, 26]. We assume that susceptibles are sedentary. This assumption is not motivated by properties of a specific disease (cf. the rabies model of Källén [17]) but derives from the insight that, at a low level of prevalence of the disease, migration of susceptibles does not essentially contribute to spread of the disease. On the other hand, one can consider epidemic models where *only* the susceptibles move [15]. Although such models may even exhibit traveling wave solutions, they are artificial; in these models a localized epidemic cannot spread in a susceptible population distributed in space (see [9]).

In our model, the parameters β_1 and β_2 are the infection rates for sedentary and migrating infected, respectively. As before α is the recovery rate. The migration operator of the form $Kv = \delta(kv - v)$ satisfies $k * 1 = 1, \delta > 0$. The rate γ_1 governs the transition from the sedentary to the migrating state, and γ_2 the reverse transition. Then the basic model reads as

$$\begin{aligned} u_t &= -u(\beta_1 w + \beta_2 z), \\ z_t &= \delta K z - \alpha z + \gamma_1 w - \gamma_2 z, \\ w_t &= u(\beta_1 w + \beta_2 z) - \alpha w - \gamma_1 w + \gamma_2 z. \end{aligned} \tag{9.8}$$

A sedentary susceptible can be infected by either an infected w residing at exactly the same spot or by a passing migrating infected z. We iterate that w and z denote the infective individuals in the population distinguished only by their momentary state of motion, either sedentary or migrating. The variables w and z do not denote successive stages or groups as in a model with latently infected or in a core group model [10, 11].

Before entering the main topic related to a *large* switching rate γ_2, we consider the special case where $\gamma_2 = 0$ and also $\beta_1 = 0$. Then the system reduces to (we exchange the second and the third equation)

$$\begin{aligned} u_t &= -u\,\beta_2 z, \\ w_t &= u\,\beta_2 z - \alpha w - \gamma_1 w, \\ z_t &= \delta K z - \alpha z + \gamma_1 w. \end{aligned} \tag{9.9}$$

Now w can be seen as a latent state. Susceptibles u become infected and enter the latent state w and then leave with rate γ_1 for the infectious state. Latent individuals do not move, but infectious individuals migrate. As Mollison [23] has pointed out, the model (9.9) is the nucleus of a rabies model by Murray. Although (9.9) is, strictly mathematically, a special case of (9.8), our biological interpretation and our range of parameters are different.

Now we return to the study of the system (9.8).

Case 1: Fast switching between the sedentary and the migrating state. We assume that γ_1 and γ_2 are both large and apply the scaling, with $\epsilon > 0$ small:

$$\gamma_1 = \rho_1/\epsilon, \qquad \gamma_2 = \rho_2/\epsilon. \tag{9.10}$$

Then, after a transient period, we have, from the second and third equations of (9.8), letting $\epsilon \to 0$ (and assuming the functions and their time derivatives stay bounded),

$$\rho_2 z = \rho_1 w. \tag{9.11}$$

With sufficient regularity (of K) we can expect that this relation carries over to the time derivatives $\rho_2 z_t = \rho_1 w_t$. Then we find

$$\begin{aligned} w_t + z_t &= w_t + \frac{\rho_1}{\rho_2} w_t \\ &= \delta K \frac{\rho_1}{\rho_2} w + u\big(\beta_1 w + \beta_2 \frac{\rho_1}{\rho_2} w\big) - \alpha\big(w + \frac{\rho_1}{\rho_2} w\big); \end{aligned} \tag{9.12}$$

hence

$$\frac{\rho_1 + \rho_2}{\rho_2} w_t = \frac{\rho_1}{\rho_2} \delta K w + u \frac{\beta_1 \rho_2 + \beta_2 \rho_1}{\rho_2} w - \alpha \frac{\rho_1 + \rho_2}{\rho_2} w, \tag{9.13}$$

and finally

$$w_t = \frac{\rho_1}{\rho_1 + \rho_2} \delta K w + \frac{\beta_1 \rho_2 + \beta_2 \rho_1}{\rho_1 + \rho_2} u w - \alpha w. \tag{9.14}$$

This equation reflects the fact that the processes of migration and infection by migrants act during a proportion $\rho_1/(\rho_1 + \rho_2)$ of the full time, while the process of direct infection acts a proportion $\rho_2/(\rho_1 + \rho_2)$ of the full time. Similarly, we get for z

$$z_t = \frac{\rho_1}{\rho_1 + \rho_2} \delta K z + \frac{\beta_1 \rho_2 + \beta_2 \rho_1}{\rho_1 + \rho_2} u z - \alpha z. \tag{9.15}$$

Notice that (9.14) and (9.15) are identical equations linear in w and z, respectively. Now it appears as if something has gone wrong since the terms giving the number of newly infected in (9.14), (9.15), and the first equation of (9.8) look different. But observe that according to our assumption (9.11) the total number of infected is

$$v = w + z = w + \frac{\rho_1}{\rho_2} w = \frac{\rho_1 + \rho_2}{\rho_2} w \tag{9.16}$$

and hence

$$u_t = -u\big(\beta_1 w + \beta_2 \frac{\rho_1}{\rho_2} w\big) = -u \frac{\rho_2 \beta_1 + \rho_1 \beta_2}{\rho_1 + \rho_2} v.$$

On the other hand, (9.14) and (9.15) are linear in w and z, respectively, and hence the total number of infected satisfies the same equation. Hence we have the following reaction-diffusion system for the susceptibles u and the total number of infected v:

$$\begin{aligned} u_t &= -\beta_m u v, \\ v_t &= \beta_m u v - \alpha v + D_m K v, \end{aligned} \tag{9.17}$$

where

$$\beta_m = \frac{\beta_1\rho_2 + \beta_2\rho_1}{\rho_1 + \rho_2} = \frac{\beta_1\gamma_2 + \beta_2\gamma_1}{\gamma_1 + \gamma_2} \tag{9.18}$$

is the *effective* infection rate and

$$D_m = \frac{\rho_1}{\rho_1 + \rho_2}\delta = \frac{\gamma_1}{\gamma_1 + \gamma_2}\delta \tag{9.19}$$

is the *effective* diffusion rate.

The effective infection rate is the arithmetic mean of β_1 and β_2, and the effective diffusion rate is the arithmetic mean of 0 and δ, both with weights γ_2 and γ_1. We conclude that the assumption of fast switching between the sedentary state and the migratory state with rates of about equal order of magnitude leads to a reaction-diffusion system.

Case 2: Rapid excursions of highly infectious individuals. We start again from the system (9.8). Now we introduce the following scaling:

$$\beta_2 = \bar{\beta}_2/\epsilon, \qquad \gamma_2 = \bar{\gamma}_2/\epsilon, \qquad \delta = \bar{\delta}/\epsilon \tag{9.20}$$

with $\epsilon > 0$ small. The assumptions say that the migrating infected move very fast (δ large) but each time just for a short period (the return rate γ_2 is also large). Hence they make rapid excursions, but they are highly infective (the infection rate β_2 for migrating infected is large). Because of the large return rate there are only few migrating infected at any given moment. The size of the migrating class z scales like ϵw in comparison to the sedentary class. Therefore, we put

$$z = \epsilon\tilde{z}, \tag{9.21}$$

and we get the equivalent rescaled system:

$$\begin{aligned} u_t &= -u(\beta_1 w + \bar{\beta}_2\tilde{z}),\\ \epsilon\tilde{z}_t &= \bar{\delta}K\tilde{z} - \epsilon\alpha\tilde{z} - \bar{\gamma}_2\tilde{z} + \gamma_1 w,\\ w_t &= u(\beta_1 w + \bar{\beta}_2\tilde{z}) - \alpha w + \bar{\gamma}_2\tilde{z} - \gamma_1 w. \end{aligned} \tag{9.22}$$

As in Case 1 we use the small parameter ϵ to get a approximately stationary relation between z and w. From the second equation of (9.22) we see that, after a transient time has passed, the following equation approximately holds:

$$\bar{\gamma}_2\tilde{z} - \bar{\delta}K\tilde{z} = \gamma_1 w. \tag{9.23}$$

The fact that δ is assumed large (and hence $\bar{\delta}$ is not of order ϵ) causes a mathematical difficulty, (9.23) being an operator equation rather than just a constant coupling as in (9.11). We solve for $\tilde{z}$ in terms of w,

$$\tilde{z} = \tilde{K}w, \tag{9.24}$$

where $\tilde{K}$ is the "resolvent operator":

$$\tilde{K} = \gamma_1(\bar{\gamma}_2 I - \bar{\delta}K)^{-1}. \tag{9.25}$$

Recall that the operator K has the form $Kv = kv - v$ with a convolution operator k. The operator $\tilde{K}$ is, up to factors, the resolvent of K. Hence it has again the form of a multiple of the identity and a convolution operator. Notice that the operator $\tilde{K}$ depends in a nonlinear way on the parameters $\bar{\gamma}_2$ and $\bar{\delta}$. In this case the total number of infected is $v = w + \epsilon \tilde{z} \approx w$. Hence we get from (9.22) a system for the susceptibles and the total number of infected:

$$\begin{aligned} u_t &= -u(\beta_1 v + \bar{\beta}_2 \tilde{K} v), \\ v_t &= u(\beta_1 v + \bar{\beta}_2 \tilde{K} v) - \alpha v - \gamma_1 v + \bar{\gamma}_2 \tilde{K} v. \end{aligned} \tag{9.26}$$

Now we take a closer look at the integral operator $\tilde{K}$. By assumption we have $K = k - I$. By simple computations we find

$$\tilde{K} = \frac{\gamma_1}{\bar{\gamma}_2 + \bar{\delta}} I + \frac{\gamma_1}{\bar{\gamma}_2 + \bar{\delta}} \frac{\bar{\delta}}{\bar{\gamma}_2} \hat{k}, \tag{9.27}$$

where the convolution operator $\hat{k}$ is given by

$$\hat{k} = \frac{\bar{\gamma}_2}{\bar{\delta}} \sum_{i=1}^{\infty} \left(\frac{\bar{\delta}}{\bar{\gamma}_2 + \bar{\delta}} \right)^i k^i \tag{9.28}$$

with kernel normalized to $\hat{k} * 1 = 1$. With this notation we get a standard form for the system for susceptibles u and infected v:

$$\begin{aligned} u_t &= -u\Big(\big(\beta_1 + \frac{\bar{\beta}_2 \gamma_1}{\bar{\gamma}_2 + \bar{\delta}}\big) v + \frac{\bar{\beta}_2 \gamma_1 \bar{\delta}}{(\bar{\gamma}_2 + \bar{\delta}) \bar{\gamma}_2} \hat{k} v\Big), \\ v_t &= u\Big(\big(\beta_1 + \frac{\bar{\beta}_2 \gamma_1}{\bar{\gamma}_2 + \bar{\delta}}\big) v + \frac{\bar{\beta}_2 \gamma_1 \bar{\delta}}{(\bar{\gamma}_2 + \bar{\delta}) \bar{\gamma}_2} \hat{k} v\Big) - \alpha v + \frac{\gamma_1 \bar{\delta}}{\bar{\gamma}_2 + \bar{\delta}} (\hat{k} v - v). \end{aligned} \tag{9.29}$$

We see that after taking the inverse in (9.25) the last term in the second equation still has the correct form of a migration operator $\hat{k} - I$.

With the present scaling (9.20) we have arrived at a system (9.29) for susceptibles and infected where the motion of infected is described by a migration term, $\hat{k} v - v$, and the infection process is described by an infection term with a contact distribution. The contact distribution has a local term (a "point mass") and a term for transmission at a distance (a continuous part with $\hat{k}$ entering as a kernel). The infection rate β_2 of the relatively few migrating infected enters not only in the continuous part but also in the local part.

Apparently these features makes sense. We have compounded both sedentary and migrating infected into one group. Although the migrating infected form only a small proportion of this group, they contribute considerably to the overall movement and the infective force in this group.

The system (9.29) gives a meaning to modeling approaches which use a migration term and a contact distribution together. From (9.29) we see that the diffusion term $\hat{k} v - v$ and the contact term $\hat{k} v$ are coupled via several parameters, notably the kernel $\hat{k}$ itself.

Now we can ask whether there is a further scaling that rids the model completely of the migration term and leads to Kendall's model. Looking at (9.29) we see that we must make γ_1 small but keep $\bar{\beta}_2 \gamma_1$ large. We can choose $\bar{\beta}_2$ large and γ_1 small such that the product stays constant, with a small parameter ε (different from the ϵ we have used in (9.20)):

$$\bar{\beta}_2 = \hat{\beta}_2 / \varepsilon, \quad \gamma_1 = \varepsilon \hat{\gamma}_1. \tag{9.30}$$

This scaling could have been done right away in (9.20) but performing both scalings in one step would be somewhat confusing. Notice that $\bar{\beta}_2$ and γ_1 did not enter into the definition of

$\hat{k}$. If we let ε go to zero, then the migration term vanishes and we get a pure contact model,

$$\begin{aligned} u_t &= -u(\beta_c v + D_c \hat{k} v), \\ v_t &= u(\beta_c v + D_c \hat{k} v) - \alpha v, \end{aligned} \tag{9.31}$$

where

$$\beta_c = \beta_1 + \frac{\hat{\beta}_2 \hat{\gamma}_1}{\bar{\gamma}_2 + \bar{\delta}} = \beta_1 + \frac{\beta_2 \gamma_1}{\gamma_2 + \delta} \tag{9.32}$$

is the point mass at the position of the susceptible and

$$D_c = \frac{\hat{\beta}_2 \hat{\gamma}_1 \bar{\delta}}{(\bar{\gamma}_2 + \bar{\delta})\bar{\gamma}_2} = \frac{\beta_2 \gamma_1 \delta}{(\gamma_2 + \delta)\gamma_2} \tag{9.33}$$

is the coefficient of the continuous part of the contact distribution (recall $\hat{k} * 1 = 1$).

In the expressions for β_c and D_c we have returned to the original parameters γ_i, β_i, and δ, which is possible because of the homogeneous dependence on ϵ. We discuss the dependence of the resulting expressions on the original parameters.

The infection rate β_c depends on the local infection rate β_1 in a simple additive way. At a fixed space position there is a local population and the spread within that local population is governed by β_1. Even if $\beta_1 = 0$, an epidemic can occur if β_2 is sufficiently large. On the other hand, if β_2 vanishes (a hypothetical situation because $\beta_2 = 0$ would make our scaling invalid), then the disease cannot spread even if β_1 were large. The rate D_c does not depend on β_1 at all, which just says that β_1 acts indeed locally only. Both β_c and D_c are monotonically increasing in β_2, saying that increasing β_2 enhances infections between distant individuals but also, indirectly, between individuals at the same position. Both expressions increase with γ_1 and decrease with γ_2, saying that infection acts stronger if the infectious individuals stay longer in the migrating compartment.

The rate D_c is increasing in the diffusion rate δ. Hence increasing the diffusion rate of migrating infected individuals increases the contact distribution everywhere.

From (9.25) and also from (9.27) it is obvious that $\tilde{K}$ is increasing in $\bar{\delta}$. But the local indirect infection term (second term in β_c) decreases with δ. The formula (9.27) shows that $\bar{\delta}$ acts like $\bar{\gamma}_2$. At a fixed position it removes migrating infected from the total number of infected.

9.3 Speed of Epidemic Waves

A traveling front of any of the models (9.17) or (9.31) is a solution u, v of the form $u(x - ct)$, $v(x - ct)$ with constant profile u, v and constant speed $c > 0$ which satisfies appropriate boundary conditions at $\pm\infty$. If we assume that the front travels from $-\infty$ to $+\infty$, then at $+\infty$ we have the uninfected equilibrium $(u, v) = (1, 0)$, whereas at $-\infty$ we have the equilibrium of a population where the epidemic has passed through, i.e., $(u, v) = (u_\infty, 0)$.

The speed of epidemic fronts can be studied by formulating a complete model and then performing a traveling wave analysis, either by determining the so-called spread number [1, 30] or by actually proving the existence of fronts and determining the speed. In complex models, only a linear analysis at the edge of the front, i.e., at the uninfected equilibrium, may be feasible. The linear approach can be justified in many, but not in all, cases [7, 30].

On the other hand, one can base the study of the spread of an epidemic on a linear model using a reproduction and dispersal kernel (RD kernel). Van den Bosch, Metz, and Diekmann

[27] study wave speeds in terms of RD kernels, and Mollison [23] derives RD kernels for various models using stochastic interpretations following earlier papers on stochastic epidemics [21, 22]. It is evident that a given RD kernel can be derived from several models with rather different interpretations (see [5, 23]) and also that deriving the kernel from a given model may be a formidable task. Metz and van den Bosch [20] give a detailed review on spatial epidemics with particular attention to pests of crops.

For the contact model with diffusion term (9.7) and for the migration model with diffusion term, the speed of the traveling front is explicitly known in terms of the parameters; see [18, 7] for the contact case and [17, 8] for the reaction-diffusion case.

We make use of this information in comparing the systems (9.17) and (9.31). Hence we use the diffusion approximation (9.5) in *all* systems. Then the system (9.8) becomes

$$\begin{aligned} u_t &= -u(\beta_1 w + \beta_2 z), \\ z_t &= \delta\sigma^2 z_{xx} - \alpha z + \gamma_1 w - \gamma_2 z, \\ w_t &= u(\beta_1 w + \beta_2 z) - \alpha w - \gamma_1 w + \gamma_2 z. \end{aligned} \tag{9.34}$$

In order to make an exact comparison we should relate the second moment of $\hat{k}$ to the moments of k. If we assume that δ is small as compared to γ_2, then $\bar{\delta}$ is small as compared to $\bar{\gamma}_2$ by (9.10), and then approximately $\hat{k} = k$ in view of (9.28).

Then it makes sense to compare the diffusion problem

$$\begin{aligned} u_t &= -\beta_m uv, \\ v_t &= \beta_m uv - \alpha v + \sigma^2 D_m v_{xx} \end{aligned} \tag{9.35}$$

and the contact problem

$$\begin{aligned} u_t &= -u(\beta_0 v + D_0 v_{xx}), \\ v_t &= u(\beta_0 v + D_0 v_{xx}) - \alpha v \end{aligned} \tag{9.36}$$

with

$$\beta_0 = \beta_c + D_c = \frac{\beta_1\gamma_2 + \beta_2\gamma_1}{\gamma_2}, \tag{9.37}$$

$$D_0 = \sigma^2 D_c, \tag{9.38}$$

and β_m, D_m, β_c, D_c given by (9.18), (9.19), (9.32), (9.33).

From the explicit formulas we find that in all expressions the scaling parameter ϵ and also ε enter in a homogeneous way and thus cancel. Hence we can express the speed in terms of the original parameters. In the migration (reaction-diffusion) case (9.35) one gets the minimal speed

$$c_m = 2\sigma\sqrt{D_m(\beta_m - \alpha)} = 2\sigma\sqrt{\frac{\delta\gamma_1}{\gamma_1+\gamma_2}\left(\frac{\beta_1\gamma_2 + \beta_2\gamma_1}{\gamma_1+\gamma_2} - \alpha\right)}. \tag{9.39}$$

In the contact case, (9.36), one gets the minimal speed

$$c_c = 2\sqrt{D_0(\beta_0 - \alpha)} = 2\sigma\sqrt{\frac{\delta\beta_2\gamma_1}{(\gamma_2+\delta)\gamma_2}\left(\frac{\beta_1\gamma_2 + \beta_2\gamma_1}{\gamma_2} - \alpha\right)}. \tag{9.40}$$

We remark in passing that for the general system (9.34) a scaling argument shows that the speed of the epidemic front is proportional to $\sqrt{\delta}$. This proportionality is destroyed by the singular perturbation transition from (9.22) to (9.26): c_m is homogeneous in $\sqrt{\delta}$, but c_c is not. This phenomenon might appear as invalidating the result but it is quite standard in singular perturbation approaches like (9.23).

One can determine the speed of epidemic waves for the general system (9.34) (assuming that the so-called linear conjecture does hold [30]). A linear analysis near the uninfected equilibrium [27] or a stochastic approach [23] leads to a cubic, which then can be discussed numerically.

Now we try to interpret the two expressions for the minimal speeds c_m and c_c. In the reaction-diffusion (migration) limit the effective infection rate does not depend on δ or any properties of the kernel. The speed c_m is proportional to $\sqrt{\delta}$ and does not depend on the infection rates other than through β_m, which is the arithmetic mean of the rates β_2 and β_1. In the contact case, (9.36), the first factor in the speed c_c also depends on δ but it is not proportional to $\sqrt{\delta}$ because δ occurs also in the denominator. The most important feature is the dependence on β_2. The first factor in (9.40) is proportional to β_2, and hence for large β_2 the speed increases essentially linearly with β_2. This statement cannot be carried to the extremes because in scaling (9.30) we had assumed that $\beta_2\gamma_1$ is constant in ε. In any case we can state that short and fast excursions of very few highly infective individuals lead to a very fast epidemic wave as compared to a balanced migration behavior of the total population. Indeed, in (9.39) the dominant factor in D is δ, and in (9.40) the dominant factor is β_2.

9.4 Conclusion

A reaction-diffusion model for epidemic spread with a sedentary and a migrating compartment is considered. Both the contact model and the migration model can be obtained as limiting cases by appropriate scaling of parameters whereby the limits to be taken reveal the appropriate epidemiological scenario. A reaction-diffusion system for susceptibles and all infected is obtained as a limiting case for large and roughly equal switching rates. In this case the speed of traveling fronts is largely determined by the migration process. If the switching parameters are scaled in an asymmetrical fashion, then one arrives at models with diffusion terms and contact distributions together and eventually at pure contact models. In the contact limit, the speed of propagation is strongly influenced by the infection process: Few highly infectious individuals which make rapid excursions lead to fast propagation of the epidemic front. The speed of the front depends linearly on the infection rate as compared to a square root dependence in the reaction-diffusion case.

The reaction-diffusion system and the contact system are both valid models but apply to different scenarios. The diffusion system is indeed well suited for the situations it was designed for, such as rabies in foxes or for the spread of plague in times of slow transport media as in [24]. The contact model seems better suited for modelling spread of disease by frequent travels of relatively few (as compared to the total infected population) highly infectious individuals.

Acknowledgment

This work was performed in part while the author was visiting the Mathematical Biology Centre with a grant from the EFF Distinguished Visitor Fund of the University of Alberta, and was also supported by DFG Priority Research Program ANumE.

Bibliography

[1] D. ARONSON, *The asymptotic speed of a simple epidemic*, in Nonlinear Diffusion, Res. Notes Math. 14, W. E. Fitzgibbon and H. F. Walker, eds., Pitman, London, 1977, pp. 1–23.

[2] C. ATKINSON AND G. E. REUTER, *Deterministic epidemic waves*, Math. Proc. Cambridge Philos. Soc., 80 (1976), pp. 315–330.

[3] F. BALL, D. MOLLISON, AND G. SCALIA-TOMBA, *Epidemics with two levels of mixing*, Ann. Appl. Probab., 7 (1997), pp. 46–89.

[4] J. COOK, *Dispersive Variability and Invasion Wave Speeds*, unpublished manuscript, 1993.

[5] O. DIEKMANN, *Thresholds and travelling waves for the geographical spread of infection, travelling waves and the asymptotic speed of propagation*, J. Math. Biol., 6 (1978), pp. 109–130.

[6] O. DIEKMANN AND H. HEESTERBEEK, *Mathematical Epidemiology of Infectious Diseases: Model Building, Analysis and Interpretation*, Wiley, New York, 1999.

[7] K. P. HADELER AND F. ROTHE, *Travelling fronts in nonlinear diffusion equations*, J. Math. Biol., 2 (1975), pp. 251–263.

[8] K. P. HADELER, *Travelling epidemic waves and correlated random walks*, in Differential Equations and Applications to Biology and Industry, M. Martelli, K. Cooke, E. Cumberbatch, B. Tang, and H. Thiene, eds., World Scientific, Singapore, 1996, pp. 145–156.

[9] K. P. HADELER, *Spatial epidemic spread by correlated random walk with slow infectives*, in Ordinary and Partial Differential Equations, Vol. V, P. D. Smith and R. J. Jarvis, eds, Pitman Res. Notes Math. Ser. 370, Longman, Harlow, UK, 1997, pp. 18–32.

[10] K. P. HADELER AND C. CASTILLO-CHAVEZ, *A core group model for disease transmission*, Math. Biosci., 128 (1995), pp. 41–55.

[11] K. P. HADELER AND P. VAN DEN DRIESSCHE, *Backward bifurcation in epidemic control*, Math. Biosci., 146 (1997), pp. 15–35.

[12] K. P. HADELER AND M. A. LEWIS, *Spatial dynamics of the diffusive logistic equation with sedentary compartment*, in preparation.

[13] H. W. HETHCOTE AND H. R. THIEME, *Stability of the endemic equilibrium in epidemic models with subpopulations*, Math. Biosci., 75 (1985), pp. 205–227.

[14] T. HILLEN, *Transport equations with resting phases*, European J. Appl. Math., to appear.

[15] Y. HOSONO AND B. ILYAS, *Existence of traveling waves with any positive speed for a diffusive epidemic model*, Nonlinear World, 1 (1995), pp. 277–290.

[16] Y. HOSONO AND B. ILYAS, *Traveling waves for a simple diffusive epidemic model*, Math. Models Methods Appl. Sci., 5 (1995), pp. 935–966.

[17] A. KÄLLÉN, *Thresholds and travelling waves in an epidemic model for rabies*, Nonlinear Anal., 8 (1984), pp. 851–856.

[18] D. KENDALL, *Mathematical models of the spread of infections*, in Mathematics and Computer Science in Biology and Medicine, Medical Research Council, London, 1965, pp. 213–225.

[19] M. A. LEWIS AND G. SCHMITZ, *Biological invasion of an organism with separate mobile and stationary states: Modeling and analysis*, Forma, 11 (1996), pp. 1–25.

[20] J. A. J. METZ AND F. VAN DEN BOSCH, *Velocities of epidemic spread*, in Epidemic Models: Their Structure and Relation to Data, Publ. Newton Inst. 5, D. Mollison, ed., Cambridge University Press, Cambridge, UK, 1995, pp. 150–186.

[21] D. MOLLISON, *Possible velocities for a simple epidemic*, Adv. Appl. Probab., 4 (1972), pp. 233–257.

[22] D. MOLLISON, *The rate of spatial propagation of simple epidemics*, in Proceedings of the 6th Berkeley Symposium on Mathematical Statisitics and Probability, Vol. III, L. M. Le Cam, J. Neyman, and E. L. Scott, eds., University of California Press, Berkeley, CA, 1972, pp. 579–614.

[23] D. MOLLISON, *Dependence of epidemic and population velocities on basic parameters*, Math. Biosci., 107 (1991), pp. 255–287.

[24] J. V. NOBLE, *Geographical and temporal development of plagues*, Nature, 250 (1974), pp. 762–768.

[25] J. RADCLIFFE AND L. RASS, *The asymptotic speed of propagation of the deterministic non-reducible n-type epidemic*, J. Math. Biol., 23 (1986), pp. 341–359.

[26] L. SATTENSPIEL AND K. DIETZ, *A structured epidemic model incorporating geographic mobility among regions*, Math. Biosci., 128 (1995), pp. 71–91.

[27] F. VAN DEN BOSCH, J. A. J. METZ, AND O. DIEKMANN, *The velocity of spatial population expansions*, J. Math. Biol., 28 (1990), pp. 529–556.

[28] M. O. VLAD AND B. SCHÖNFISCH, *Mass action law versus contagion dynamics. A mean-field statistical approach with application to the theory of epidemics*, J. Phys. A, 29 (1996), pp. 4895–4913.

[29] M. O. VLAD, B. SCHÖNFISCH, AND C. LACOURSIÈRE, *Statistical-mechanical analogies for space-dependent epidemics*, Phys. A, 229 (1996), pp. 365–401.

[30] H. F. WEINBERGER, M. A. LEWIS, AND BINGTUAN LI, *Analysis of linear determinacy for spread in cooperative models*, J. Math. Biol., 45 (2002), pp. 219–233.

Chapter 10

Modeling the Spread of Influenza among Cities

James M. Hyman and Tara LaForce†*

10.1 Introduction

Every year influenza kills thousands of Americans, and millions are stricken ill. Despite attempts to vaccinate high-risk populations the cost of influenza in terms of lost life and lost productivity is still high. Mathematical models can provide insight into how influenza spreads between individual people within a community and across the United States. This insight can potentially help guide health care workers anticipate the number of influenza cases each year and identify anomalies that might foretell an unexpected new or stronger epidemic.

Weekly data on the mortality and morbidity attributed to influenza and pneumonia (P/I) is collected by the Centers for Disease Control and Prevention (CDC) Epidemiology Branch Office for each of 122 cities in the United States. We analyze this data search for correlations between the number of cases in a city and other epidemiological parameters, such as the population of the city, the infectiousness of the disease, and the transmission from city to city during an epidemic. We also searched for possible precursor cities that might herald the start of a new season.

The simplest susceptible-infected-recovered (SIR) model, developed in 1927 by Kermack and McKendrick [9], assumes random mixing of the population. Models for the initial spread of infectious agents through nonrandom mixing populations where people move between groups in a single population is needed to predict the early spread of a disease [13]. However, after the short initial growth of an epidemic in a city, the random mixing assumption can be used for the spread of diseases, such as influenza, within a city.

The mixing between the cities still needs to be modeled to predict the spread of a disease between cities when their populations are unevenly spread over a large geographic area. In 1985, Rvachev and Longini published the first of a series of papers modeling the spread of influenza from city to city around the world during the 1968–1969 pandemic [8, 12].

*Theoretical Division, MS-B284, Center for Nonlinear Studies, Los Alamos National Laboratory, Los Alamos, NM 87545.

†Petroleum and Geosystems Engineering, University of Texas at Austin, Austin, TX 78712.

They developed a multicity model in which air travel was used to approximate the spread of the pandemic from its (assumed) origins in Hong Kong to 51 other major populations centers worldwide. The structure of our model is similar to the one developed by Longini and Rvachev.

We model the spread of the disease in each city by a system of deterministic differential equations. The susceptible and infected people are assumed to mix randomly within a city. We assume that the nonrandom mixing of the population among the cities between the major cities is captured in the model by the air travel. We also assume that people continue to travel when they are infectious. This assumption restricts the current model to specific diseases, such as influenza, where people are asymptomatic or only mildly ill while infected. Our model could be extended to account for people who change their travel plans due to the severity of the disease by adding another infection stage in the model.

The model parameters for the transmission and disease progression are estimated from the literature [14, 1]. The rate that people move between the cities is based on airline data, as was done in [12]. We were unable to estimate two parameters from the literature: the average number of contacts a typical city person has during a day that could result in transmitting the disease, and the seasonal change in the infectivity of influenza. We assumed that these two parameters were the same for every city and estimated them by an L1 fit to the CDC data. We fit the parameters on several different cities. The fits on the cities with the best data provided similar estimates for these parameter values.

After defining the model and parameters, we compare the results of the model with the data collected by the 122 Cities Mortality Reporting System and reported to the CDC and conclude that this model approximates the magnitude and fluctuation of the influenza seasons for 1996–2001. This simple model did remarkably well at predicting the yearly influenza epidemic.

The SIR transmission model for a single city is then extended to predict the spread of the virus among multiple cities. We provide details on how we estimated the model parameters and data for the population and traffic between the cities. We provide a simple analysis for the reproductive number and the sensitivity of the model predictions to small changes in the parameter estimates. Finally, we provide a comparison of the model predictions with the CDC data for several representative cities.

10.2 Multicity Transmission

We first define and analyze the single-city model, then add a term to account for the return to susceptibility of an infected person and an emigration/migration term. Next, we generalize the model to a multicity transmission model where the epidemic is spread through the network of cities by the migration of infected people.

10.2.1 Single-City Transmission Model

By dividing a fixed population into susceptible (S), infected (I), and recovered (R) individuals, the simple SIR model

$$dS/dt = -\lambda S, \tag{10.1a}$$

$$dI/dt = \lambda S - \alpha I, \tag{10.1b}$$

$$dR/dt = \alpha I \tag{10.1c}$$

can predict the spread of the single outbreak of a disease. Here the susceptible population is infected at a rate $\lambda > 0$, and the infected population recovers from illness at the rate $\alpha > 0$. For a randomly mixing population, the infection rate, λ, can be estimated by the product of three terms:

$$\lambda = r\beta\frac{I}{N}. \tag{10.2}$$

The average number of contacts per individual per unit time, $r > 0$, is assumed to be constant for large populations. A contact is defined as an encounter between people that could transmit the disease. When defining r, we have assumed that the contacts in the population are random and the number of contacts per unit time is independent of the population size. This is an appropriate assumption for large cities, but for smaller populations or villages it might be more appropriate for r to be proportional to the total population, $N = S + I + R$, of the city. That is, if the population of the village increased by 10%, then a typical person would have 10% more contacts. This is not true for large cities.

The next factor in λ accounts for the probability of transmitting the disease in a contact. This factor, $\beta > 0$, is proportional to the average infectiousness of an infected person times the average susceptibility of a susceptible person. Influenza is more likely to spread in the winter than the summer [17, 18, 19, 20]. This may be caused by an increased infectiousness of the disease, an increased susceptibility of people, or an increased number contacts with others that might result in transmitting the infection during the winter. For example, people may spend more time indoors. Because r and β only occur as a product in the model, we can approximate both of these effects by allowing the product $r\beta$ to vary through the year:

$$r\beta = \hat{\beta}\hat{r}[1 + \epsilon \sin(2\pi t/365)]. \tag{10.3}$$

Here $\epsilon < 1$ is the fluctuation in infectivity between seasons and will be determined by a fitting ϵ, the seasonal variation of the model to CDC influenza data. We assume that ϵ is the same in all of the cities.

The final factor (I/N) in λ reflects that the probability a randomly chosen contact infected is equal to the fraction of the population that is infected.

10.2.2 Loss of Immunity

Once entering the recovered state, people slowly return to being susceptible to the currently circulating strain of the virus. This can happen because the small mutations in genetic code in the surface antigens allow influenza to drift in ways that eventually allow it to evade the immune system defenses from an earlier infection. We partially account for the loss of immunity by including a new state between the recovered state and the susceptible state in which people have partial immunity to the current strain of the disease. We define the constant rates η^R for return to partial immunity and η^P for a full return to susceptibility. We include the state by modifying the SIR model, $S \to I \to R$, to an SIRP model, $S \to I \to R \to P \to S$, where P is the stage of partial immunity. To fully account for the drift of the virus requires a multistrain model, and is beyond the scope of this study.

We also extend the model to account for people entering and leaving the population either through birth and death or by leaving the network of cities in the model. We define the migration term in a form where S^0 is the stable susceptible population in the absence of infection. In this simple model, we do not explicitly include a term to account for the increase in mortality due to influenza.

The resulting single-city transmission SIPR model is

$$dS/dt = -\lambda^S S + \eta^P P + \mu(S^0 - S), \tag{10.4a}$$
$$dI/dt = \lambda^S S + \lambda^P P - \alpha I - \mu I, \tag{10.4b}$$
$$dR/dt = \alpha I - \eta^R R - \mu R, \tag{10.4c}$$
$$dP/dt = \eta^R R - \eta^P P - \lambda^P P - \mu P. \tag{10.4d}$$

Both susceptible and partially immune people can be infected. As before, the infection rate for the fully susceptible population is $\lambda^S = \beta^S r(I/N)$, where β^S is the susceptibility of a person in S. Similarly, partially immune people are infected with the rate $\lambda^P = \beta^P r(I/N)$. Because people in the partially immune state are less likely to be infected in a single contact, we assume $\beta^S > \beta^P$. The total population is now defined as $N = S + I + R + P$.

10.2.3 Multicity Transmission Model

We begin by modeling two cities (city 1 and city 2) with people traveling between them. The epidemic is modeled within each city by the single-city model. We use subscripts to indicate the city number and define m_{ij} to be the number of people traveling from city i to city j per unit time. The equations for city 1 and city 2 are

$$dS_1/dt = -\lambda_1^S S_1 + \eta^P P_1 + \mu(S_1^0 - S_1) + m_{21}\frac{S_2}{N_2} - m_{12}\frac{S_1}{N_1}, \tag{10.5a}$$
$$dI_1/dt = \lambda_1^S S_1 + \lambda_1^P P_1 - \alpha I_1 - \mu I_1 + m_{21}\frac{I_2}{N_2} - m_{12}\frac{I_1}{N_1}, \tag{10.5b}$$
$$dR_1/dt = \alpha I_1 - \eta^R R_1 - \mu R_1 + m_{21}\frac{R_2}{N_2} - m_{12}\frac{R_1}{N_1}, \tag{10.5c}$$
$$dP_1/dt = \eta^R R_1 - \eta^P P_1 - \lambda_1^P P_1 - \mu P_1 + m_{21}\frac{P_2}{N_2} - m_{12}\frac{P_1}{N_1}, \tag{10.5d}$$

$$dS_2/dt = -\lambda_2^S S_2 + \eta^P P_2 + \mu(S_2^0 - S_2) + m_{12}\frac{S_1}{N_1} - m_{21}\frac{S_2}{N_2}, \tag{10.6a}$$
$$dI_2/dt = \lambda_2^S S_2 + \lambda_2^P P_2 - \alpha I_2 - \mu I_2 + m_{12}\frac{I_1}{N_1} - m_{21}\frac{I_2}{N_2}, \tag{10.6b}$$
$$dR_2/dt = \alpha I_2 - \eta^R R_2 - \mu R_2 + m_{12}\frac{R_1}{N_1} - m_{21}\frac{R_2}{N_2}, \tag{10.6c}$$
$$dP_2/dt = \eta^R R_2 - \eta^P P_2 - \lambda_2^P P_2 - \mu P_2 + m_{12}\frac{P_1}{N_1} - m_{21}\frac{P_2}{N_2}. \tag{10.6d}$$

Now $\lambda_j^S = \beta^S r \frac{I_j}{N_j}$ and $\lambda_j^P = \beta^P r \frac{I_j}{N_j}$ for $j =$ cities 1, 2 and $i = S, P$.

The populations of the two cities are constant if $m_{21} = m_{12}$. In this simple model with only one infection stage, we assume that infected people continue to travel with the same rate as the uninfected population. Thus the fraction of people traveling from city 2 to city 1 that are infected is equal to the fraction of people in city 2 who are infected, (I_2/N_2). Consequently, the number of infected people entering city 2 from city 1 is $I_1\ m_{21}(I_2/N_2)$.

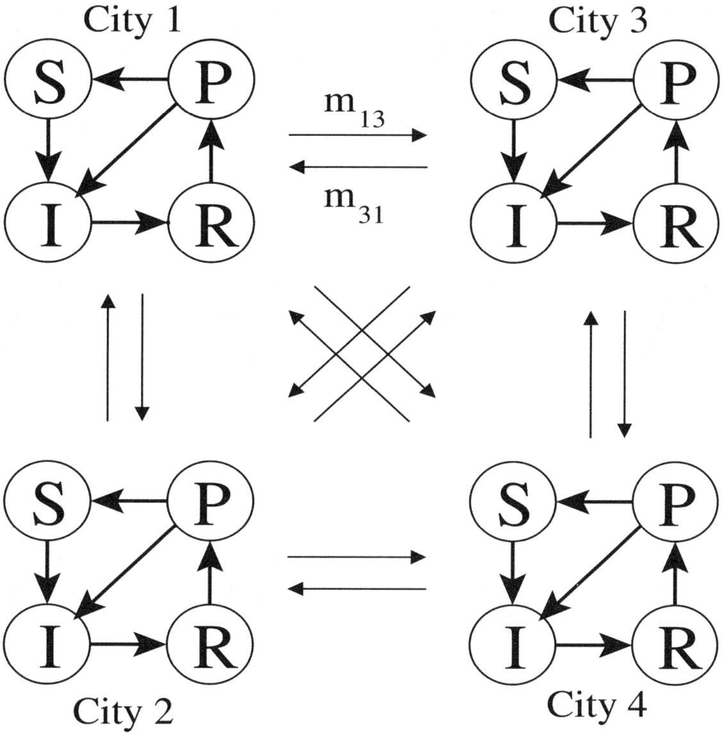

Figure 10.1. *A schematic of how people travel through the stages of the illness and between the cities for a four-city system. S—susceptible, I—infected, R—recovered (immune), P—partially susceptible.* $m_{i,j}$*—the number of people traveling from city i to city j per day.*

Figure 10.1 is a schematic of how people move through the stages of illness in a four-city model. To generalize the two-city model to a multicity model with n cities we first generalize the migration matrix M for the number of people traveling between cities:

$$M = \begin{pmatrix} m_{1,1} & m_{1,2} & \cdots & m_{1,n} \\ m_{2,1} & m_{2,2} & \cdots & m_{2,n} \\ \vdots & \vdots & \ddots & \vdots \\ m_{n,1} & m_{n,2} & \cdots & m_{n,n} \end{pmatrix}.$$

Here m_{ij} is the number of people per unit time who travel from city i to city j. We have assumed that the simulation will be used for a short enough time that the migration matrix is constant. If the model is modified to allow the populations of the cities to change, then the migration matrix must be recast as a function of the city populations. In our simulations, M is symmetric, $m_{ij} = m_{ji}$, and therefore the population of each city remains constant. The diagonal terms, m_{ii}, account for all the nontravelers that do not explicitly appear in the equations. The number of susceptible people entering city k is $\sum_{j=1}^{n} m_{jk}\frac{S_j}{N_j}$, and the number of susceptible people leaving city k is $\sum_{j=1}^{n} m_{kj}\frac{S_k}{N_k}$.

The resulting multicity model is

$$dS_k/dt = -\lambda_k^S S_k + \eta^P P_k + \mu(S_{0k} - S_k) + \sum_{j=1}^{n} m_{jk}\frac{S_j}{N_j} - \sum_{j=1}^{n} m_{kj}\frac{S_k}{N_k}, \quad (10.7a)$$

$$dI_k/dt = \lambda_k^S S_k + \lambda_k^P P_k - \alpha I_k - \mu I_k + \sum_{j=1}^{n} m_{jk}\frac{I_j}{N_j} - \sum_{j=1}^{n} m_{kj}\frac{I_k}{N_k}, \quad (10.7b)$$

$$dR_k/dt = \alpha I_k - \eta^R R_k - \mu R_k + \sum_{j=1}^{n} m_{jk}\frac{R_j}{N_j} - \sum_{j=1}^{n} m_{kj}\frac{R_k}{N_k}, \quad (10.7c)$$

$$dP_k/dt = \eta^R R_k - \eta^P P_k - \lambda_k^P P_k - \mu P_k + \sum_{j=1}^{n} m_{jk}\frac{P_j}{N_j} - \sum_{j=1}^{n} m_{kj}\frac{P_k}{N_k}. \quad (10.7d)$$

Where rates of infection are $\lambda_k^S = \beta^S r \frac{I_k}{N_k}$ and $\lambda_k^P = \beta^P r \frac{I_k}{N_k}$ for each city $k = 1, \ldots, n$. This is a generalization of an earlier model studied by Hyman and LaForce [6].

10.3 Model Parameters

We establish model parameters appropriate for influenza virus in this section. The values we discuss here are summarized in Table 10.1 below. Where possible, the parameters were obtained for strains of H3N2, the dominant strain of flu for the majority of our data set. The average illness lasts $1/\alpha$ days. We assume that rate of recovery from illness, α per day, is the same for all cities, regardless of location or season. We fix the duration of the infectious period to be 4.1 days; therefore $\alpha = 0.2439$ [1].

We assume that the infectiousness of an average infected person, the susceptibility of an average susceptible person, and the average number of contacts are the same in all of the cities. All infected people are equally infectious and all fully susceptible people are equally likely to contract the illness. In a single person, infectivity is also assumed to be constant through the course of the illness. The infectiousness of an individual is higher in the winter than in the summer, regardless of the climate of the city. This may be due to increased infectiousness of the disease (β) or a change in the contact pattern of the average individual. This is reflected in the model by allowing the product $r\beta$ to be dependent on the season.

We define $\hat{\beta}^S$ as the mean infectivity per contact. Stilianakis et al. estimate the transmission rate as $\beta \approx 6 \times 10^{-4}$ in a randomly mixing population in the absence of drug resistance for a subclinical infection in [14]. We will use this as our baseline infectivity parameter. Subclinical infection is defined as an infection where a person is asymptomatic but still infectious, and it accounts for approximately 75% of all H3N2 infections [10].

Including a mechanism for previously infected people to return to susceptibility is a simple mechanism to account for the loss of immunity caused by the natural genetic drift of the virus away from the strain with which a person was infected, as opposed to an actual loss of immunity to a specific strain of the virus within a person. That is, we assume that the genetic strain of virus drifts at a constant rate and that previously infected people slowly become susceptible to the currently circulating strain. The rates of return to partial and full susceptibility after illness are η^R and η^P. Frank et al. [4] reported no recorded cases of people becoming reinfected with H3N2 in the same season as a primary infection. Thus $\eta^R \leq 1/365 days$. That is, people are fully immune for at least a year.

Table 10.1. *Model parameters, dimensions, baseline values used in the simulations, and estimated ranges of validity. Most of these parameters are assigned values estimated from the epidemiology literature. The two parameters r and ϵ, whose values are unknown, were determined by a least squares fit so the model best matches the influenza data. The ranges for the average number of adequate contacts per day to transmit the disease, $r\beta$, and ϵ was chosen to reflect the differences between in the least squares fit for different cities. The range for r was determined by dividing $r\beta$ by β.*

meaning	parameter	baseline	suitable range
rate of recovery $(1/days)$ [1]	α	0.2439	$[0.07, 0.5]$
mean transmission probability per contact(fully susceptible) $(1/contacts)$ [14]	$\hat{\beta}^S$	6×10^{-4}	$[5 \times 10^{-4}, 6 \times 10^{-3}]$
mean transmission probability per contact (partially susceptible) $(1/contacts)$	$\hat{\beta}^P = \sigma\hat{\beta}^S$	$\sigma = 0.55$	$[0.3, 0.7]$
seasonal fluctuation of transmission probability (dimensionless)	ϵ	0.0210	$[0.007, 0.026]$
number of contacts per unit time $(contacts/day)$	r	410.38	$[42, 625]$
number of adequate contacts per unit time $(contacts/day)$	$r\beta$	0.246	$[0.24, 0.26]$
removal rate of people from population in the absence of infection $(1/days)$	μ	0.0002740	$[0.000027, 0.25]$
rate of return to partial susceptibility $(1/days)$ [4]	η^R	0.00274	$[0.00137, 0.00274]$
rate of return to full susceptibility$(1/days)$	η^P	0.00137	$[0, 0.00549]$

Partially immune people are infected at a rate that is a fraction of the rate at which fully susceptible people are infected. Following full immunity, we allow a person to become partially susceptible to similar circulating strains of influenza. A person may have a partial immunity to the circulating strain of influenza for several years, or none at all, depending on the rate of drift of the virus. We define the time a person becomes partially immune when their susceptibility is the ratio $\sigma = (\beta^P/\beta^S)$ less than the susceptibility of people who have never been infected. In our simulations, we used a threshold of $\sigma = 0.55$ to define partially immune population. The average length of time a person is partially immune is unknown and we arbitrarily assign the period of partial immunity to be $\eta^P = 1/720 days$. We also assume that the length of time a person is infected is the same for both susceptible and partially immune people. In section 10.5.1 we will show that the model is not sensitive to either of these parameters.

The model is insensitive to the rate μ at which people leave a city by either moving away or dying. We assigned the value $\mu = 0.0003 days^{-1}$, meaning that people live in a single city for approximately 10 years before moving.

10.3.1 Estimation of $r\beta$ and ϵ

Two parameters must be estimated by fitting the available data. The number of adequate contacts a person has per day sufficient to transmit the disease, $r\beta$, and the amount which infectivity varies with the seasons, ϵ, are assumed to be the same for all of the cities. We fix $\beta = 6 \times 10^{-4}$ (the baseline value) and simultaneously estimate ϵ and r to fit the CDC data.

The CDC cities weekly mortality data set required some additional analysis before it could be used to estimate r and ϵ. We will also describe some of the variations and anomalies that had to be accounted for based on individual cities. To account for missing data in weeks where it was not reported, we estimated the number of cases for nonreporting weeks as the average value neighboring weeks that did report cases. For the purpose of fitting parameters, a weighting system w_i was used in which nonreporting weeks are assigned the value $w_i = 0$ and all other weeks are assigned the value $w_i = 1$.

Finally, the data are presented as weekly mortality, while the model predicts daily morbidity. In order to make a comparison between the two, the mortality is assumed to be 1% of the morbidity and the number of deaths per week is divided up into an even number of cases every day. This assumption is a simple scaling parameter that can be accounted for by a simple multiplicative factor in the model predictions.

We define model_t as the value predicted by the model at time t and data_t is the CDC data as defined in the last section. The residual

$$\text{error} = \sum_{t=1}^{n} w_t |\text{data}_t - \text{model}_t| \tag{10.8}$$

was then minimized in the l_1 norm using an alternating line search over ϵ and r until r converged to five significant figures of accuracy and ϵ converged to within three significant figures of accuracy. The l_1 norm is chosen because it minimizes the effect of outliers as compared with the l_2 norm.

To reduce the effect of outliers, we fit the data after smoothing it with a Hamming filter, $d_i \leftarrow (d_{i-1} + 2d_i + d_{i+1})/4$, three times. We observed that the minimizing values for ϵ and r, shown in Table 10.2, were not sensitive to the filtering process. We believe that the insensitivity to prefiltering the data may be caused by the smoothing properties already inherent in the l_1 norm minimization procedures.

Because of uncertainties in the data, we chose to minimize the number of free parameters in the model and elected to find one value for ϵ and r for the entire data set. In order to choose these average values, the cities Denver, San Francisco, Portland, Kansas City, and Cincinnati were used because they had the smallest residuals. The model was then fitted over these five cities simultaneously to establish $r = 410.38$ and $\epsilon = 0.0210$ values for the entire data set.

Because in the model r always appears as a product with β, fitting r to data is equivalent to fitting the baseline value for the average number of adequate contacts per day to transmit the disease, $r\beta$. Therefore, in the process of fitting the model to r, we also were correcting for uncertainty in β. The best fit for r is based on dividing the best fit for $r\beta$ by the baseline value for $\beta = 6 \times 10^{-4}$ [14]. This gives the baseline value $r = 410.38$, which seems to be high. If $\beta = 6 \times 10^{-3}$, then $r = 41.038$ average contacts per day.

The reproductive number is linearly related to the product $r\beta$. Thus this fit indirectly establishes a reproductive number consistent with the data. The model is more sensitive to the transmission rate $r\beta$ than to ϵ. Figure 10.2 illustrates the sensitivity of the least squares fit for r and ϵ. This will be discussed more in section 10.5.1.

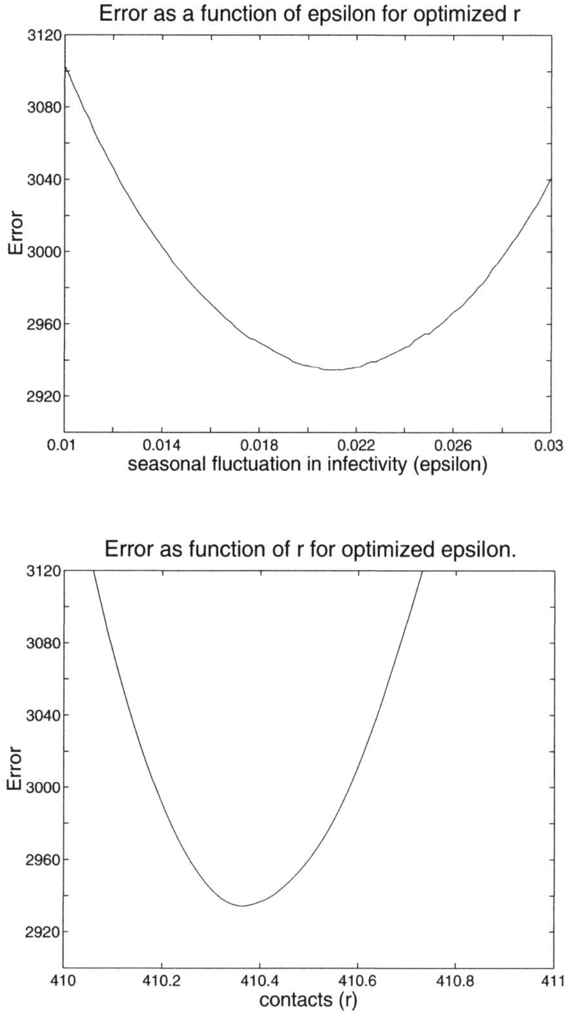

Figure 10.2. *The residual in the l_1 norm over five cities for* $\epsilon = [0.01, 0.03]$ *with* $r = 410.38$ *and* $\beta = 6 \times 10^{-4}$. *The residual for* $r = [410, 411]$ *is plotted for with* $\epsilon = 0.0210$ *and* $\beta = 6 \times 10^{-4}$. *Note that if* $\beta = 6 \times 10^{-3}$, *then the best fit range for the number of contacts per day would be* $r = [41.0, 41.1]$.

10.3.2 Population and Mobility

The population of each of the cities in Table 10.2 is based on the Census 2000 data [16]. In some instances, we combined the two major cities to account for the population in the area covered by the reporting station. There is some inconsistency in our population estimates because the population reporting to a CDC reporting station in a major city may be smaller than the population that uses the airport in the same city.

Migration between cities is approximated by airline flight data for the third quarter of 2000 as posted on the US Department of Transportation website [11] in the Air Travel

Table 10.2. *Populations of 33 largest cities in the US based on the 2000 census data and optimal r and ϵ fit for each city.*

min over		entire	time
City	Population	r	ϵ
New York, NY	9314235	411.05	0.0154
Los Angeles, CA	9519338	409.89	0.0189
Chicago, IL	8272768	410.49	0.0093
Philadelphia, PA	5100931	410.49	0.0106
Washington, DC**	4923153	N/A	N/A
Detroit, MI**	4441551	N/A	N/A
Houston, TX	4177646	412.230	0.0119
San Francisco–Oakland, CA	4123740	410.12	0.0185
Atlanta, GA**	4112198	N/A	N/A
Dallas, TX	3519176	410.49	0.0200
Boston, MA	3406829	410.52	0.0216
Phoenix–Mesa, AZ	3251876	409.98	0.0236
Minneapolis–St Paul, MN	2968806	412.21	0.0240
Cleveland–Akron, OH	2945831	409.89	0.0251
San Diego, CA	2813833	411.57	0.0126
St. Louis, MO**	2603607	N/A	N/A
Baltimore, MD	2552994	412.17	0.0203
Seattle–Tacoma, WA	2414616	410.49	0.0129
Tampa–St. Petersburg, FL	2395997	412.76	0.0168
Pittsburgh, PA**	2358695	N/A	N/A
Miami, FL	2253362	410.49	0.0092
Denver	2109282	410.84	0.0207
Portland–Vancouver, OR	1918009	410.44	0.0269
Kansas City, MO–KS	1776062	409.89	0.0202
Cincinnati, OH–KY–IN	1646395	411.79	0.0147
Orlando*, FL	1644561	N/A	N/A
Sacramento, CA	1628197	415.89	0.0180
Fort Lauderdale*, FL	1623018	N/A	N/A
Indianapolis, IN	1607486	412.39	0.0069
San Antonio, TX	1592383	413.53	0.0249
Las Vegas, NV– AZ	1563282	412.60	0.0257
Columbus, OH	1540157	413.89	0.0179
Milwaukee–Waukesha, WI	1500741	411.64	0.0173
Totals:	107620755	411.1965	0.01368

*This city does not participate in the 122 Cities Mortality Reporting.

**This city's data were too noisy to obtain a parameter fit [16].

Consumer Report. The estimates are for the average number of one-way passenger trips per day between two cities. Thus for a city with more than one airport, such as New York, the flight statistics are for all of the airports combined. Some of the airline data are in Table 10.3.

Since the flight data are the total flights per day, half the flights are assumed to be going in each direction in the migration matrix in the model. This leads to a symmetric migration matrix where the same number of people enter and leave the city each day and there is a constant population for each city.

10.4 The 1996–2001 Influenza Seasons

The model assumption that there is only one strain of influenza circulating in the population is justified for the 1996–1997 influenza season through the 1999–2000 season, when the

Table 10.3. *Daily flights between 33 major US cities* [11].

	New York, NY	Los Angeles, CA	Chicago, IL	Philadelphia, PA–NJ	Washington, DC	Detroit, MI	Houston, TX	San Francisco–Oakland, CA	Atlanta, GA	Dallas, TX	Boston, MA–NH	Phoenix–Mesa, AZ	Minneapolis–St. Paul,	Cleveland–Akron, OH	San Diego, CA	St. Louis, MO–IL	Baltimore, MD	Seattle, WA	Tampa–St. Petersburg, FL	Pittsburgh, PA	Miami, FL	Denver, CO	Portland–Vancouver, OR	Kansas City, MO–KS	Cincinnati, OH–KY–IN	Orlando, FL	Sacramento, CA	Fort Lauderdale, FL	Indianapolis, IN	San Antonio, TX	Las Vegas, NV–AZ	Columbus, OH	Milwaukee–Waukesha, WI
New York, NY		7324	7637	0*	5765	2953	2302	6055	6885	3382	6566	1318	2103	1136	1338	1117	398	1590	3485	1377	3342	2560	737	939	994	7658	302	7675	849	411	4414	856	826
Los Angeles, CA			3986	1289	1929	1559	1363	8220	1611	1576	1993	3162	1323	647	311	747	1582	2997	542	424	816	2075	1662	735	395	1258	2229	527	826	419	5324	462	522
Chicago, IL				2313	2783	3673	1943	3468	3711	3140	2579	2265	4049	2591	1144	2735	2933	1649	1836	965	918	2820	786	2555	762	2657	475	1566	867	542	3084	1291	0*
Philadelphia, PA–NJ					0*	848	719	1268	2553	1054	1561	548	634	322	505	402	0*	529	1040	750	461	629	175	319	358	1882	0*	1058	370	0*	1332	278	323
Washington, DC						618	881	1426	3221	1919	3307	321	1242	304	610	420	0*	749	990	319	1016	1415	248	584	414	1807	185	887	318	361	381	443	456
Detroit, MI							693	927	1699	992	661	853	1095	127	450	1196	951	675	946	219	369	616	291	534	225	1550	0*	562	517	189	1700	0*	466
Houston, TX								1081	1645	5652	751	598	448	479	465	793	863	606	440	255	418	930	274	533	195	790	191	365	341	944	1053	269	0*
San Francisco–Oakland, CA									1313	1530	1951	2427	1577	505	5041	451	1570	4615	326	471	543	2320	2653	551	378	643	0*	264	395	174	4310	341	614
Atlanta, GA										2855	2552	618	1307	596	560	653	1320	785	1914	765	1564	1249	420	856	493	2060	262	1774	437	391	865	496	396
Dallas, TX											1327	939	1244	382	710	821	529	939	768	354	470	1989	473	1187	314	1143	257	588	407	2228	1429	378	397
Boston, MA–NH												422	1312	445	574	317	759	695	1006	648	540	1358	282	371	425	1627	132	1332	321	0*	822	450	389
Phoenix–Mesa, AZ													1186	458	2351	568	928	1444	235	233	133	1353	1027	564	206	414	797	217	492	291	2578	448	377
Minneapolis–St. Paul,														346	625	539	311	1353	321	375	242	1770	429	860	405	986	226	181	291	316	1123	227	773
Cleveland–Akron, OH															260	639	1449	282	410	0*	179	342	0*	293	0*	651	0*	340	182	0*	714	0*	220
San Diego, CA																363	899	1614	204	192	0*	1103	929	371	0*	324	2023	202	244	182	2136	230	161
St. Louis, MO–IL																	1170	501	496	157	131	481	242	1033	176	716	0*	310	339	298	489	459	0*
Baltimore, MD																		796	1127	0*	313	818	463	537	177	1436	324	911	502	445	1093	32	0*
Seattle, WA																			291	266	172	1438	554	431	245	535	1258	238	294	230	2021	323	341
Tampa–St. Petersburg, FL																				457	328	459	0*	348	284	0*	0*	1244	582	0*	646	664	204
Pittsburgh, PA																					150	390	0*	211	0*	582	0*	309	0*	0*	348	0*	0*
Miami, FL																						328	0*	156	172	308	0*	0*	109	0*	512	118	101
Denver, CO																							870	1249	435	766	401	205	351	263	1248	246	385
Portland–Vancouver, OR																								262	0*	298	1197	0*	0*	0*	1296	173	0*
Kansas City, MO–KS																									195	699	228	156	322	204	729	219	216
Cincinnati, OH–KY–IN																										376	0*	216	0*	0*	246	0	0*
Orlando, FL																											190	798	975	264	660	818	623
Sacramento, CA																												0*	0*	0*	1143	0*	0*
Fort Lauderdale, FL																													337	0*	561	281	124
Indianapolis, IN																														0*	860	0*	0*
San Antonio, TX																															588	0*	0*
Las Vegas, NV– AZ																																937	792
Columbus, OH																																	0*
Milwaukee–Waukesha, WI																																	

*Market not large enough to be listed in top 1000 US markets.

influenza A subtype H3N2 dominated. However, the 1995–1996 and the 2000–2001 seasons were not decisively dominated by any one strain. In the 1997–2000 influenza seasons the epidemic threshold was exceeded for at least six weeks [2, 17, 18, 19]. In the 2000–2001 season the epidemic threshold was never reached, although this may be because the epidemic threshold was adjusted upward before this season [20]. The CDC data are summarized in Table 10.4.

The 122 Cities Mortality Reporting System is a volunteer system run by the participating cities. The cities each use their own system to count the P/I deaths and report the data directly to the CDC. There are inconsistencies in the data set due to changes in the volunteer staff and because there are often insufficient people to keep up with reporting during the peak of the influenza season [5]. The data also shows that some cities are more thorough in reporting than others. Also, there is uncertainty in the undercounting and what portion of P/I deaths may be attributed to influenza [5]. For the purposes of the model we assume that all influenza deaths are recorded and that all P/I deaths are attributable to influenza.

The data contains weeks in which no cases were reported. Frequently the unreported cases in one week are accounted for cumulatively in a single later report. Also, there is a lag time between the actual death dates and the report sent to the CDC. This lag time can be several weeks and tends to be longer in the winter than in the summer [5]. The CDC reports indicate that the office visits for influenza and P/I deaths peak with the mortality between 1 and 4 weeks after the morbidity peaks. The mean time between the morbidity peak and the mortality peak is 3 weeks. The reporting delay accounts for some of the time lag between the recorded peak of the epidemic as estimated through the physicians surveillance network and the peak of the CDC 122 Cities influenza mortality epidemic [2, 17, 18, 19, 20]. We did not account for the time lag in our simulations and assumed that it was the same for all the reporting cities. If the reporting accuracy is estimated, then this could be used to adjust the data before defining the parameters. For example, to account for an estimated three-week time lag, a first-order correction of the model is obtained by shifting the predictions back by the same time.

Table 10.4. *Summary of the data reported by the CDC in their Surveillance for Influenza reports for the* 1995–1996 *through* 2000–2001 *influenza seasons. Note that the influenza mortality peaks about three weeks after influenza activity* [2, 17, 18, 19, 20].

Influenza Season	Week influenza morbidity peaked	Week influenza mortality peaked	Predominant strain	Percent which are predominant strain	Number of weeks epidemic threshold exceeded
1995–1996	Jan 6–13	Jan 20	A(H1N1)	50	6
1996–1997	Dec 28	Jan 25	A(H3N2)	78	10
1997–1998	Feb 7	Feb 28*	A(H3N2)	99.5	11
1998–1999	Feb 6–27	March 13	A(H3N2)	76	12
1999–2000	Jan 1	Jan 22	A(H3N2)	96.5	22
2000–2001	Feb 3–10	Feb 24*	A(H1N1)	52	0

*The week in which influenza mortality peaked was not provided in the report, so the week in which influenza activity peaked + three weeks is used.

10.5 Model Threshold Conditions

The reproductive number, R_0, is the number of secondary infections that result from a single primary infection in a fully susceptible population. In a simple SIR model (as outlined in section 10.2.1) with no migration, no time dependence on transmission probability, or no return to susceptibility, we have $R_0 = r\beta/\alpha$. If $R_0 > 1$, then the epidemic spreads within the population to a stable equilibrium value. If $R_0 < 1$, then the only stable equilibrium is the zero equilibrium and the epidemic dies out.

For more complex models, the reproductive number is determined by the dominant eigenvalue of the Jacobian matrix at the infection free equilibrium [7]. The reproductive number for the network of n cities can be found by solving the eigenvalues of the $4n \times 4n$ Jacobian matrix for the multicity model (10.7). As the parameters change, the eigenvalues must be recalculated for each case.

Often a simple analytic formula for the threshold conditions (T_0) estimating when the epidemic will take off and when it will die out can give more insight into the behavior of an epidemic than computing the reproductive number for a specific set of parameters. We found it useful to analytically calculate the reproductive number for each city, assuming that it is isolated, and then to use this as a guide to estimate a threshold reproductive number for the entire population.

An upper bound for R_{0k} for the kth city is limited by the maximum of the time-dependent infectivity, $\beta = \beta_{\max}$. That is, $R_0 = \max[r\beta_k/(\phi_k + \alpha)]$. Here $1/(\phi_k + \alpha)$ is the average time a person is infected and remains in the kth city. To account for the natural removal rate, μ, and the migration of people from the population in a city in addition to the fraction of people who leave city k for other cities per unit time, we define $\phi_k = \mu + \sum_{i=1}^{n} D_{k,i}$.

We define an upper bound for threshold condition for the multicity system by defining it to be the maximum reproductive number of any of the cities, because if the epidemic spreads in one city, then it will persist in the entire population. That is,

$$T_0 = \max(R_{0k}) \quad \text{for } k = 1, \dots, n.$$

While only a threshold value, not a reproductive number, T_0 provides an accurate indication of whether an epidemic in a multicity population will persist or die out. For the baseline parameters in Table 10.1, the $T_0 = 1.02$ and occurs in Pittsburgh. Because $R_0 \approx T_0 \approx 1$, the model is sensitive to small changes in βr. If $\epsilon > 0.02$, then effective reproductive number of the system falls below one in the summer seasons, indicating that influenza cannot persist long under summer conditions. We were surprised at how closely our model and estimated parameters resulted in predicting that the annual influenza epidemic is perched precariously close to being able to sustain itself.

Because T_0 and R_0 depend linearly on the number of contacts per day, this implies that one of the most effective strategies to slowing the initial outbreak of an epidemic with $R_0 \approx 1$, like influenza or severe acute respiratory syndrome (SARS), would be for people to limit the number of adequate contacts with other individuals during the epidemic. When this approach is applied in the early stages of and epidemic, as the World Health Organization did for SARS in Toronto [3], then there is a good chance to contain the early spread of epidemic. The dependence on the connectivity ϕ to other cities on threshold conditions is less obvious.

10.5.1 Sensitivity Analysis

The relative sensitivity of the model prediction to small changes in each parameter can help determine the most important parameters in slowing the epidemic. The normalized sensitivity of five quantities ($I_{\max}$, $I_{\min}$, $I_{\text{cumulative}}$, $S_{\max}$, and $S_{\min}$) with respect to each of the eight parameters is approximated by

$$\frac{dQ}{dp}\frac{1}{Q_0} \approx \frac{[Q_{p(1+\delta)} - Q_{p(1-\delta)}]/Q_0}{[p(1+\delta) - p(1-\delta)]/p_0}, \tag{10.9}$$

where δ is the percent change in each of the parameters, $Q = I_{\max}, I_{\min}, I_{\text{cumulative}}, S_{\max}$, and $S_{\min}$ and $p = r, \beta^S, \beta^P, \alpha, \mu, \eta^R, \eta^P$, and ϵ. The factor $1/Q_0$ normalizes the importance of each parameter with respect to Q so the relative importance of the parameters can be compared.

The parameter α, the rate of recovery from infection, is the most important single parameter in every quantity we evaluated. For the baseline case, a 1% change in α results in a 122% change in peak of the infected population and a 1.5% change in the susceptible population.

The model is also sensitive to changes in β^S and r, the mean infection rate per contact for fully susceptible people and the number of contacts per unit time, respectively. The two quantities always appear in a product and have approximately equal importance in the model. A 1% change in either parameter yields approximately a 120% change in the peak of infected population and a 1.5% change in the susceptible population.

The infection terms $I_{\max}, I_{\min}$, and $I_{\text{cumulative}}$ are three orders of magnitude less sensitive to β_P, η^R, and η^P than they were to β^S and r. A 1% change in β_P, η^R, or η^P results a [0.53%–0.35%] change in the infected population. A 1% change in these parameters yields only a 0.002% change in the susceptible population.

The epidemic is also relatively insensitive to ϵ. A 1% change in ϵ results in a 0.3% change in $I_{\max}$ and $I_{\min}$. None of the other measurements are significantly affected by this parameter.

For simulations of a couple of years, the predictions are insensitive to changes in the migration/natural death rate, μ. These terms have such a small impact that they could be eliminated without significantly affecting the results.

In the model, the peak of the epidemic does not occur at the same time as the peak in the infectivity. The time lag follows from the seasonal variation in the infectivity and can be estimated from an approximate solution for the infected stage in a single-city model. When the multicity model is in periodic equilibrium, the same fraction of people are infected in each city. Therefore, the same number of infected people are entering a city as leaving, and the migration terms balance. When the multicity model is at equilibrium, the reduced equation for the infected stage in a city is

$$\frac{dI}{dt} = \lambda^S S + \lambda^P P - \alpha I - \mu I \tag{10.10}$$

or, in terms of the basic parameters,

$$\frac{dI}{dt} = I\left(\hat{\beta^S}\hat{r}\left(\frac{S}{N} + \sigma\frac{P}{N}\right)(1 + \epsilon\sin(\tau t)) - (\alpha + \mu)\right). \tag{10.11}$$

To simplify notation, we define

$$a(t) = \hat{\beta^S}\hat{r}\left(\frac{S(t)}{N} + \sigma\frac{P(t)}{N}\right), \tag{10.12a}$$

$$b = -(\alpha + \mu) \tag{10.12b}$$

and rewrite the equation as

$$\frac{dI}{dt} = I(a(t)(1 + \epsilon\sin(\tau t) + b). \tag{10.13}$$

After some time $\hat{t}$, the model with the baseline parameters settles down into a solution with period τ. In this periodic solution, the difference between the maxima and the minima of the number of people in the susceptible and partially susceptible populations changes by approximately 1% and $\hat{t}\ a(t) \approx a$ is constant. The simplified equation

$$\frac{dI}{dt} = I(a(1 + \epsilon \sin(\tau t) + b) \tag{10.14}$$

and solution

$$I = Ce^{(a+b)t}e^{-\frac{a\epsilon}{\tau}\cos(\tau t)} \tag{10.15}$$

account for the $\tau/4$ lag in the peak of the infection every year, provided that $e^{(a+b)t}$ is bounded.

For the baseline parameters, we verified that $a(t) \in [0.2445, 0.2447]$ for $t > \hat{t}$ days. Therefore $(a + b) \in [-7.66 \times 10^{-5}, 7.25 \times 10^{-5}]$, and

$$I \approx Ce^{-\frac{a\epsilon}{\tau}\cos(\tau t)}, \tag{10.16}$$

which is periodic with period τ. This provides an accurate estimate for time lag between the peak of the infectivity and the peak of the epidemic. This time lag might be an important factor that should be considered when studying the theory on the influence of the cold weather on the survival of the virus. That is, changes in people behavior during the winter months may give the virus a window of rapid transmission, the effects of which are not felt until much later.

10.6 Numerical Simulations

The model equations were integrated numerically with a variable order Adams–Bashforth–Moulton MATLAB solver, ode113. The parameters were set to the baseline values in Table 10.1 with the network of the 33 cities listed in Table 10.3, unless otherwise noted.

The influenza subtype H3N2 first became a dominant strain in the pandemic of 1968 [15]. We assumed that recent infections for this strain is close to a periodic equilibrium and used it as the benchmark data to be compared with the simulated model periodic equilibrium.

A small initial infection was introduced into the model and the solution was integrated until it reached periodic equilibrium. The timing for the model was set by the time that the largest fraction of people in each city is infected. In the periodic equilibrium, the same percent of people are in each stage in each city. At the peak of the epidemic, we observed that 98.76% of people were susceptible, 0.005731% of people were infected, 0.4503% of people were recovered, and 0.7854% of people were partially immune. The solution of the model, shown in Figures 10.3, 10.4, 10.5, 10.6, and 10.7, illustrate that some of the cities report data close to the model predictions, while others differ greatly. The smooth curves in the figures are the model predictions, while CDC P/I data plotted is highly variable.

The magnitude of the epidemic in Los Angeles and Kansas City is slightly overestimated by the model; however, the prediction is quite good. The data for New York City are fit quite nicely, although the number of cases is slightly underestimated by the model. The overestimates occur in the first two seasons; then for the last two influenza seasons the model fit is reasonable. The data for San Francisco/Oakland, Boston, Phoenix, Denver, and Portland are well predicted by the model. In all five cities the seasonal variation in the number of cases is greater than that predicted by the model.

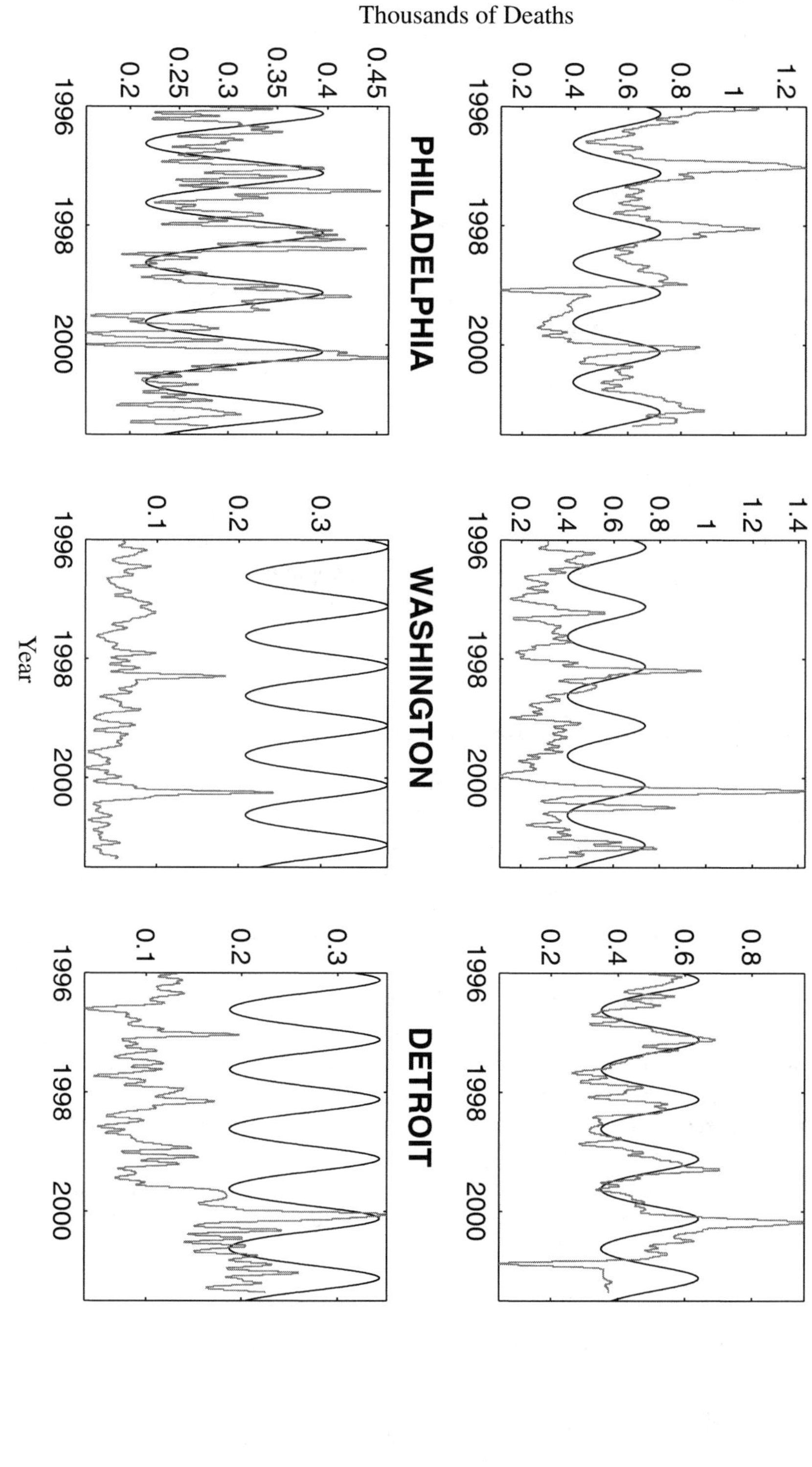

Figure 10.3. *Comparison of the results of the simulation with data from the Centers for Disease Control* [2, 17, 18, 19, 20].

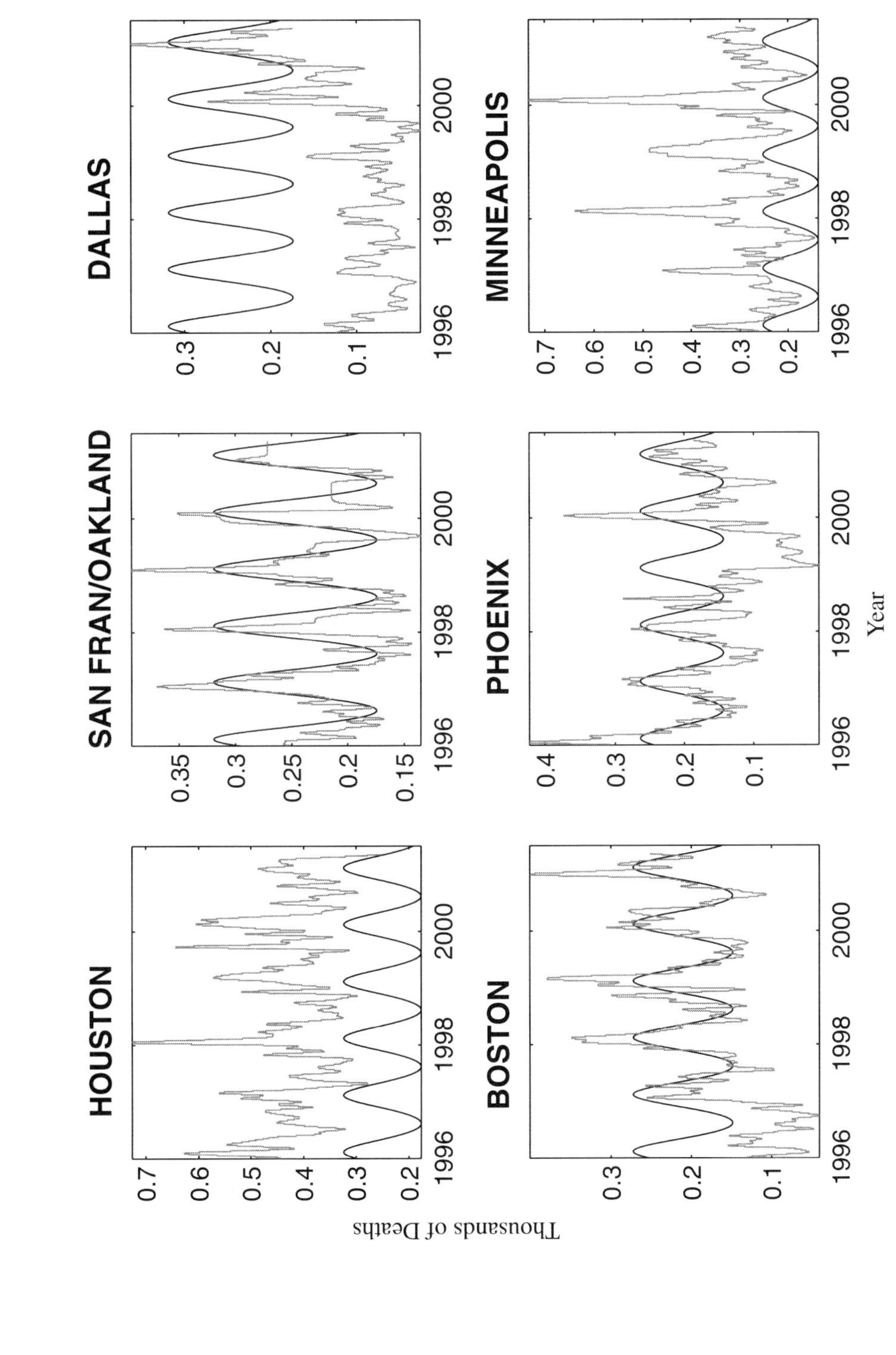

Figure 10.4. *Comparison of the results of the simulation with data from the Centers for Disease Control* [2, 17, 18, 19, 20].

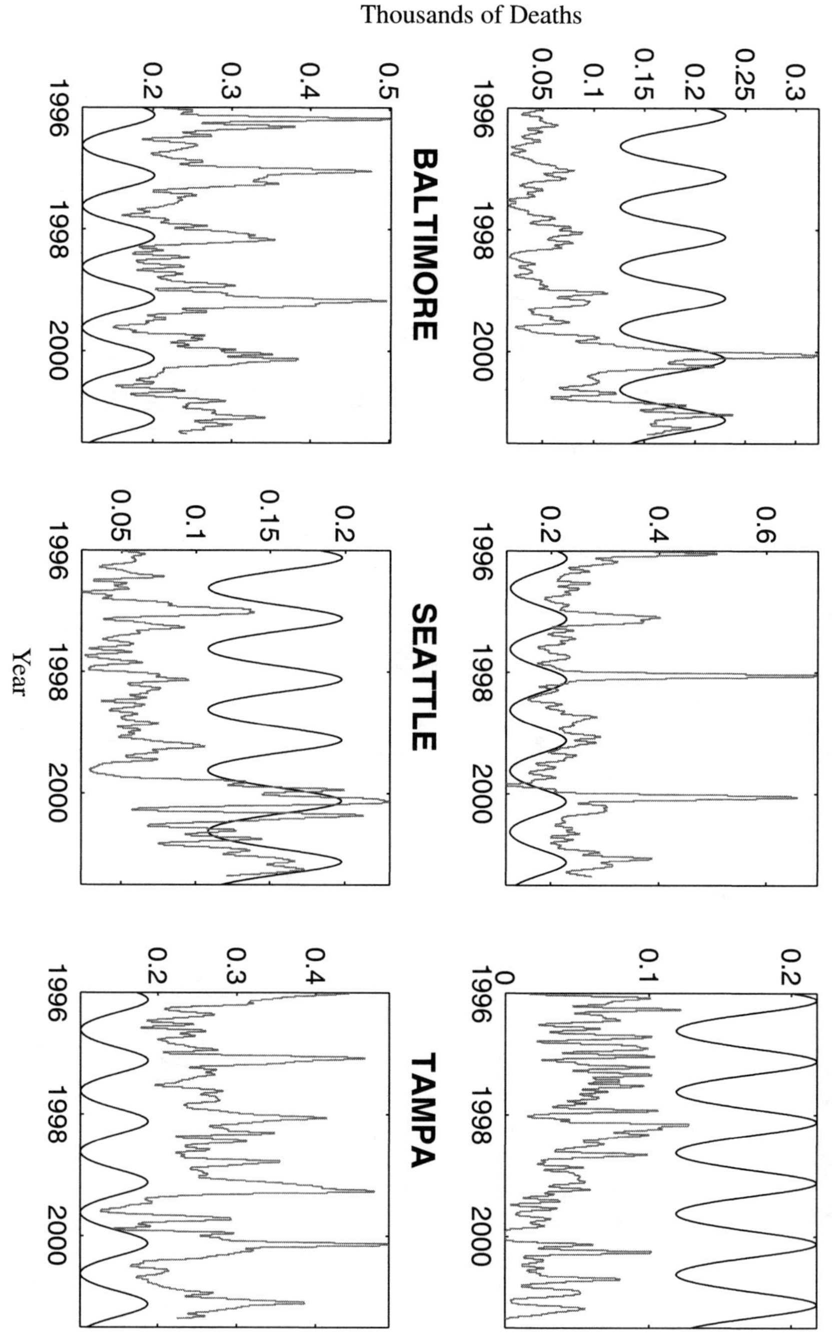

Figure 10.5. *Comparison of the results of the simulation with data from the Centers for Disease Control* [2, 17, 18, 19, 20].

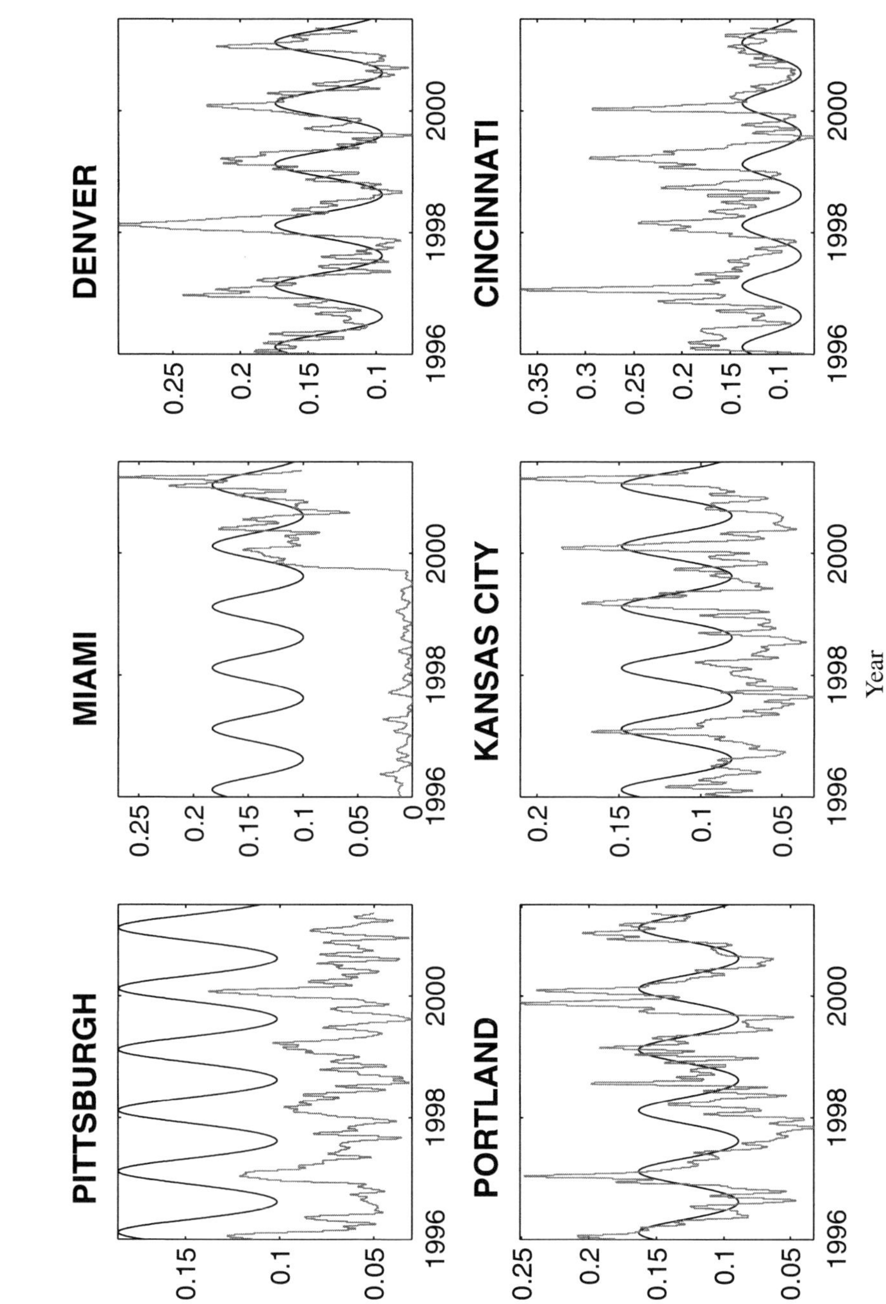

Figure 10.6. *Comparison of the results of the simulation with data from the Centers for Disease Control* [2, 17, 18, 19, 20].

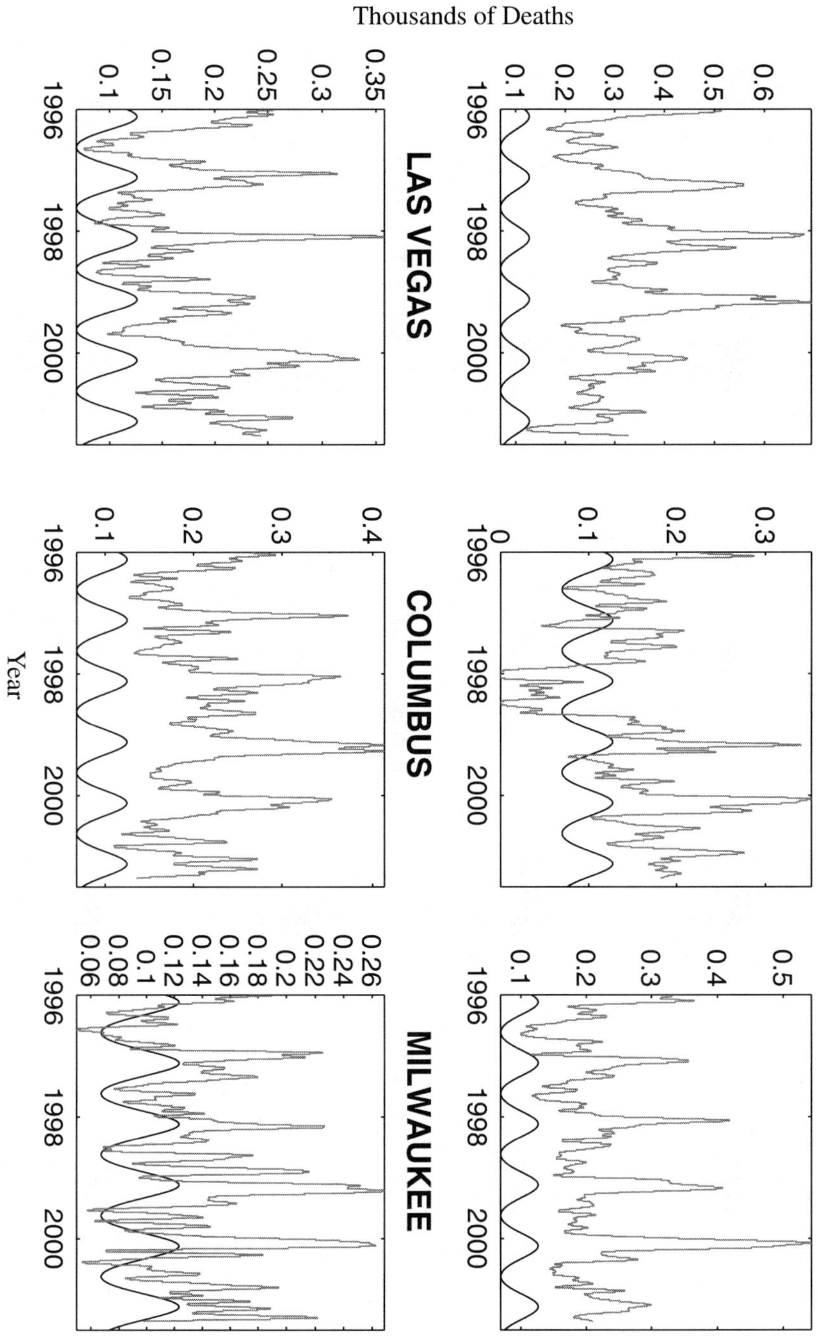

Figure 10.7. *Comparison of the results of the simulation with data from the Centers for Disease Control* [2, 17, 18, 19, 20].

Neither Chicago nor Philadelphia was used to fit the parameters in the model. Even so, they were two of the cities where the magnitude and seasonal fluctuation of the yearly epidemic were well approximated by the model.

Indianapolis, Milwaukee, and Cincinnati are all slightly underestimated by the model. Note that Indianapolis and Milwaukee seem to have very little seasonal variation in the number of P/I deaths, unlike most of the other cities posting consistent data; this is surprising as the cold weather in these cities would be expected to increase the seasonal variation in the epidemic.

The influenza mortality for Minneapolis, San Diego, and Las Vegas is underestimated by the model. This is likely because of differing (or more comprehensive) data collection techniques employed in these cities.

The model prediction of the mortality for Houston, Baltimore, Tampa, Sacramento, San Antonio, and Columbus is substantially lower than the actual data. There are a number of potential reasons for this. One is that the population base used for the data collection is of the metro area, as opposed to the city proper, as was assumed in the model. Another is that these cities collect more complete data than the other cities or that they have a more liberal criteria for classifying cause of death as P/I. We did not identify any correlation between the reported data and the climate in these cities. However, it could be significant that four of these cities are among the six smallest in the model. This might indicate that even in populations as large as the these cities, the size of the population might still affect the number of contacts. Recall that the model predictions are sensitive to the number of contacts per day for a typical individual.

The CDC mortality is inconsistent for Detroit, Dallas, Cleveland, Seattle, and Miami. In the last two years these cities reported significantly more influenza deaths. Whether this is because of a new method of data collection, a larger population base being used, or a genuine change in the number of P/I deaths is unknown. As none of the other cities showed a substantial increase in reported P/I deaths in this time frame, it seems likely that a change in data collection took place. What is significant is that in all five cases the higher estimates were similar with what the model predicted. It seems likely that the new data collection method was comparable with that of the cities which the model was fitted to; thus a good prediction was obtained.

The data for Washington, St. Louis, Atlanta (not shown), and Pittsburgh are substantially lower than predicted by the model. These four cities turned in very low estimates for the five years of the data set, compared to the other cities we studied. This could be because they were using a small subset of the population to approximate the whole, or because they had a very stringent definition of what was recorded as a P/I death. Atlanta was included in the model as the ninth largest city in the US; however, the fit is not shown because they stopped collecting data for most of 1999 and 2000.

The data does not support a correlation between climate and number of influenza deaths. It was anticipated that, since the infectivity and contact rate are constant for all of the cities, the model would underestimate the number of influenza cases in the cities with harsher climates. In fact two of the cities for which the model underestimates the number of deaths the most are Houston and Sacramento, both of which have warm climates.

10.6.1 Initial Stages of an Epidemic

In the initial outbreak the size of the network of cities has a substantial impact on the early spread of the epidemic. We formulated the model on a subset of the major cities of the United States. At the periodic equilibrium, running the reduced models yields the same results as the full model because the same percentage of people in each city are infected. Therefore the same number of infected people are leaving each city as entering. This is not true for the transient epidemic before it reaches the equilibrium. The recent SARS epidemic demonstrated how quickly an infection can spread from a local outbreak to a global problem. For these simulations the epidemic is assumed to start in a single city (New York) and spread outward from there.

In our simulation, the epidemic is started with 1000 people ill, as it would take a reasonably large number of people getting infected before an emerging epidemic is identified. We investigated how the size of the subset of cities in the network affects the spread of the epidemic. We observed that the smaller the network is, the faster the epidemic approaches equilibrium. Also, in smaller networks it spreads in the city of the initial outbreak because of the larger reproductive number resulting from fewer people migrating into and out of the city. Figure 10.8 shows the initial outbreak for an epidemic that emerged on the day of highest infectivity (in November).

10.7 Discussion

We defined a modified SIR model for the spread of influenza that accounts for nonrandom mixing among a discrete network of 33 cities. The nodes of the network were weighted by the population of the city and the bonds between the nodes represented the daily movement of people among the data as estimated by airline flight data. Data from the influenza transmission studies and the reported mortality data attributed to influenza and pneumonia were used to define the model parameters. Despite noise in the 122 Cities data from the CDC, similar estimates for fluctuation in infectivity and number of contacts (and thus reproductive numbers) are obtained for each city.

The essential features of the yearly influenza epidemic approximated the influenza and pneumonia mortality data reported by the CDC. The magnitude and fluctuation of the yearly epidemic is well matched by the prediction of the model at the periodic equilibrium. At the endemic equilibrium the travel terms in this model are not important.

We observed that the threshold reproductive number for the network of cities is very close to one and that the model predictions are most sensitive to α and $r\beta$. That is, the most effective approaches to slowing an epidemic is to treat the ill to reduce the length of the infectious stage (α), to reduce the number of contacts an infected has in a typical day (r), and to reduce the probably of transmission in a typical contact (β). These are precisely the measures taken in Toronto to contain the initial spread of SARS and were demonstrated to be highly effective [3].

There is a substantial time lag between the peak of the infectivity and the peak of the epidemic, which is accounted for by deriving an approximate solution to the differential equation of the infected stage.

We also used the model to predict the initial spread of a new infectious agent when it is introduced into one of the cities. In the initial stages of the epidemic, the number of cities in the network and the timing of the outbreak determine how the epidemic spreads within and among each of the cities.

One area of primary concern in the model is the assumption of random mixing of the populations within the cities. However, people within cities are in groups, based on age, geography, socioeconomic factors, or religious affiliation, that they mix with regularly and have fairly few contacts with people outside of these groups. Additionally, a small portion of the citizens of a city will do the majority of the traveling. This suggests the development

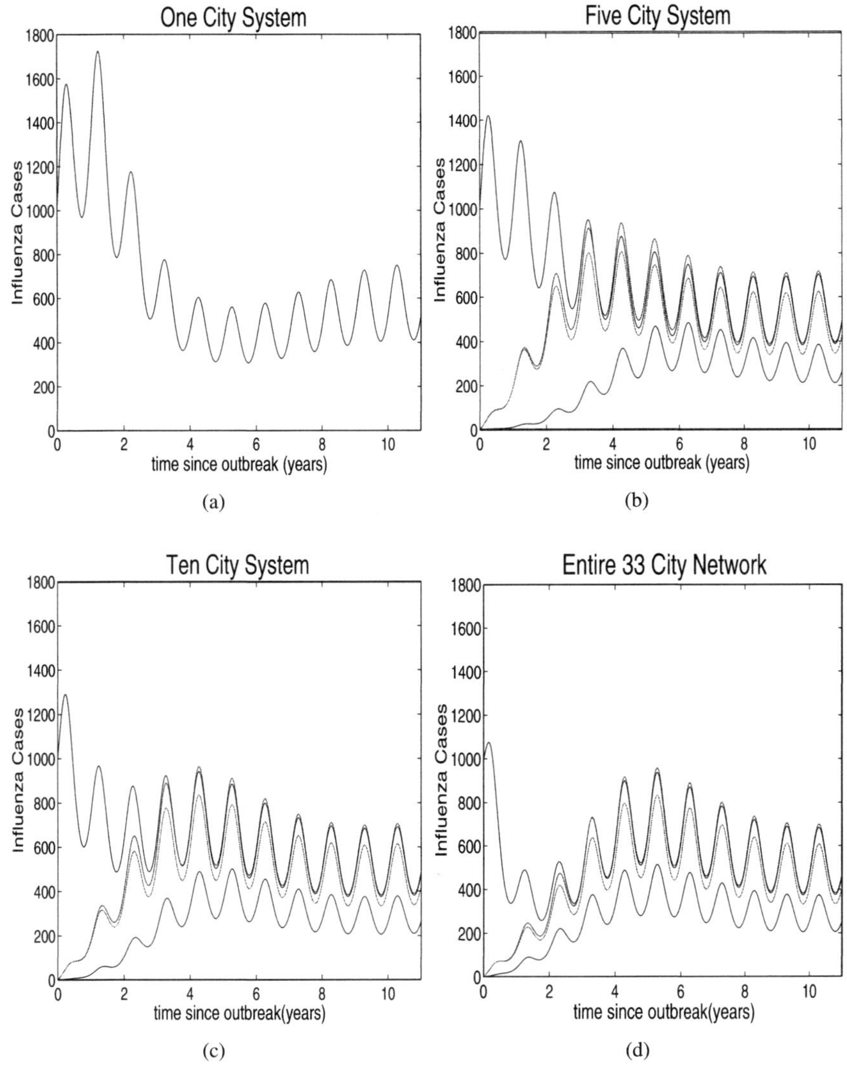

Figure 10.8. *Initial outbreaks for various size systems of cities when the outbreaks starts at the time of peak infectivity.* (a) *is for New York alone.* (b) *is for the largest four cities in the US on a network of the largest five cities in the US.* (c) *is the results of the largest four cities in the US on a network of the largest ten cities in the US.* (d) *is the results of the largest four cities in the US on the entire* 33*-city network.*

of a model which has nonrandomly mixing groups, called a small-world model, within each city in addition to the travel between cities. Some of the groups would travel more often than others. A first step in this direction would be to have a model with two groups in each city, travelers and nontravelers. We have observed in more detailed biased mixing models that after a brief startup period, the random mixing assumption is fairly accurate.

In the model the same fraction of people in each city are susceptible, infected, immune and partially susceptible to the epidemic at periodic equilibrium. This is a result of the contact rate and infectivity being the same for each city. This assumption has been used in other multicity SIR models [12], with good results. However, uniform contacts is not an appropriate assumption for the early stages of an epidemic, because of the nonrandom mixing of the population, variations in the population density, and availability of public transportation in different cities. A better approach would be to fit contact rates to the data for subsets of cities based on availability of public transportation or population density.

The populations of the cities were kept constant by not including any increased mortality due to influenza. As long as we are concerned only with the endemic equilibrium and the annual influenza epidemic, this assumption is reasonable. However, for longer term simulations or more severe epidemics, the increased mortality due to influenza must be accounted for.

The current model accounts for a slight drift in the genetic code for one strain of influenza by allowing a previously infected person to gradually lose immunity to the currently circulating strain of the virus. This is a reasonable assumption in a simulation of two or three years if the annual influenza epidemic is decisively dominated by a single strain. It will not predict the impact of a major shift in the virus. A straightforward extension of the model could account for multiple strains. However, there are few data on the strain that causes a particular infection or the variation of the susceptibility of the population to different strains needed to validate the model.

Acknowledgments

This research was supported by the Department of Energy under contracts W-7405-ENG-36 and the National Infrastructure Simulation and Analysis Center (NISAC).

We thank Lori Hutwagner, CDC Epidemiology Program, for answering questions about the 122 Cities Mortality Data; Lynette Brammer, CDC Influenza Branch, for answering questions about the Physicians Surveillance Network; and Leon Arriola for help analyzing the model. We also thank Bill Sailor, Shilpa Khatri, Thomas Park, Miriam Nuno, and Andy Perelson for their help and feedback on the paper and Mike McKay for his comments on the sensitivity analysis.

Bibliography

[1] C. L. Addy, I. M. Longini, and M. Haber, *A generalized stochastic-model for the analysis of infectious-disease final size data*, Biometrics, 47 (1991), pp. 961–974.

[2] T. L. Brammer, H. S. Izurieta, K. Fukuda, L. M. Schmeltz, H. L. Regnery, H. E. Hall, and N. J. Cox, *Surveillance for Influenza—United States,* 1994–95, 1995–96, *and* 1996–97 *Seasons*, Morbidity and Mortality Weekly Report 49(SS03);13–28, Centers for Disease Control, Atlanta, April 28, 2000; available online from http://www.cdc.gov/epo/mmwr/preview/mmwrhtml/ss4903a2.htm.

[3] G. Chowell, P. W. Fenimore, M. A. Castillo-Garsow, and C. Castillo-Chavez, *SARS Outbreaks in Ontario, Hong Kong, and Singapore: The Role of Diagnosis and Isolation as a Control Mechanism*, Report, Los Alamos National Laboratory, Los Alamos, NM, 2003.

[4] A. L. Frank and L. H. Taber, *Variation in frequency of natural reinfection with influenza-a viruses*, J. Med. Virology, 12 (1983), pp. 17–23.

[5] L. Hutwagner, CDC Epidemiology Program, *Telephone conversation about* 122 *cities mortality reporting system*, June 2001.

[6] J. M. Hyman and T. LaForce, *Multi-City SIR Epidemic Model*, Report, Los Alamos National Laboratory, Los Alamos, NM, 2001.

[7] J. M. Hyman and J. Li, *An intuitive formulation for the reproductive number for the spread of diseases in heterogeneous populations*, Math. Biosci., 167 (2000), pp. 65–86.

[8] I. M. Longini, *A mathematical-model for predicting the geographic spread of new infectious agents*, Math. Biosci., 90 (1988), pp. 367–383.

[9] D. Mollison, ed., *Epidemic Models: Their Structure and Relation to Data*, Cambridge University Press, Cambridge, UK, 1995.

[10] A. S. Monto, J. S. Koopman, and I. M. Longini, Jr., *Tecumseh study of illness* XIII: *Influenza infection and disease*, Amer. J. Epidemiology, 121 (1985), pp. 811–822.

[11] Office of the Assistant Secretary for Aviation and International Affairs, *Domestic Airline Fares Consumer Report: Third Quarter* 1999 *Passenger and Fare Information*, US Department of Transportation, Washington, DC, May 2000; available online from http://ostpxweb.dot.gov/aviation/domfares/993web.pdf.

[12] L. A. Rvachev and I. M. Longini, *A mathematical-model for the global spread of influenza*, Math. Biosci., 75 (1985), pp. 3–23.

[13] L. Sattenspiel and C. P. Simon, *The spread and persistence of infectious-diseases in structured populations*, Math. Biosci., 90 (1988), pp. 341–366.

[14] N. I. Stilianakis, A. S. Perelson, and F. G. Hayden, *Emergence of drug resistance during an influenza epidemic: Insights from a mathematical model*, J. Infectious Diseases, 177 (1998), pp. 863–873.

[15] S. B. Thacker, *The persistence of influenza a in human-populations*, Epidemiologic Rev., 8 (1986), pp. 129–142.

[16] US Census Bureau, *Census* 2000 *PHC-T*-3 *Ranking Tables for Metropolitan Areas* 1990 *and* 2000, US Census Bureau, Washington, DC, April 2001; available online from http://blue.census.gov/population/cen2000/phc-t3/tab01.pdf.

[17] Centers for Disease Control, *Update: Influenza Activity—United States and Worldwide,* 1997–98 *Season, and Composition of the* 1998–99 *Influenza Vaccine*, Morbidity and Mortality Weekly Report 47(14);280–284, Centers for Disease Control, Atlanta, April 17, 1998; available online from http://www.cdc.gov/mmwr/preview/mmwrhtml/00052002.htm.

[18] Centers for Disease Control, *Update: Influenza Activity—United States and Worldwide,* 1998–99 *Season, and Composition of the* 1999–2000 *Influenza Vaccine*, Morbidity and Mortality Weekly Report 48(18);374–378, Centers for Disease Control, Atlanta, May 14, 1999; available online from http://www.cdc.gov/mmwr/preview/mmwrhtml/mm4818a2.htm.

[19] CENTERS FOR DISEASE CONTROL, *Update: Influenza Activity—United States and Worldwide,* 1999–2000 *Season, and Composition of the* 2000–01 *Influenza Vaccine*, Morbidity and Mortality Weekly Report 49(17);375–381, Centers for Disease Control, Atlanta, May 05, 2000; available online from http://www.cdc.gov/epo/mmwr/preview/mmwrhtml/mm4917a5.htm.

[20] CENTERS FOR DISEASE CONTROL, *Update: Influenza Activity—United States and Worldwide,* 2000–01 *Season, and Composition of the* 2001–02 *Influenza Vaccine*, Morbidity and Mortality Weekly Report 50(22);466–470, Centers for Disease Control, Atlanta, June 08, 2001; available online from http://www.cdc.gov/mmwr/preview/mmwrhtml/mm5022a4.htm.

Index